GROUNDWATER CONTAMINATION AND EMERGENCY RESPONSE GUIDE

GROUNDWATER CONTAMINATION AND EMERGENCY RESPONSE GUIDE

by

J.H. Guswa   **W.J. Lyman**

Arthur D. Little, Inc.
Cambridge, Massachusetts

A.S. Donigian, Jr., T.Y.R. Lo, E.W. Shanahan

Anderson-Nichols & Co., Inc.
Palo Alto, California

NOYES PUBLICATIONS
Park Ridge, New Jersey, USA

Copyright © 1984 by Noyes Publications
Library of Congress Catalog Card Number 84-14842
ISBN: 0-8155-0999-5
ISSN: 0090-516X
Printed in the United States

Published in the United States of America by
Noyes Publications
Mill Road, Park Ridge, New Jersey 07656

10 9 8 7 6 5 4 3 2 1

Library of Congress Cataloging in Publication Data
Main entry under title:

Groundwater contamination and emergency response guide.

    (Pollution technology review, ISSN 0090-516X ; no. 111)
    Bibliography: p.
    Includes index.
    1. Water, Underground--Pollution--Handbooks, manuals,
etc. 2. Organic water pollutants--Handbooks, manuals,
etc. I. Guswa, John H. II. Lyman, Warren J.
III. Series.
TD426.G73 1984   363.7'394   84-14842
ISBN 0-8155-0999-5

Foreword

An overview of groundwater hydrology; a technology review of equipment, methods, and field techniques; and a methodology for estimating groundwater contamination under emergency response conditions are provided in this book. It describes the state of the art of the various techniques used to identify, quantify, and respond to groundwater pollution incidents.

Interest in the causes and effects of groundwater contamination has increased significantly in the past decade as numerous incidents have brought the potential problems to public attention. Protection of our groundwater resources is of critical importance, thus making the book both timely and relevant.

Part I assesses methodology for investigating and evaluating known or suspected instances of contamination. Part II surveys groundwater fundamentals, state-of-the-art equipment, monitoring methods, and treatment and containment technologies. It will serve as a desk reference and guidance manual. The information will supply input to developmental efforts leading to more efficient, durable, and cost-effective monitoring and remedial action technologies. Part III details possible emergency response actions at toxic spill and hazardous waste disposal sites. It provides a rapid assessment methodology for performing such an evaluation within a 24-hour emergency response time frame.

The information in the book is from:

> *Groundwater Contamination Response Guide. Volume I: Methodology,* by J.H. Guswa and W.J. Lyman of Arthur D. Little, Inc., for the U.S. Air Force Engineering and Services Center, June 1983.

Groundwater Contamination Response Guide. Volume II: Desk Reference, by J.H. Guswa and W.J. Lyman of Arthur D. Little, Inc., for the U.S. Air Force Engineering and Services Center, June 1983.

Rapid Assessment of Potential Ground-Water Contamination Under Emergency Response Conditions, by A.S. Donigian, Jr.; T.Y.R. Lo; and E.W. Shanahan of Anderson-Nichols & Co., Inc., for the U.S. Environmental Protection Agency, November 1983.

The table of contents is organized in such a way as to serve as a subject index and provides easy access to the information contained in the book.

Advanced composition and production methods developed by Noyes Publications are employed to bring this durably bound book to you in a minimum of time. Special techniques are used to close the gap between "manuscript" and "completed book." In order to keep the price of the book to a reasonable level, it has been partially reproduced by photo-offset directly from the original reports and the cost saving passed on to the reader. Due to this method of publishing, certain portions of the book may be less legible than desired.

NOTICE

Contents and Subject Index

Part II
Groundwater Contamination Response Desk Reference

Part I

Groundwater Contamination Response
Methodology

The information in Part I is from *Groundwater Contamination Response Guide. Volume I: Methodology,* by J.H. Guswa and W.J. Lyman of Arthur D. Little, Inc., for the U.S. Air Force Engineering and Service Center, June 1983.

Acknowledgments

This document was prepared by Arthur D. Little, Inc., Acorn Park, Cambridge, Massachusetts, for the Engineering and Services Laboratory, Air Force Engineering and Services Center (AFESC), Tyndall Air Force Base, Florida. The AFESC/RDVW project officer was Captain Glen E. Tapio.

Several volumes of technical material were critically reviewed and compiled during preparation of this report. Expert summaries were provided by the following individuals:

J. Adams	M. Mikulis
J. Bass	R. O'Neill
A. Brecher	A. Preston
D. Brown	C. Thrun
D. Leland	

1. Introduction

Since the mid-seventies, the emphasis on understanding the causes and effects of groundwater contamination by organic chemicals has increased because a few Air Force facilities have encountered significant problems with the presence of organic contaminants under their property. During investigation and cleanup of these known incidents, it became obvious that there was no organized procedure to guide Air Force personnel in determining the location, extent, and level of groundwater contamination, or to select the most appropriate containment or treatment technology. Because Federal legislation related to contamination of groundwater resources affects the Air Force, a methodology to assess and control groundwater pollution by organic chemicals became expedient.

The general problem with protection of groundwater resources is to identify the areas and mechanisms by which contaminants enter the groundwater system, to develop reliable methods for predicting contaminant transport, to select an appropriate contaminant/treatment technology, and to ensure compliance with federal and state legislation.

For Air Force personnel, this requires:

- identification and analysis of available information to estimate the extent, nature, direction, and rate of movement of the contaminant;

- development of a field investigation program to quantify the rate and direction of contaminant movement, as well as the extent of the contaminated zone;

- selection of method(s) for containing the spread of contaminants or treating contaminated groundwater; and

- response to the appropriate federal and state agencies.

To effectively respond to groundwater contamination incidents, the Air Force is developing the capability to rapidly identify organic contaminants in groundwater, to determine pollutant pathways, and to determine the fate of organic constituents in groundwater. The results of this effort will be included in the Spill Prevention and Response Plan for each Air Force installation.

Until such time as the Spill Prevention and Response Plan can be updated, an interim solution is needed. A user-oriented field manual, based on a literature review and describing the current best practicable

3

methodology for Air Force field personnel to respond to incidents of groundwater contamination by organic chemicals, is proposed.

This Methodology and the companion Desk Reference are designed to help base level engineering personnel to address groundwater pollution problems in a logical manner. This will address such specific issues as:

- the initial response to identified contamination incidents;

- developing a strategy for determining the origin of organic contaminants;

- determining the rate and direction of movement of the pollutants and extent of the contaminated zone;

- identifying possible strategies for control, containment, and cleanup of groundwater contamination.

These volumes do not provide specific solutions for groundwater contamination problems. The data necessary to design the response for a particular contamination incident must be developed from site-specific soil and groundwater investigations. They do, however, describe an overall approach which can be followed to ensure a logical, scientifically based response to a groundwater pollution incident.

The Methodology is a summary document which describes the logical flow of action to be taken in responding to a contamination incident. The Desk Reference is based on a thorough review of the scientific and technical literature related to groundwater contamination and summarizes the state of the art of the various techniques used to identify, quantify, and respond to groundwater pollution incidents.

2. Identifying and Assessing
the Contamination Problem

Initial identification of groundwater contamination is generally unexpected; that is, there usually is no advance warning that a well or spring which has previously had good quality water is going to show evidence of contamination. The complex flow paths which can exist in groundwater systems, the wide variety of contamination sources, and the fact that groundwater flow is not directly observable all contribute to this "surprise factor." When initially informed of a potential groundwater contamination incident, the questions of most immediate concern are usually:

- What is the nature of the contamination?
- What is the source of the contaminant?
- How extensive is the contamination? and
- What is an appropriate remedial response?

1. INITIAL ASSESSMENT

The initial indication of a contamination problem may be water with an unusual taste, odor, or physical appearance, an indication of vegetative or wildlife stress, or it may be noted during routine water quality testing. The initial indication may provide some information regarding the nature of the contaminant, but it will not usually provide information regarding the source, extent, or severity of the problem. These concerns need to be addressed by a problem-specific investigation and analysis program. Table 1 lists the major information categories and specific data elements that usually need to be evaluated during the investigation and analysis program. All of the data elements in Table 1 may not be required for every problem, but they should be considered during the initial problem assessment and definition.

The physical framework includes all the geologic and topographic information which describes the environment through which groundwater and, hence, the contaminant flows. This includes the thickness and areal extent of various geologic units, as well as maps of the spatial variability of water transmitting and storage properties. The hydrologic system is defined by those properties which control water movement through the physical framework. These data include water level information, identification of natural and human-induced recharge and discharge locations, the hydraulic connection between groundwater and surface water, spatial variability of water quality, and other factors which define boundary conditions to the flow system. Site information includes a description of present and past site uses which may provide information regarding the nature and source of contamination, identification of existing monitoring points which may be used in the problem investigation, and construction information, such as the location of buried utilities, which is important for safety reasons, as

5

TABLE 1.   PRINCIPAL DATA REQUIREMENTS FOR GROUNDWATER CONTAMINATION
ASSESSMENT

Physical Framework	Hydrogeologic maps showing extent and boundaries of all geologic units
	Topographic maps showing surface water bodies and landforms (including springs and seeps)
	Water table, potentiometric, bedrock configuration, and saturated thickness maps
	Maps showing variations in water-transmitting properties
	Maps showing variations in storage coefficient
Hydrologic System	Water levels and water level changes (maps and hydrographs)
	Depth-to-water map (for evapotranspiration estimates, selection of sampling method)
	Type and extent of recharge areas (irrigated areas, recharge basins, recharge wells, etc.)
	Groundwater inflow and outflow
	Groundwater pumpage (temporal and spatial distribution)
	Climatologic information
	Surface water diversions

TABLE 1. (CONCLUDED)

Stream flow quality (temporal and spatial
distribution)

Temporal and spatial distribution of
groundwater quality

Relation of surface water bodies (hydraulic
connection) to aquifers

Stream flow variation (including gain and
loss measurements)

Site Information   Previous site use (system operations,
                                materials handled, safety considerations)

                                Potential sources of contamination (on and
                                off site)

                                Location of buried utilities (contamination
                                source, safety considerations, affect on
                                groundwater flow)

                                Location of established monitoring points
                                (including complete construction details for
                                monitoring wells)

well as for identifying possible sources of contamination and
pertubations on the local groundwater flow system.

Each of these types of data can provide useful information in the
identification and evaluation of the contamination incident. Some may
already exist and be available at the Air Force base or from other
sources, while some may have to be collected, and some may not be
required in response to the contamination incident. The types and level
of data detail necessary to address a groundwater contamination incident
are problem-specific and can only be determined during the problem
assessment. However, a methodology does exist which can be used to
select an appropriate response to almost all groundwater contamination
incidents. This methodology, summarized in Figure 1, is described in
detail in the following paragraphs.

The initial notification of the potential groundwater contamination
will provide the first information regarding the nature and extent of
contamination. The notification would probably include at least
qualitative information such as taste or odor, for example, and specfic
location information. This information will be the base for developing
an initial strategy for defining the nature and extent of contamination
and selecting an appropriate remedial response.

The first step is to gather the available information and assess
the problem. Because it is unlikely that all of the information listed
in Table 1 would be available, the initial assessments must be based on
interpretations from whatever information is available.

The available geologic and hydrologic information may include
published maps of a general nature or it may include data collected as
part of other investigations. If data from the specific site are not
available, then representative values from similar materials or terrains
can be used as an initial approximation. Some of the likely sources of
geologic and hydrologic data have been identified in Part II of
this report.

The geologic information should be summarized to prepare a
three-dimensional representation of the site. This is done to identify
the relationships between the various hydrologic unit(s) and to identify
those units most likely to be affected by the contaminant. Maps showing
the areal distribution and variations in thickness and
water-transmitting properties of the different units and geologic
sections showing their vertical relationships are most useful for this
purpose.

The available water level information is interpreted to determine
groundwater flow directions. This is done by plotting water level
information on maps and preparing water table or potentiometric contour
maps. Flow directions are determined on the basis of the contour
patterns. The estimated groundwater flow directions and
water-transmitting properties of the geologic units are used to estimate
the rate and direction of contaminant transport and the size of the

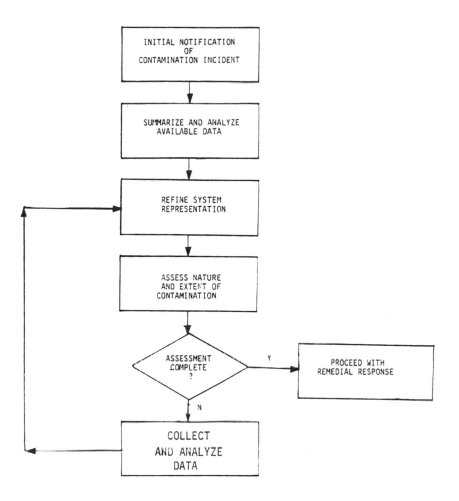

Figure 1.   Schematic Representation of Interaction Between Data Analysis, Problem Assessment, and Data Collection Leading to Selection of Remedial Response

contaminant plume. Table 2 includes equations which can be used to calculate transport velocities and contaminant concentrations. These equations are based on simplifying assumptions about the groundwater flow system and contaminant transport, but can be used to estimate contaminant distribution. Input parameter values would be determined from available site data, or representative values for similar materials or terrains could be used to make the initial estimates. The greater the uncertainty for any input parameter, the wider the range of values which should be used in the analyses. These estimates would be evaluated during the subsequent data collection program, and revised according to the new information. Some hydrologic principles which need to be recognized when dealing with contaminant plumes in groundwater are:

- the contaminant plume is not diluted with the entire body of groundwater, but tends to remain as an intact body with only slight dispersion and diffusion along the edges;

- the contaminant actually moves faster than the average groundwater velocity because of hydrodynamic dispersion;

- the path of a soluble contaminant plume will generally follow the direction of groundwater flow. Diversions in flow direction from induced changes in gradient (e.g., a pumped well) will also divert the contaminant plume;

- the flow direction of a water-immiscible contaminant will be affected by the groundwater flow direction, but they do not necessarily coincide;

- hydraulic and lithologic conditions and fluid density determine the vertical depth to which the contaminant will migrate into the aquifer. The thickness of the plume in the aquifer will probably increase with distance downgradient from the source;

- the extent and movement of various constituents in the contaminant plume will vary depending on attenuation from the various chemical and biochemical reactions.

The initial contamination assessment should attempt to identify the particular chemical or contaminant and identify all possible sources of the contamination. Initial assessments should also be used for screening the remedial alternatives to identify those likely to be effective.

TABLE 2. EQUATIONS FOR ESTIMATING CONTAMINANT TRANSPORT VELOCITIES AND CONTAMINANT CONCENTRATIONS FOR IDEALIZED FLOW SYSTEMS

	POROUS MEDIA	FRACTURED MEDIA
BULK FLOW	$\bar{V} = (K/n)(\Delta h/L)$	$\bar{V} = (K/Nb)(\Delta h/L)$
WITH RETARDATION	$V_c = \bar{V}/(1 + (\rho b \cdot Kd)/n)$	$V_c = \bar{V}/(1 + (2\ Ka/b))$
WITH DISPERSION	$(C/C_o) = (0.5)\ (erfc\ (L-Vt)/(2\sqrt{D_L t}))$	NA

where

V is the average linear velocity of the fluid

K is the hydraulic conductivity

n is the effective porosity

$\Delta h/L$ is the headgradient between two points

N is the number of fractures per unit distance

b is the width of the fractures

V_c is the velocity of the 0.5 point on the concentration profile of the retarded species

ρb is the bulk mass density of the porous medium

Kd, Ka are the distribution coefficients of the contaminant for porous and fractured media, respectively

TABLE 2. (CONCLUDED)

C is the contaminant concentration at a point X_1

C_o is the source concentration of the contaminant

1 is the distance along the flow path from the source to the point where C is calculated

t is the time since the contaminant entered the groundwater system

D_1 is the coefficient of longitudinal hydrodynamic dispersion

2.   ADDITIONAL DATA COLLECTION AND ASSESSMENT

    a.   Field Investigations

    After analysis of available data and estimates of the probable
source, nature and extent of contamination, a field investigation
program is implemented to evaluate initial estimates, as well as gather
additional hydrogeologic information about the site.   Table 3 lists
field investigation activities which might be undertaken to improve the
information base.   Selection of field investigation methods is based
upon the type and amount of additional data needed to supplement the
available information.   These methods have been described in the Desk
Reference (Part II) and are summarized below.

    TABLE 3.   FIELD INVESTIGATION ACTIVITIES WHICH CAN BE UNDERTAKEN TO
             IMPROVE THE INFORMATION BASE FOR CONTAMINATION ASSESSMENT

                    Field Mapping
                    Surface Geophysical Surveys
                    Test Drilling and Sampling
                    Monitor Well Installation
                    Borehole Geophysical Surveys
                    Hydraulic Testing
                    Water Quality Sampling

        (1)   Field Mapping

    Field mapping can be done on a topographic or aerial photographic
base.   The maps and photos are used to accurately identify the locations
of various surface features and the different geologic units.
Identifying the location of surface features, such as springs, seeps,
streams, sampling points, and cultural features can facilitate problem
identification.   Comparisons of present and past aerial photographs may
provide information regarding changes in land use or base operations
which may help to identify the source of contamination.

        (2)   Surface Geophysical Surveys

    The two most common types of surface geophysical surveys are
electrical earth resistivity and seismic refraction surveys.   Both of
these methods provide geologic interpretation based on indirect
measurements of physical characteristics.   They can provide subsurface
geologic information much faster and cheaper than drilling, but they
must be calibrated against more direct measurements.   They can also be
used in areas which may not be accessible to a drilling rig.

    Earth resistivity surveys are commonly used to define subsurface
geology and, occasionally, zones of contaminated groundwater.   In
complex geologic environments or in the vicinity of some manmade

structures, such as buried pipelines and fences, the results of a resistivity survey are inconclusive. Seismic surveys are generally used to provide information regarding the depths and thicknesses of different geologic units, as well as depth to water. The results of these surveys can also be difficult to interpret in complex environments. Other surface geophysical surveys which provide more specialized information have been described in the Desk Reference.

(3)   Test Drilling and Sampling

A test drilling and sampling program is often necessary to describe the local hydrogeologic environment. This includes the type, thickness and depth of the various geologic units, their water-bearing and chemical characteristics, and depth to water. Samples may be collected to provide visual identification of the materials encountered, or they may be collected for specialized laboratory tests. The drilling and sampling method used depends upon the type and depth of material to be sampled. Several drilling and sampling methods have been described in Section IV of the Desk Reference. Drilling methods vary on a regional basis because of the large scale variation in geologic conditions. Discussions with local drillers can provide information regarding the drilling techniques used locally.

Samples of geologic materials can also be obtained when they are exposed at land surface, such as quarries, sand and gravel pits and bedrock outcrops. Because of weathering, however, the samples collected from land surface may not be representative of conditions below ground, particularly for bedrock materials.

(4)   Monitor Well Installation

It is frequently desirable to install monitoring wells during the test drilling and sampling program. Monitoring well locations are usually chosen after analysis of available information. The expense associated with well construction materials and installation, maintenance, and operation of the monitoring network, necessitates careful selection of monitoring well locations.

Monitoring sites are usually chosen to provide information regarding temporal changes in water levels or quality, to document the presence or absence of a contaminant or to provide early warning of an unexpected change in direction of movement or size of the contaminant plume. The specific data that should be evaluated in designing the monitoring network include:

- groundwater flow direction;

- distribution of geologic and hydrologic characteristics of various units

- background water quality;

- present or future effects of groundwater withdrawals on the flow system;

- the type and frequency of measurements to be made at the monitoring site, as well as the expected temporal variation in those parameters.

The information provided by a monitoring well represents a small portion of the geologic unit being sampled. Interpolation of the information collected from the well to the geologic material in general is frequently limited by the heterogeneity of the material. The greater the geologic variability, the larger the number of sampling points necessary to adequately define the subsurface environment.

(5) Borehole Geophysical Surveys

The most commonly used borehole geophysical surveys in groundwater contamination assessments are resistivity and natural gamma logging. These surveys generally provide qualitative information regarding the variations in geologic materials (resistivity and natural gamma logging) and water quality (resistivity logging). These surveys are usually used to supplement the driller's and geologists' log of the test drilling operation.

Resistivity surveys can only be made in uncased boreholes and, therefore, may not be possible for all test holes. Natural gamma surveys do not have that restriction and are particularly useful for interpreting lithologic information from previously drilled wells for which this information is not available.

(6) Hydraulic Testing

Hydraulic testing is usually done to determine in-situ hydraulic properties. Tests can be done using single or multiple wells or piezometers. These field methods are based on analyzing water level changes in wells or piezometers in response to a sudden introduction or removal of a known volume of water or to an instantaneous pressure pulse. Single well tests provide in-situ values of hydraulic properties which represent a small volume of the aquifer, while multiple well tests provide values that represent a larger portion of the aquifer. The hydraulic properties are frequently determined by comparing observed water test level changes with those calculated for idealized aquifer geometries.

In-situ tests may provide information regarding the hydraulic properties of the geologic media in the immediate vicinity of the contamination problem. A disadvantage to this field technique is that the analysis of the water level change data is usually not straightforward. In particular, observed water level change data are affected by well contruction and aquifer geometry and heterogeneity. Misrepresentation of either of these parameters will yield erroneous results.

(7)   Water Quality Sampling

Water samples are collected to obtain information regarding natural variations in water quality, as well as to determine areas which are contaminated.   Samples may be collected from surface water bodies or from the groundwater system.   Groundwater samples can be collected from existing wells or springs, during test drilling activities, or from monitoring wells installed as part of the field investigation.

The objective of the water quality sampling program should be to collect and preserve water samples so that the water quality of the sample is representative of the environment from which it was collected. This is not a trivial task because the techniques which can be used to obtain a sample are often limited by the ease of access to the sampling point.   For groundwater samples, the major limitations are commonly the depth to water and well diameter.   Table 4 lists some of the common sampling methods used for various well diameters and depths to water. Sampling methods are described in Section 3 (4) of the Desk Reference.

    b.   Chemical Analysis Methods

    (1) Overview

During the analysis of groundwater and sediment samples for organic and/or inorganic constituents, it is necessary to follow some steps fundamental to the analytical process.   These steps are as follows:

●   obtain a representative sample;

●   prepare the sample for analysis;

●   separate constituent(s) that interfere;

●   identify/measure the constituent(s) of interest in the sample;

●   calculate the results including, as appropriate, precision, accuracy and detection limits of numerical results.

The purpose for obtaining a representative sample in the field was discussed in paragraph b(1), Section 1, of this report.   Once a field sample has been received in the laboratory, it is usually necessary to obtain a representative aliquot of that field sample for subsequent analysis.   Representative aliquots from groundwater samples are typically obtained by constructing a composite field sample using homogenization (a blender) or shaking (by hand), and quickly withdrawing an appropriately sized aliquot from the composite field sample.

Successfully implementing steps 2, 3, and 4 of the analytical process depends upon selecting appropriate analytical techniques.   In turn, selection of those techniques is based on what is known about the

TABLE 4. PUMPING EQUIPMENT SELECTION

Diameter Casing	Bailer	Peristaltic Pump	Vacuum Pump	Airlift	Diaphragm "Trash" Pump	Submersible Diaphragm Pump	Submersible Electric Pump	Submersible Electric Pump w/Packer
1.25-inch								
Water level < 20 ft		X	X	X	X			
Water level > 25 ft								
2-inch								
Water level < 20 ft	X	X	X	X	X	X		
Water level > 25 ft	X			X	X			
4-inch								
Water level < 20 ft	X	X	X	X	X	X	X	X
Water level > 25 ft	X			X		X	X	X
6-inch								
Water level < 20 ft				X	X		X	X
Water level > 25 ft				X			X	X
8-inch								
Water level < 20 ft				X	X		X	X
Water level > 20 ft				X			X	X

15

source of the sample and what the ultimate use of the data will be. Selection of analytical techniques is discussed later in this section. Data gathered during a sampling and analysis program may vary in quality. For the purposes of this report, quality refers to the validity, reliability and, more specifically, the precision and accuracy of the data.

To ensure the generation of high quality data for groundwater samples, a Quality Assurance/Quality Control (QA/QC) program should be implemented throughout a sampling and analysis program. In addition, the user of data generated during a groundwater sampling and analysis program should have sufficient information about the original method of analysis and the data quality to assess whether or not the data meet the purposes of the program. Use of a QA/QC program ensures that the quality of data is documented in a way that permits users of the data to make independent assessments. The basic elements of a QA/QC program are discussed later in this section.

(2)   General Approach

Before analysis of field samples, it is necessary to prepare an analytical plan directed towards solving the specific problem. Development of this plan with appropriate selection of analytical methods requires review of information concerning the intended purpose of the data, and previously obtained data.

The methods selected for the analysis of groundwater and sediment samples must be appropriate for the purpose of the chemical analysis data. The following criteria for the analytical method should be used: adequate sensitivity, detection limits, selectivity, precision and accuracy. Other characteristics that should be considered include: dynamic measurement range; ease of operation; multiconstituent applicability; low cost; ruggedness; portability. For example, when chemical analysis data is to be used to demonstrate compliance with regulatory standards, these standards may require that certain analytical techniques be used, and that particular organic compounds be analyzed at or below specified limits of detection.

Further, available information concerning the samples and the source of the samples must be reviewed in order to provide further input into selecting analytical methods. Preparation and analysis of a sample depends upon the type of sample (groundwater, sediment, interstitial water), the organics being analyzed, and the potential interferences to be dealt with. The chemical composition of the samples may be available through previous analyses, known chemical disposal and management practices, obvious odors, or other means. Chemical composition information may be used to determine what organics are of interest to the study and what interferences are expected to be a problem. The planning stage for selecting an analytical approach is summarized in Figure 2.

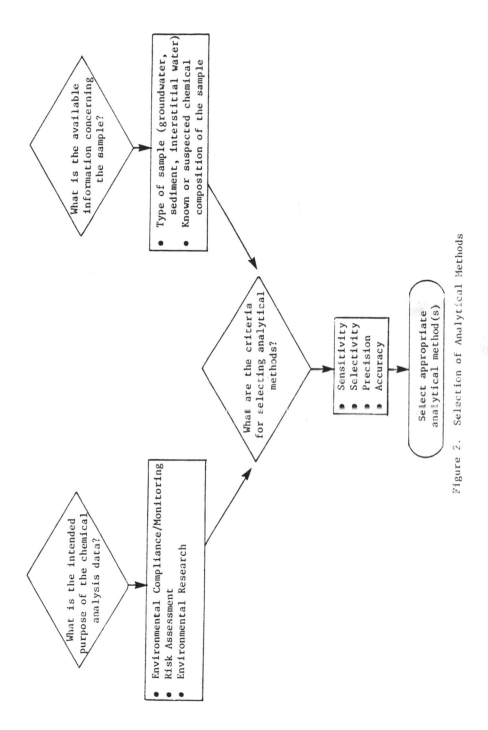

Figure 2. Selection of Analytical Methods

When there is uncertainty as to the most appropriate analytical methods, one set or subset of samples should be analyzed and the appropriate statistics calculated to determine the applicability of the methods. If the data from one approach are within the original criteria, then analysis of the balance of the samples may proceed. If the initial data are determined to be unsuitable, then it is necessary to either change or modify the analytical procedures. This process for determining the applicability of the analytical plan is shown in Figure 3.

(3) Methods for Organic Chemicals

The organic content of groundwater, sediment, and interstitial water may be approximated through analysis of the classical parameters given in Figure 4. The data obtained from these analyses do not provide the identity and concentration of specific organic compounds. However, such information can be useful since these parameters have been extensively used as indicators of water quality. There is, therefore, a large data base available for comparison. Of the conventional parameters, total organic carbon (TOC) and total organic halogens (TOH) are the most useful. TOC and TOH are rapid, cost-effective measurements which provide an assessment of organics in the samples. They are useful in the assessment of the level of contamination of a sample, and the subsequent determination of the procedures necessary for sample preparation and measurement of specific organic compounds. For example, if a TOC value is high, it may be desirable to dilute the sample prior to a particular measurement to prevent overloading the detector. The results of TOC and/or TOH analyses should not be used to determine whether or not a sample is hazardous, since some organic chemicals may be present at extremely low concentrations (below the detection limits for TOC and TOH) and the data from TOC and TOH analyses will not indicate their presence.

(4) Other Classical Parameters

After review of the classical water parameter tests, the next step in selection of a method is the determination of the organic species of interest. A directed analysis may be performed for a particular species, or if the organics present have not been identified, a screening analysis may be performed. Directed analyses are designed to provide qualitative confirmation of presence and compound identities, as well as quantitative data of known quality for each of the specified organics of interest. A screening analysis is designed to provide an overall description of the major types and approximate quantities of organics present in the sample. The results of a survey analysis may lead to subsequent directed analyses.

A comprehensive scheme for directed and/or survey analyses is given in Figure 5. This scheme does not include all the possible analytical techniques which may be used when characterizing a sample, however, it does include the more commonly utilized approaches. A more complete description of analytical methods is included in Part II.

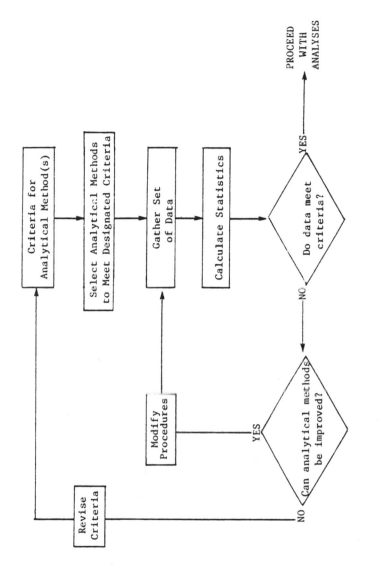

Figure 3. Applicability of the Analytical Plan

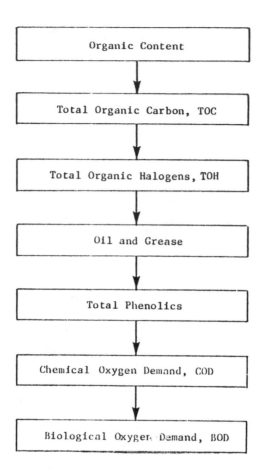

Figure 4.   Analyses Parameters

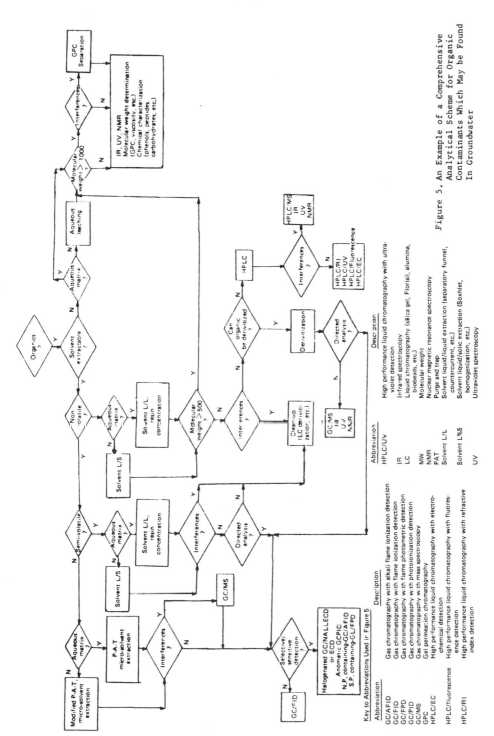

Figure 5. An Example of a Comprehensive Analytical Scheme for Organic Contaminants Which May be Found In Groundwater

(5)   Quality Assurance/Quality Control

The implementation of a Quality Assurance/Quality Control (QA/QC) Program during a sampling and analysis program is critical to providing reliable analytical results.   A QA/QC program provides procedures and guidelines to:

- ensure data of the highest quality possible;

- maintain the quality of data within predetermined (tolerance) limits and to provide specific guidelines for activities to be taken where those predetermined limits are exceeded;

- document the quality (accuracy and precision) of generated data.

A QA/QC program addresses several areas.   These include:

- Personnel responsibilities
- Documentation
- Data and procedures reviews
- Audits
- Maintenance of facilities and equipment
- Training
- Sample preservation
- Standards and reagents
- Chemical analysis methods
- Quality control samples
- Quality control data

c.   Integrating Data Collection and Analysis

The initial data analysis and problem assessment will provide information regarding the types of data that may need to be collected during the field investigation program.   These data needs will reflect the uncertainties in the interpretation of the available data as well as the short and long term goals of assessing the degree of contamination and selecting a remedial response plan.   It should not be expected, however, that the data needs required for complete problem assessment can be determined during the initial problem analysis.   Some of the data needs may be satisfied with a limited amount of additional data collection, but it should be anticipated that the data collection activities may identify new uncertainties or data needs which require more data collection.   The data collection effort should be planned, therefore, to provide for periods of data analysis in order to redefine and establish priorities for data collection.   After the data needs have been redefined the data collection program should be redone to provide for collecting the most critical data in a timely manner.   Figure 6 illustrates the sequence of steps to be followed when integrating the data collection and analysis efforts.

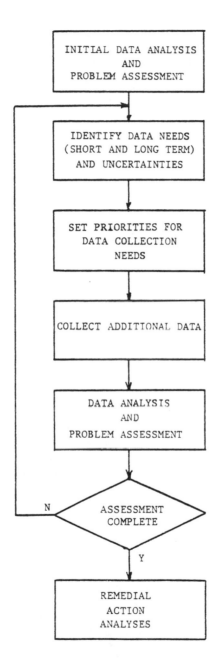

Figure 6. Schematic Illustration of the Interaction Between Data
Collection and Analysis

3. Developing Decision Parameters

1. ULTIMATE FATE OF ORGANICS IN GROUNDWATER

This section describes a two-step procedure to obtain information on the ultimate fate of organic chemicals in the soil/groundwater system. It identifies the information needs and provides direction for obtaining such information and using it for an assessment of ultimate fate.

A key aspect of this approach is an initial (preliminary) assessment making use of readily available data, supplemented with estimates and/or surrogates (Figure 7). The purpose of this initial step is twofold: (1) to provide a more rapid, preliminary analysis - avoiding the time and expense of laboratory and field studies - so that timely decisions and plans can be made, e.g., on response actions; and (2) to provide a sharper focus on just what additional laboratory and field tests - if any - need to be done. While the use of such a procedure should save both time and money, there is no formal requirement for its use, and some conditions may warrant a different approach.

Key sections of Part II that will be referred to are:

●   Fate of organic groundwater pollutants (Section 2(f); and

●   Physical, chemical, and biological parameters and constants applicable to organic contaminants and physical systems of concern (Section 4(a)).

a.   Step I - Preliminary Assessment

(1) Information Requirements

Section 4(1) of Part II provides a discussion of important chemical-specific (cf. Table 23) and environment-specific (cf. Table 24) properties.

Perhaps the chemical-specific properties are more important to evaluate in the preliminary assessment. However, without detailed knowledge of certain environment-specific properties, it may not be possible to determine the correct value(s) of some chemical-specific properties within an order of magnitude. For example, the soil adsorption coefficient will vary with soil type (especially soil organic carbon content) and other parameters. An order-of-magnitude uncertainty may be quite acceptable in a preliminary assessment considering (as described in Section 2(f) of Part II) that many of these properties range over at least seven orders of magnitude and that the importance of a fate process may be associated with a wide range of a property.

26

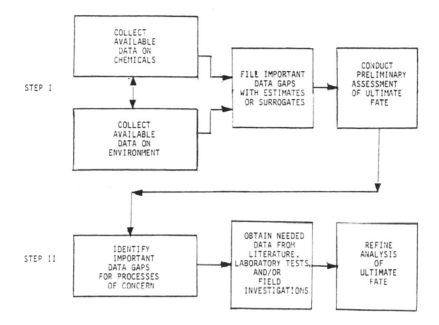

Figure 7.   Two-Step Process for Assessing the Ultimate
Fate of Organics in Groundwater

Measured values of the chemical-specific properties are much preferred over estimates; however, data have been published for only a relatively small fraction of the more common pesticides, solvents, fuel constituents, and other synthetic chemicals. This is especially true for persistence-related properties (e.g., rate of hydrolysis, rate of biodegradation). Because much of the important literature is widely scattered (i.e., in different books, journals, government reports, and unpublished reports), often confusing (due to the use of different test methods), and of variable quality, a literature search by a qualified, experienced environmental scientist is required. Such searches cannot rely on computer searches of bibliographic data bases because of the nature of the sources containing the data and the manner in which they are abstracted. Computerized data bases of physical and chemical properties are just starting to become available, but, at the present, have not proven their worth.

If measured values of the chemical-specific properties are not available, reasonable estimates may frequently be derived. (See Lyman et al., 1982.)

Environment-specific properties may be available in literature describing the geohydrological and meteorological conditions near the site. Possible sources of additional information are described later in this report.

As described in the overview, it is suggested that a preliminary assessment proceed without recourse to special, often expensive, laboratory or field tests to fill all data gaps.

(2) Assessment of Fate

The first 'law' of environmental pollution states that: "Everything must go somewhere." An assessment of the ultimate fate of a groundwater pollutant should answer two questions following logically from this law: a. "Where does it go?" and b. "How fast does it get there?" Both of these questions may reasonably be asked with regard to three types of processes:

- partitioning of the chemical between the three phases (soil, water, air) of the 'soil';

- degradation of the chemical by such processes as hydrolysis, biodegradation, and oxidation; and

- transport of the chemical, either in the vapor phase to the atmosphere or in solution with the groundwater.

The answers for the third type of process (transport) usually require modeling which may be beyond the resources and data availability associated with a preliminary assessment.

For partitioning, the preliminary assessment should determine (predict) how the chemical partitions between the soil, water, and air phases of the groundwater system. Section 2(b) of Part II provides a technical discussion of the methodology. This answers the question "Where does it go?" and provides important information on the mobility of the chemical. The question of "how fast?" is seldom important for partitioning since the time scales of groundwater movement are usually much longer than the time required for equilibrium partitioning to be achieved.

For degradation, the question of "how fast?" should come first. This refers to assessing the rates at which the chemical is transformed (degraded) from its original form to some other compound, or series of compounds, by the processes mentioned. The answers will be in the form of rate constants and will be environment-specific (i.e., will depend on such properties as temperature and pH). The question of "Where does it go?" is translated in this case to "What are the products of degradation?" The answer to this question will include a list of "intermediate" and "final" (stable) chemicals which will also be environment-specific in many cases. Information on the degradation products is important for assessments of potential human health effects and for monitoring programs. Section 2(3) of Part II provides background information on this subject.

b.   Step II - Revised Assessment

The preliminary assessment should have provided an identification, and possibly even a semiquantitative description, of the important fate processes acting on the chemicals of concern. It will also have, almost certainly, identified a number of important data gaps for both chemical- and environment-specific properties. The revised assessment will require more detailed knowledge about the key chemical- and environment-specific properties and the factors affecting their values. Although some of these data may be found after specialized literature searches, laboratory and/or field tests will usually be required. These tests will often require considerable time (weeks to months) and expense.

A hypothetical example may help to illustrate the process. Assume that a preliminary assessment for a chemical indicated that: (1) only water-soil partitioning was important, but the soil adsorption coefficient estimated for the soils at the site was uncertain by a factor of 10; (2) hydrolysis was the only important degradation pathway, but the rate constant, extrapolated from laboratory data obtained under much different conditions, was uncertain by a factor of 100; and (3) only partial information on the hydrolysis reaction products was available. In this case, laboratory tests - using site-specific conditions (soils, water, temperature, etc.) - could provide measured values of the adsorption coefficient and hydrolysis rate constant (as a function of key environmental variables) whose uncertainties were closer to 10 percent. Details on reaction pathways and products would also be available, and a revised fate assessment could be made with confidence.

2.   TYPES OF REMEDIAL RESPONSE

The National Oil and Hazardous Substances Contingency Plan (NCP) (40CFR Part 300) identifies three general categories of remedial response. These are:

- initial remedial measures;
- source control remedial actions; and
- offsite remedial action.

Initial remedial measures are actions which are "feasible and necessary to limit exposure or threat of exposure to a significant health or environmental hazard and . . . are cost effective . . . and should begin before final selection of an appropriate remedial action," (40CFR 300.68(e)(1)). They are, in short, cost-effective measures to protect the public and the environment while long term solutions are being sought.

Source control remedial actions are appropriate when "the threat can be mitigated and minimized by controlling the source of the contamination at or near the area where the hazardous substances were originally located," (40CFR 300.68(f)). Removal or repair of a leaky underground fuel storage tank are examples of source control remedial action.

Offsite remedial actions are appropriate when "the hazardous substances have migrated from the area of their original location," (40CFR 300.68(f)). An impermeable barrier, such as a slurry wall, placed underground to contain a contaminated plume while it is being pumped for treatment is an example of offsite remedial action.

Technologies available for remedial action of groundwater contamination may be useful in any of these three categories, depending on site-specific conditions. Groundwater pumping, for example, may be used to protect drinking water supply wells (initial), remove contaminated groundwater at or near the site (source control), or purge a contaminated plume downgradient from the site (offsite). Remedial measures applicable to the treatment, containment, and control of groundwater contamination are discussed in Section 5 of Part II. A comprehensive list of technologies more generally applicable to all types of remedial action is given in the NCP (40CFR 300.70).

4. Selecting a Remedial Response

Part 300.68 of the NCP details the thought process recommended for selecting an appropriate remedial response. This process, diagrammed in Figure 8, involves four basic steps:

- preliminary assessment;
- development of alternatives;
- analysis of alternatives; and
- selection of appropriate response.

Each of these steps is discussed below with respect to response to groundwater contamination at Air Force facilities.

1. PRELIMINARY ASSESSMENT

Preliminary assessment of the remedial response includes scoping, determination of appropriate type or types of response, and the remedial investigation.

Scoping can be considered the initial assessment of the magnitude of the problem based on available information. Its purpose is to determine expected funding requirements and types of remedial action necessary, and to provide a starting point for remedial investigation. Factors used in scoping are given in Part 300.68(e) of the NCP and summarized in Table 3. The problem assessment phase should be planned so that information required to address these factors is collected.

The remedial investigation is a more detailed analysis of the conclusions of the scoping process. According to the NCP, the remedial investigation has two purposes (40CFR 300.68(f)):

- "to determine the nature and extent of the problem presented by the release," and

- to gather "sufficient information to determine the necessity for and proposed extent of remedial action."

In the context of this report, the first purpose is fulfilled in the problem assessment phase discussed earlier. The remedial investigation considered here fulfills the second purpose. The fact that these two purposes are considered together in the NCP, however, again indicates the importance of considering remedial response information requirements in the problem assessment phase.

Information gathered in the remedial investigation is used to evaluate the conclusions of the scoping process based on the factors given in Table 5. The process is therefore iterative, repeating the

31

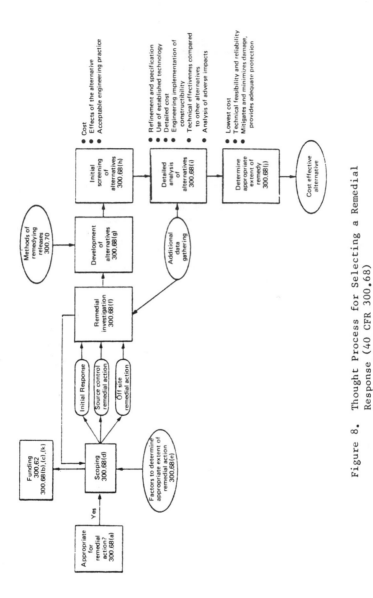

Figure 8.   Thought Process for Selecting a Remedial
Response (40 CFR 300.68)

Source: Arthur D. Little, Inc. 1982

TABLE 5. REMEDIAL ACTION CONSIDERATION (40 CFR 300.68(e))

Type of Action

Initial Remedial Measures

Factors to Consider

- Actual or potential direct contact with hazardous substances by nearby population.

- Absence of an effective drainage control system (with an emphasis on run-on control).

- Contaminated drinking water at the tap. (Measures might include the temporary provision of an alternative water supply.)

- Hazardous substances in drums, barrels, tanks, or other bulk storage containers, above surface posing a serious threat to public health or the environment. (Measures might include transport of drums offsite.)

- Highly contaminated soils largely at or near surface, posing a serious threat to public health or the environment. (Measures might include temporary capping or removal of highly contaminated soils from drainage areas.)

- Serious threat of fire or explosion or other serious threat to public health or the environment. (Measures might include security or drum removal.)

- Weather conditions that may cause substances to migrate and to pose a serious threat to public health or the environment.

Source Control Remedial Action

- The extent to which substances pose a danger to public health, welfare, or the environment. Factors which should be considered in assessing this danger include:

TABLE 5. (CONTINUED)

Type of Action	Factors to Consider
	(a) Population at risk; (b) Amount and form of the substance present; (c) Hazardous properties of the substances; (d) Hydrogeological factors (e.g., soil permeability depth to saturated zone, hydrologic gradients, proximity to a drinking water aquifer); and (e) Climate (rainfall, etc.).
	• The extent to which substances have migrated or are contained by either natural or manmade barriers.
	• The experiences and approaches used in similar situations by State and Federal agencies and private parties.
	• Environmental effects and welfare concerns.
Offsite Remedial Action	• Contribution of the contamination to an air, land, or water pollution problem.
	• The extent to which the substances have migrated or are expected to migrate from the area of their original location and whether continued migration may pose a danger to public health, welfare, or environment.
	• The extent to which natural or manmade barriers currently contain the hazardous substances and the adequacy of the barriers.
	• The extent to which substances pose a danger to public health, welfare, or the environment. Factors which should be considered in assessing this danger include: (a) Population at risk; (b) Amount and form of the substance present;

TABLE 5. (CONCLUDED)

Type of Action   Factors to Consider

                                 (c) Hazardous properties of the
            substances;
  (d) Hydrogeological factors (e.g.,
     soil permeability depth to
     saturated zone, hydrologic
     gradients, proximity to a
     drinking water aquifer); and
  (e) Climate (rainfall, etc.).

•   The experiences and approaches used
     in similar situations by State and
     Federal agencies and private
     parties.

•   Environmental effects and welfare
     concerns.

scoping process until its conclusions are consistent with information gathered in the remedial investigation.

The outcome of the preliminary assessment is a request for funding (as appropriate), a decision on the types of remedial response required, and information necessary to develop alternatives for action. In addition, initial remedial measures may be implemented at this stage.

2.   DEVELOPMENT OF ALTERNATIVES

Development of alternatives for remedial action involves selecting a limited number of alternatives "for either source control or offsite remedial actions (or both), depending upon the type of response that has been identified [in the preliminary assessment]" (40CFR 300.68(g)). These alternatives can be selected from the discussion provided in Section 5 of Part II. In addition, a "no-action" alternative may be assessed. "No-action alternatives are appropriate, for example, when response action may cause a greater environmental or health danger than no action," (40CFR 300.68(g)). The no-action alternative was considered by the Air Force in Case History (c) in Part II. The outcome of this stage is a list of potential alternatives for remedial action to be considered for use at the site.

3.   ANALYSIS OF ALTERNATIVES

Once a list of remedial alternatives has been developed, the alternatives must be analyzed so that the most appropriate alternative may be selected. This analysis involves two basic steps:

- initial screening; and
- detailed analysis.

Initial screening of alternatives is designed to eliminate alternatives which are clearly inappropriate to the given situation or are clearly inferior to other alternatives. It is based primarily on three factors:

- cost;
- effects of the alternative; and
- acceptable engineering practice.

Alternatives may be eliminated from consideration on the basis of cost if the alternative "far exceeds (e.g., by an order of magnitude) the costs of other alternatives evaluated and . . . does not provide substantially greater public health or environmental benefit" than other alternatives, (40CFR 300.68(h)(1)).

An alternative can also be eliminated from consideration at this stage if the effect of the "alternative itself or its implementation has any adverse environmental effects" or if the alternative is not "likely to achieve adequate control of source material . . .[nor] effectively mitigate and minimize the threat of harm to public health, welfare, or the environment," (40CFR 300.68(h)(2)). Groundwater pumping, for example, would not be considered appropriate if pumping would change hydrologic conditions causing contamination of adjacent aquifers.

An alternative may be eliminated from consideration on the basis of acceptable engineering practice if the alternative is not "feasible for the location and condition of release, applicable to the problem, . . . [or does not] represent a reliable means of addressing the problem." Chlorination of groundwater contaminated with organic waste, for example, would not be considered acceptable engineering practice.

Alternatives which remain after initial screening should be evaluated in more detail. This detailed analysis of each alternative should include (40CFR 300.68(i)(2)):

- refinement and specification of the alternative in detail;

- detailed cost estimation, including distribution of costs over time;

- determination of engineering constructability;

- assessment of technical effectiveness; and

- detailed analysis of adverse environmental impacts and methods (with costs) for mitigating these impacts.

Additional data gathering may be required to complete this analysis. In addition, laboratory or pilot scale studies may be required at this stage, particularly for treatment technologies.

4.   SELECTING AN APPROPRIATE REMEDIAL ACTION

Based on the results of the detailed analysis, the appropriate alternative(s) for remedial action may be selected. The NCP considers the most appropriate alternative to be "the lowest cost alternative that is technologically feasible and reliable, and which effectively mitigates and minimizes damage to and provides adequate protection of public health, welfare, or the environment," (40CFR 300.68(j)).

The result of this stage is the selection of appropriate cost-effective remedial actions to be implemented at the site. At any step in this process, as new information or data are manifested, it may be necessary to go back to previous steps and consider new types of response and new alternatives for action. This process, however, provides an effective way of approaching groundwater contamination problems to determine the appropriate extent and method of remedial action.

5.   RESPONDING TO REGULATORY REQUIREMENTS

Federal regulations for response to groundwater contamination are based primarily on the Resource Conservation and Recovery Act of 1976 (RCRA), and the Comprehensive Emergency Response, Compensation and Liability Act of 1980 (CERCLA or Superfund). These statutes, as well as others which apply to groundwater, are described in Section 2.4 of the Appendix. State and local regulations for response vary from state to state, and from municipality to municipality. Notification requirements based on these regulations are depicted in Figure 9.

Once groundwater contamination is discovered at a site, the source, extent, and other parameters of the contamination should be investigated. Procedures for discovering, investigating, and characterizing groundwater contamination are discussed elsewhere in the report. Notification requirements depend primarily on the location and source of the contamination. Appropriate state and local agencies (e.g., Board of Health, Public Health Department) should be notified if the contamination presents a threat to the local community, or if required by state or local regulation. Contamination discovered at a facility permitted under RCRA requires special notification procedures, dependent on whether the contamination is found upgradient or downgradient of the site. Contamination discovered at a facility not permitted under RCRA may require notification under CERCLA if a "reportable quantity" of waste, as specified in 40CFR 117.3, is determined to have been released. In this case, the National Response Center or designated alternate officials should be contacted. If a quantity less than the reportable quantity has been released, or if the source of contamination is unknown, the National Response Center should be notified to determine appropriate action. These notification procedures are discussed in more detail in Section 1(d) of Part II.

Response procedures for groundwater contamination at Air Force bases may change because of shifting regulatory policies. In particular, the EPA is currently preparing to give the states the lead role in groundwater protection and is considering giving defense facilities special status with respect to environmental regulations. The RCRA/Superfund Hotline (800-424-9346) can be called for information on new regulations pertaining to RCRA or CERCLA, or to answer questions about recommended response procedures.

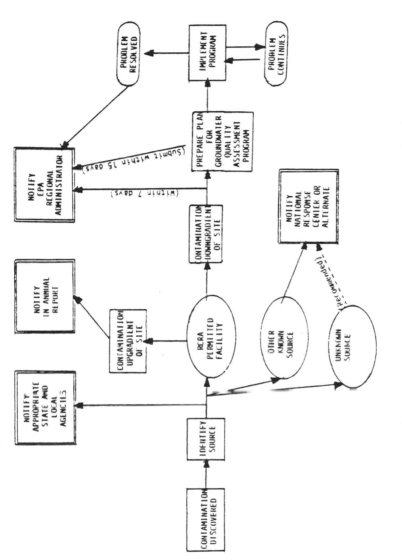

Figure 9. Regulatory Response to Groundwater Contamination

References

Arthur D. Little, Inc., "User Guide for Evaluating Remedial Action Technologies," 68-01-5949. Draft for EPA Municipal Environmental Research Laboratory, Cincinnati, OH., June 1982.

Lyman, W.J., W.F. Reehl and D.H. Rosenblatt (eds.), <u>Handbook of Chemical Property Estimation Methods</u>, McGraw-Hill, New York, 1982.

Part II

Groundwater Contamination Response Desk Reference

The information in Part II is from *Groundwater Contamination Response Guide. Volume II: Desk Reference,* by J.H. Guswa and W.J. Lyman of Arthur D. Little, Inc., for the U.S. Air Force Engineering and Services Center, June 1983.

Acknowledgments

This document was prepared by Arthur D. Little, Inc., Acorn Park, Cambridge, Massachusetts, for the Engineering and Services Laboratory, Air Force Engineering and Services Center (AFESC), Tyndall Air Force Base, Florida. The AFESC/RDVS project officer was Captain Glen E. Tapio.

Several volumes of technical material were critically reviewed and compiled during preparation of this report. Expert summaries were provided by the following individuals:

J. Adams	M. Mikulis
J. Bass	R. O'Neill
A. Brecher	A. Preston
D. Brown	C. Thrun
D. Leland	

1. General Description
of Groundwater Contamination

1.   OVERVIEW OF GROUNDWATER HYDROLOGY AND CONTAMINATION

   a. Groundwater Hydrology

   Groundwater is the name given to water moving through the
land-based portion of the hydrologic cycle (Figure 1). Water in the
ground fills pores in sediment or cracks in rocks and usually moves
slowly along indirect paths around each particle. About 4 percent of
the total world volume of water is groundwater; however, groundwater
constitutes between 68 percent and 99 percent (depending on the sources)
of all useable fresh water (Table 1). At present, the United States is
using 25-percent groundwater (water recovered from permeable aquifers)
and 75-percent surface water for its water supply (Figure 2). The
Western United States constitutes the bulk of groundwater use with 45-
percent groundwater and 55-percent surface water (Freeze and Cherry,
1979). The increasing need for clean sources of fresh water and the
long return period for replenishment or cleaning of groundwater explain
the recent emphasis on protecting the purity of groundwater.

   In the subsurface all gradations exist between freely flowing water
and water firmly fixed in the crystal structure of minerals. Figure 3
is a schematic of these gradations; however, there are no sharp
boundaries between the various water types. Soil water, readily
evaporated on a hot day, grades into intermediate vadose or suspended
water. Intermediate vadose water, in turn, grades slowly to capillary
water in silts and clays although there is a distinct boundary between
these water types in coarser grained sediments. The boundary between
vadose water and groundwater is known as the water table. Davis and
DeWiest (1966) define the water table as follows:

>   "The most common definitions of the water
> table state that it is the surface
> separating the capillary fringe from the
> 'zone of saturation,' or that it is the
> surface defined by the water levels in wells
> which tap an unconfined saturated material.
> A more exact definition states that the water
> table is the surface in unconfined material
> along which the hydrostatic pressure is equal
> to the atmospheric pressure."

The water below the water table is known as groundwater, the saturated
zone or phreatic water. Because the lower portions of the vadose zone
may be saturated with capillary water, phreatic water is the better
synonym for groundwater (Figure 3). Figure 4 shows the major water
zones superimposed on a topographic profile (Bear, 1979).

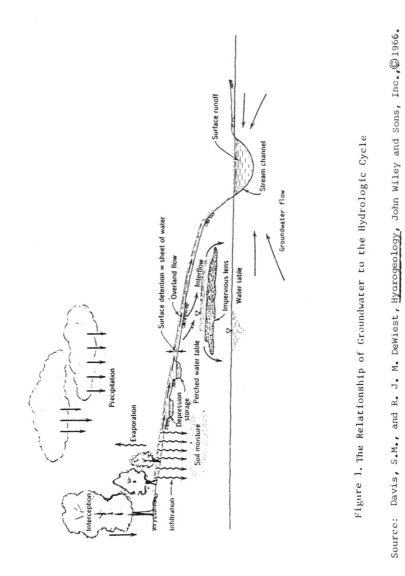

Figure 1. The Relationship of Groundwater to the Hydrologic Cycle

Source: Davis, S.M., and R. J. M. DeWiest, Hydrogeology, John Wiley and Sons, Inc.,©1966.

TABLE 1. ESTIMATE OF THE WATER BALANCE OF THE WORLD

Parameter	Surface area $(km^2) \times 10^6$	Volume $(km^3) \times 10^6$	Volume (%)	Equivalent depth (m)*	Residence time
Oceans and seas	361	1370	94	2500	~4000 years
Lakes and reservoirs	1.55	0.13	<0.01	0.25	~10 years
Swamps	<0.1	<0.01	<0.01	0.007	1–10 years
River channels	<0.1	<0.01	<0.01	0.003	~2 weeks
Soil moisture	130	0.07	<0.01	0.13	2 weeks–1 year
Groundwater	130	60	4	120	2 weeks–10,000 years
Icecaps and glaciers	17.8	30	2	60	10–1000 years
Atmospheric water	504	0.01	<0.01	0.025	~10 days
Biospheric water	<0.1	<0.01	<0.01	0.001	~1 week

*Computed as though storage were uniformly distributed over the entire surface of the earth.

Source: Freeze, R.A. and J. A. Cherry, Groundwater, Prentice-Hall, Inc., © 1979.

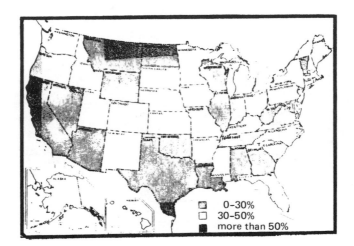

Figure 2.   Percent of Population Served by Groundwater

Source: Bartlet, R.E., "State Groundwater Protection Programs --
        A National Summary," <u>Ground Water</u>, Volume 17, Number 1,
        pp. 89-93,© 1979.

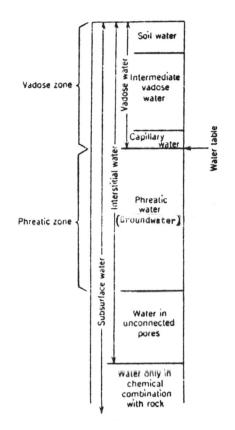

Figure 3. Classification of Subsurface Water

Source: Davis, S.M., and R.J.M. DeWiest, Hydrogeology,
         John Wiley and Sons, Inc.,© 1966.

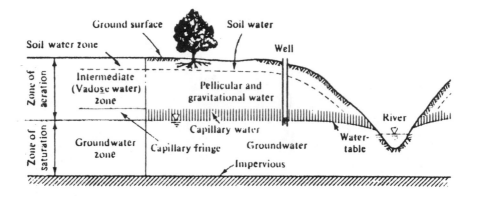

Figure 4.   The Distribution of Subsurface Water

Source: Bear, J., Hydraulics of Groundwater, McGraw-Hill Book
           Company, New York,© 1979.

Within the groundwater system, water occurs and moves through the void spaces of the geologic materials. These void spaces may be the intergranular porosity of unconsolidated materials or the fractures and solution cavities commonly found in consolidated material. The volume of water contained within any particular geologic unit and the ease with which it moves through that unit depends upon the frequency, size, and degree of interconnection between the void spaces. The term aquifer is applied to the geologic units through which water generally moves easily, and the term confining bed is applied to those units through which water movement is generally restricted. The term "leakage" is commonly used to describe the exchange of water between aquifers and confining beds.

Aquifers are subdivided into two general categories: confined or unconfined. Confined aquifers are commonly referred to as artesian aquifers and unconfined aquifers are commonly referred to as water table aquifers. The classification of confined or unconfined aquifers is based on the absence or presence of a water table or free water surface. An unconfined aquifer is one in which the water table acts as the upper surface of the zone of saturation. Confined aquifers are under pressure greater than atmospheric and bounded above and below by confining beds. The water level of a well which penetrates a confined aquifer will rise above the base of the overlying confining bed. The water level in the well is referred to as the piezometric or potentiometric level and corresponds to the hydrostatic pressure level of water in the aquifer. If the piezometric level is above ground level, a flowing well results.

Water moves through the ground from areas of recharge to areas of discharge. For unconfined aquifers, the source of water recharge that infiltrates the unsaturated zone is a portion of the precipitation that falls on land surface. For a confined aquifer, the source of water is commonly leakage through an overlying or underlying confining bed and from surface infiltration where the confined aquifer is exposed at land surface. Groundwater discharge areas are commonly surface water bodies, such as streams, marshes, and oceans.

As water moves downward through the unsaturated zone, it may encounter zones where its rate of downward movement is slowed. This can result in localized zones where all the void spaces are saturated with water. Such a condition is referred to as a perched water table condition. Figure 5 illustrates the types of aquifers and the relationship between aquifer and confining beds and recharge and discharge areas.

Aquifers are characterized by their ability to conduct water under a hydraulic gradient and by their ability to store water. Hydraulic conductivity (K) is a measure of the aquifer's ability to conduct water under a hydraulic gradient. It is a property of the geologic medium and the fluid flowing through it. Aquifer transmissivity (T) is a measure of the ability of an aquifer to transmit water through its entire thickness and is equal to the product of the conductivity and aquifer thickness (T=Kb). The storage coefficient (S) is a measure of the

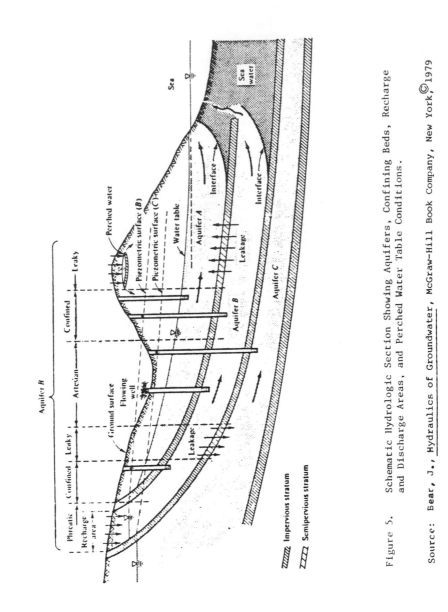

Figure 5.   Schematic Hydrologic Section Showing Aquifers, Confining Beds, Recharge
and Discharge Areas, and Perched Water Table Conditions.

Source: Bear, J., Hydraulics of Groundwater, McGraw-Hill Book Company, New York, ©1979

volume of water than an aquifer releases from or takes into storage in response to a change in piezometric head. For confined aquifers, the storage coefficient ranges from 5×10^{-3} to 5 to 10^{-5} and is a measure of the fluid and rock compressibility. For unconfined aquifers, the storage coefficient ranges from 0.1 to 0.3 and is an indicator of the aquifer pore space volume.

Groundwater movement can be predicted using Darcy's Law, which can be expressed in the following way:

$$q = \frac{Q}{A} = -K \frac{\Delta h}{\Delta X} \tag{1}$$

where

q is the Darcy velocity (L/T)
Q is the flow rate (L^3/T)
A is the cross sectional area perpendicular to the flow direction (L^2)
K is the hydraulic conductivity (L/T)
Δh is the total potential head change (L)
ΔX is the distance across which the head change occurs (L)

The Darcy velocity, q, is essentially unidirectional, is based on the total cross-sectional area, and does not represent the true velocity of the individual water particles that follow irregular paths around individual grains (Figure 6). The true velocity or seepage velocity is determined by dividing the Darcy velocity by the effective porosity of the geologic medium. If, for example, the effective porosity were 0.20, then the seepage velocity would be five times faster than the Darcy velocity. This is not the true velocity of every water particle because it does not consider the flow path followed but it is a good estimate of the average velocity.

Prediction of the rate and direction of movement of contaminants in groundwater, begins with evaluation of the simple flow system, specifically, the variables in Equation 1. The next step is to analyze the effect that the local geology has on the flow system. The CEQ (1981) states that:

> "The degree of threat to groundwater
> [becoming contaminated] depends on the
> material underlying the surface site and the
> particular geologic and hydrologic
> conditions. For example, a dump sited on top
> of a thick layer of impermeable clay poses
> little threat to an aquifer beneath it, but a
> landfill on permeable material is a serious
> threat."

As the critical variables are defined at a particular site, the simple flow law (Darcy's law) is expanded upon to better model the given

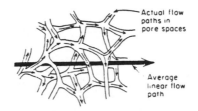

Figure 6.   Average Linear Velocity Compared
             With Actual Flow Paths of
             Groundwater Around Grains

Source: Freeze, R.A. and J.A..Cherry, <u>Groundwater</u>, Prentice-Hall,
          Inc., © 1979.

situation. With every increase in required input data, more extensive field methods must also be used to fully evaluate the site.

The subsequent sections of this report are designed to show how each of the preceding variables as well as variations in properties of the contaminants themselves can affect the overall transport of contaminants in a groundwater system. Defining and evaluating the most critical variables at a given site are the most important steps in defining the extent of contamination and predicting future movement of contaminants.

b. Groundwater Contamination

Because groundwater is becoming so widely used for drinking water, preventing and detecting its contamination has become increasingly important. Although filtration and chemical reactions between contaminants and sediment or rock cleanse many potential pollutants from percolating groundwater, larger concentrations and more inert chemicals often preclude this self-cleansing mechanism (CEQ, 1981). Soluble contaminants may move more quickly in groundwater than the average water flow thus causing a large volume of groundwater to be polluted. Insoluble or immiscible fluids often do not move as rapidly as the average water flow, and may, therefore, persist as in contaminants in an area for a long time.

The normal components of clean groundwater may include any of the long list of possible inorganic ions listed in Table 2. The levels of any specific chemicals considered objectionable depend on the use of the water. Tables 3 and 4 list concentration limits for various inorganic and organic water constituents. The list for drinking water standards (Table 3) is longer and has smaller recommended concentrations than Table 4, the standard for agricultural water. Water with concentrations exceeding these limits would be considered polluted for that use. The severity of the problem of groundwater pollution is highlighted by the rapid increase in development of manmade organic compounds to a total number now near 2 million (Freeze and Cherry, 1979).

Organic chemicals make their way to the land surface as potential groundwater contaminants as a result of the use of pesticides, the use of land for sewage disposal, the use of sanitary landfills or refuse dumps for disposal of organic compounds, burial of containers with organic compounds at special burial sites, leakage from liquid waste storage ponds, and accidental spills along highways, or storage and handling areas. Figure 7 shows the interactions between sources of organic chemicals and the hydrologic cycle.

TABLE 2.   CLASSIFICATION OF DISSOLVED INORGANIC
CONSTITUENTS IN GROUNDWATER

Major constituents (greater than 5 mg/ℓ)

Bicarbonate	Silicon
Calcium	Sodium
Chloride	Sulfate
Magnesium	Carbonic acid

Minor constituents (0.01–10.0 mg/ℓ)

Boron	Nitrate
Carbonate	Potassium
Fluoride	Strontium
Iron	

Trace constituents (less than 0.1 mg/ℓ)

Aluminum	Molybdenum
Antimony	Nickel
Arsenic	Niobium
Barium	Phosphate
Beryllium	Platinum
Bismuth	Radium
Bromide	Rubidium
Cadmium	Ruthenium
Cerium	Scandium
Cesium	Selenium
Chromium	Silver
Cobalt	Thallium
Copper	Thorium
Gallium	Tin
Germanium	Titanium
Gold	Tungsten
Indium	Uranium
Iodide	Vanadium
Lanthanum	Ytterbium
Lead	Yttrium
Lithium	Zinc
Manganese	Zirconium

Source: Freeze, R.A. and J.A. Cherry, Groundwater, Prentice-Hall, Inc., © 1979.

TABLE 3   STANDARDS FOR DRINKING WATER

Constituent	Recommended concentration limit* (mg/ℓ)
Inorganic	
Total dissolved solids	500
Chloride (Cl)	250
Sulfate (SO$_4^{2-}$)	250
Nitrate (NO$_3$)	45†
Iron (Fe)	0.3
Manganese (Mn)	0.05
Copper (Cu)	1.0
Zinc (Zn)	5.0
Boron (B)	1.0
Hydrogen sulfide (H$_2$S)	0.05
	Maximum permissible concentration‡
Arsenic (As)	0.05
Barium (Ba)	1.0
Cadmium (Cd)	0.01
Chromium (Cr^{VI})	0.05
Selenium	0.01
Antimony (Sb)	0.01
Lead (Pb)	0.05
Mercury (Hg)	0.002
Silver (Ag)	0.05
Fluoride (F)	1.4–2.4§
Organic	
Cyanide	0.05
Endrine	0.0002
Lindane	0.004
Methoxychlor	0.1
Toxaphene	0.005
2,4 D	0.1
2,4,5-TP silvex	0.01
Phenols	0.001
Carbon chloroform extract	0.2
Synthetic detergents	0.5
Radionuclides and radioactivity	Maximum permissible activity (pCi/ℓ)
Radium 226	5
Strontium 90	10
Plutonium	50,000
Gross beta activity	30
Gross alpha activity	3
Bacteriological	
Total coliform bacteria	1 per 100 mℓ

SOURCES: U.S. Environmental Protection Agency, 1975 and World Health Organization, European Standards, 1970.

*Recommended concentration limits for these constituents are mainly to provide acceptable esthetic and taste characteristics.

†Limit for NO$_3$ expressed as N is 10 mg/ℓ according to U.S. and Canadian standards; according to WHO European standards, it is 11.3 mg/ℓ as N and 50 mg/ℓ as NO$_3$.

Source: Freeze, R.A. and J.A. Cherry, Groundwater, Prentice-Hall, Inc., © 1979.

TABLE 4.   RECOMMENDED CONCENTRATION LIMITS FOR WATER USED
FOR LIVESTOCK AND IRRIGATION CROP PRODUCTION

	Livestock: Recommended limits (mg/ℓ)	Irrigation crops: Recommended limits (mg/ℓ)
Total dissolved solids		
Small animals	3000	700
Poultry	5000	
Other animals	7000	
Nitrate	45	—
Arsenic	0.2	0.1
Boron	5	0.75
Cadmium	0.05	0.01
Chromium	1	0.1
Fluoride	2	1
Lead	0.1	5
Mercury	0.01	—
Selenium	0.05	0.02

Source: Freeze, R.A. and J.A. Cherry, Groundwater, Prentice-Hall,
Inc.,© 1979.

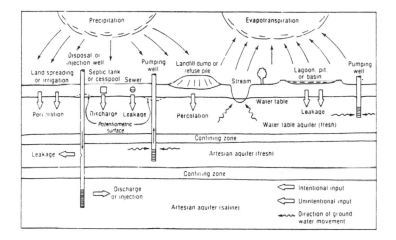

Figure 7.   Sources of Groundwater Contamination

Source: CEQ, 1981.

2.   CAUSES AND PREVENTION OF GROUNDWATER CONTAMINATION

   a. Introduction

   Groundwater is contaminated by the movement of pollutants through
an aquifer recharge zone into the aquifer. Pollutants percolate through
a recharge zone which usually includes a soil layer. The soil layer may
partially cleanse percolating water of contaminants by biologic
degradation, adsorption, and ion exchange processes. However, soils are
not capable of removing many synthetic organic compounds such as
chlorinated solvents. In fact, a study in New Jersey found that
groundwater toxic organic contamination paralleled surface water toxic
organic contamination (Page, 1981). Groundwater contamination can also
be caused by the following: contact between contaminated surface waters
and groundwater (i.e., wetlands); disposal of contaminants below the
high groundwater table (i.e., deep-well disposal); and subsurface
discharges from leaky pipes, storage tanks, etc. (Freeze and Cherry,
1979).

   Contaminants in a groundwater system will slowly disperse in and
move with the groundwater. Typically groundwater moves at a rate of 5
to 500 feet per year (Environmental Science and Technology, 1980).
Groundwater can be pristine within a few hundred feet of contaminated
water because the laminar flow characteristics of groundwater are not
conducive to mixing and dilution of contaminants (CEQ, 1981). Although
groundwater moves slowly, contaminants may travel long distances over
long periods of time since little degradation or dilution takes place in
the anaerobic groundwater environment.

   Landfills and chemical dump sites have received the most attention
as a sources of groundwater contaminants. The disposal of industrial
wastes in impoundments and solid waste sites was identified in a 1977
EPA report as the major cause of groundwater contamination (U.S. EPA,
1977a). Other significant sources of contamination include:

   - surface impoundments (e.g., wastewater lagoons);
   - mines;
   - septic systems;
   - underground petroleum storage tanks;
   - spills;
   - the intentional application of pesticides; and
   - runoff from paved areas and piles (e.g., coal,
     salt).

A number of these sources are depicted in Figure 7.

   Prevention of groundwater contamination requires that contaminants
not be allowed to travel through an aquifer recharge zone and that
spread of contaminants be contained. Preventive measures include:

- removal of contaminants from the recharge zones;

- restriction or diversion of water movement with barriers to isolate the contaminants; and

- collection of the contaminated leachate.

In this section we describe, in general terms, the causes and prevention of groundwater contamination. Prevention of groundwater contamination due to past disposal practices is the focus rather than protection strategies at new disposal facilities. Examples are cited wherever possible.

An extensive bibliography of reports, manuals, and books on spill control and groundwater contamination prevention is included at the end of this report.

b. Sources

(1) Land Disposal

(a) Causes

Groundwater may be contaminated by migration of surface water associated with the land disposal site into the aquifer and/or percolation of liquid waste disposed at the site into the aquifer (Figure 7). An unlined land disposal facility situated over permeable geologic material in an aquifer recharge zone is likely to cause groundwater contamination. Many land disposal facilities sited prior to the Resource and Recovery Act (RCRA) may be in locations where the hydrogeologic environment is conducive to transport of contaminants to the groundwater. A study of 50 industrial disposal sites indicated that a large number of the approximately 50,650 active and inactive land disposal sites may be contaminating groundwater (CEQ, 1981). Groundwater at 40 of the 50 sites surveyed during the study contained organic solvents, benzene, and chlorinated phenols.

(b) Prevention

Containment of contaminants above the groundwater table or isolation of the contaminants from the groundwater flow are prevention methods. The location of the contaminants in relation to the groundwater table is an important containment consideration. If the contaminants are above the water table, the reduction of surface infiltration and/or collection of leachate and other percolating liquids can be used as prevention techniques. Available technologies to reduce surface infiltration include surface seals, surface water diversions and graded surfaces.

Lowering of the water table and groundwater diversion techniques can prevent contamination below the groundwater table. Groundwater may be adjusted to below the contamination level by either pumping or subsurface drains. It is necessary to ensure contaminants do not percolate down to the lowered water table level. Diversion techniques include upgradient and downgradient diversions or barriers. Upgradient diversions or barriers prevent groundwater from contacting the contaminants while downgradient diversions contain groundwater which has contacted the contaminants. Available diversion and barrier technologies include slurry walls, grout curtains, and sheet pile barriers. Diversion techniques may also be applicable to contaminants located above the groundwater table.

(2) Surface Impoundments

(a) Causes

Leakage of contaminants through the bottom of surface impoundments is the principal cause of groundwater contamination associated with this source. An estimated 26,000 industrial surface impoundments are currently in operation (Wyss, 1980). A survey of over 8000 sites indicated that: 50 percent of the impoundments may contain hazardous contaminants; and 10 percent are unlined, overlie permeable material, and are within 1 mile of a water supply well (CEQ, 1981).

(b) Prevention

Lining surface impoundments to prevent leakage is a relatively recent engineering practice. Low permeability clays or synthetic materials are typically used for liners. If a liner begins leaking it may be possible to repair; however, liner repair feasibility has not yet been demonstrated. Therefore, lined surface impoundments which leak should be treated as unlined facilities in most cases.

The first step toward preventing a leaking surface impoundment from contaminating groundwater is generally to drain the impoundment. After dewatering, prevention of further groundwater contamination will require the same technologies as those cited in the land disposal section.

(3) Mines

(a) Causes

Mines have been used for the disposal of hazardous materials in the past. Groundwater flows to subsurface mine walls, through the mine caverns where it can entrain contaminants, and back to a surface or groundwater system.

(b) Prevention

Groundwater barriers may be an effective technique, but implementation would require an extensive hydrogeologic survey because of the complexity of water movement in mines. Lowering the water table adjustment may be a more cost-effective prevention alternative.

(4) Septic Systems

(a) Causes

Many homeowners clean septic systems with fluids which contain such chemicals as trichloroethylene (TCE), benzene, or methylene chloride. The fluids dissolve sludge in the septic system so it is transported through the soil absorption field. Both the dissolved sludge and the cleaning fluid can percolate down to the groundwater levels along with other mobile components of the leachate. Each homeowner may use less than a gallon per cleaning but the cumulative use by 19.5 million owners of onsite disposal   systems (CEQ, 1980) can be significant. For example, the aggregate annual use of septic system cleaning fluids on Long Island, New York was on the order of 400,000 gallons (CEQ, 1980).

(b) Prevention

Septic system cleaning fluids pose a significant groundwater contamination prevention problem because of the great number of small sources which must be controlled. Eliminating the use of septic cleaners which contain potential groundwater contaminants is probably the most cost-effective prevention measure which can be taken. The State Legislation of New York has considered banning the use of septic cleaners which contain certain organic solvents (NYDEC, 1979).

(5) Underground Petroleum Tanks

(a) Causes

Leaking underground petroleum tanks pose a groundwater contamination threat. Gasoline contains aromatic hydrocarbons and lead which can be hazardous to humans if consumed. For example, in Nassau County, New York, 36 leaking gasoline storage tanks were identified in 1979 (NYDEC, 1979).

(b) Prevention

Groundwater pumping to collect and purify the groundwater near a leaky tank is the usual course of action (NYDEC, 1979). The tank itself may be excavated and repaired or replaced to prevent a chronic groundwater contamination.

(6) Spills

(a) Causes

Spills of toxic materials generally occur during transportation or storage of the materials. A spill in a hydrogeologic area favorable for quick migration to a groundwater system can pose a major problem.

(b) Prevention

When spills are reported promptly, many techniques exist to clean them up. Cleanup technologies include removal and in-situ treatment. The U.S. Environmental Protection Agency maintains a technical assistance team to respond to spill cleanup needs.

3. CASE HISTORIES

The following three case histories provide examples of how groundwater contamination incidents were identified, the problems assessed, and remedial actions proposed or undertaken at three Air Force Bases. They represent contamination by trichloroethylene (TCE), benzene, and petroleum, oil, lubricant (POL).

a. Case Study #1: TCE Contamination

(1) Background

Trichloroethylene (TCE) contamination of groundwater was discovered in 1977, at a midwestern Air Force base. The Base overlies a sand and gravel aquifer which overlies a thick clay layer at an average depth of 65 feet. The water table ranges from 10 feet below the land surface at the western part of the base to 25 feet at the eastern part. The TCE contaminated groundwater flows northeast under the influence of the natural groundwater gradient and base water supply wells toward a nearby lake.

TCE is a degreasing solvent and is a suspected carcinogen. The U.S. Environmental Protection Agency (EPA) estimates that the incremental increase of cancer risk over an individual's lifetime is 10^{-6} when drinking water concentrations of TCE are 2.7 micrograms per liter (USEPA, 1980). According to information supplied by the Air Force, the EPA is currently considering a TCE drinking water standard of 4.5 micrograms per liter, and the National Academy of Science suggests a maximum concentration of 270 micrograms per liter in drinking water at USAF bases.

(2) Problem Identification

In October 1977, the Base Civil Engineering Squadron received complaints about the peculiar odor and taste of the drinking water at the base. Analysis of faucet water showed a TCE concentration of 1,100 micrograms per liter. In addition, analysis of water taken directly from a supply well in the eastern part of the base showed a TCE concentration of 6,700 micrograms per liter.

The suspected source of contamination was an underground 500-gallon tank used for temporary storage of waste TCE. Approximately 5,000 gallons of TCE had been added to the tank, and an unspecified amount had been pumped out of the tank since 1962. Upon discovery of contamination, the tank was excavated and a leak discovered where the filler pipe connected the tank. Further investigation of base water supply wells indicated TCE contamination at various locations throughout the base. Analyses of offbase water supplies showed no contamination except for one well located 1,000 feet east of the base. These analyses and the discovery of a leak in the TCE storage tank pointed to the tank as the major source of TCE contamination.

Following the discovery of TCE contamination, use of water supply wells in the eastern portion of the base was discontinued in favor of wells in the southern portion. In January 1978, two of these wells were again put into use when they were found to have only trace amounts of TCE contamination. Later in 1978, TCE was detected in the southern wells, and their use was also discontinued. Only one other water supply well was affected by TCE.

(3) Problem Assessment

Late in 1979, the U.S. Geological Survey (USGS) began collecting data with the installation of wells in areas of known contamination. Over the study period, the USGS installed 165 wells -- 116 four-inch diameter wells used for pumping tests, water level measurements, and collection of water samples for analysis; and 49 wells 1 1/4 inches in diameter, used only to measure water level. Information from the wells and other sources of data was used to:

1) determine the rate and direction of groundwater movement;

2) determine the horizontal and vertical extent of TCE contamination;

3) investigate all suspected sources of TCE, past and present; and

4) develop and calibrate mathematical models to predict groundwater and TCE movement and provide information in deciding how to remove the contamination.

This allowed the USGS to develop an effective remedial program.

The Air Force and USGS investigations also revealed other contamination problems at the base which were unrelated to the TCE storage tank leak. These included TCE contamination from other sources, dichloroethylene contamination from an unknown source, and contamination of groundwater by a fuel substance. Although these problems were discovered in the course of the investigation, they will not be further discussed.

(4)   Remedial Action

Initial efforts to control TCE contamination were made prior to the involvement of the USGS. In March 1978, two of the eastern water supply wells were pumped to remove (or purge) TCE-contaminated water. Purged water was treated in aeration reservoirs near the water supply treatment plant. In addition, three new purge wells were added in May 1978, and three more in August 1979. These wells had removed approximately 535 gallons of TCE by June 1980. Their effectiveness, however, had greatly decreased. While 215 gallons of TCE were purged from September 1978 to August 1979, only 75 gallons were purged between September 1979 and August 1980.

Based on USGS analyses, therefore, three new purge well sites were identified. Groundwater models were used to determine the optimal pumping rates for the purge well system and predict its effect on TCE contamination. The models indicated that the central part of the plume would be lowered 15 feet and that water north, east, and south of the most highly contaminated zone would be drawn into the wells. TCE, therefore, would not escape eastward toward the lake. The three new purge wells have recently been constructed and have been operating since May 1982. No information on their effectiveness is yet available.

b.   Case Study #2: Benzene Contamination

(1)   Background.

Benzene contamination of groundwater was discovered in the late 1970s at an Air Force base in the northern United States. Benzene is used in the synthesis of organic chemicals and as a solvent and degreasing agent. It is a component of aviation fuels, gasoline, lacquers, and paints. The EPA estimates that the incremental increase of cancer risk over an individual's lifetime is 10^{-6} when drinking water concentrations of benzene are .66 micrograms per liter (USEPA, 1980a). Benzene contamination is considered significant in this case because it is thought to indicate the presence of JP-4 fuel or gasoline in the groundwater.

(2)   Problem Identification

Benzene concentrations exceeding 1000 micrograms per liter were discovered in a well during the investigation of another groundwater problem at the base. Water from an adjacent deeper well had benzene concentrations from 96 to 197 micrograms per liter. This indicated local contamination caused by surface storage. Laboratory analyses showed that the water not only contained benzene, but toluene and other unidentified organic compounds as well.

The suspected source of contamination was the base motor pool which has surface tanks for bulk storage of JP-4 fuel. Samples of JP-4 fuel

and gasoline from the motor pool were analyzed to try to verify the
source of contamination. These results were inconclusive because of the
decomposition of the organic compounds in the groundwater. Comparison
of gas chromatograph spectra, however, suggested JP-4 fuel as the most
likely contaminant.

(3) Problem Assessment

Current data available are insufficient to predict the distribution
and movement of contaminated groundwater. Only the general direction of
groundwater flow is known. Additional work is required to adequately
assess the contamination problem at this site. This will include
installing deep and shallow wells in the path of groundwater flow and
developing a mathematical flow model. These will be used both to
predict the distribution and movement of contaminated groundwater and to
design and evaluate purge pumping schemes.

(4) Remedial Action

No remedial action activities have been carried out at the site to
date. However, purge pumping wells may eventually be installed to
remove the contaminated groundwater.

c. Case Study #3: POL Contamination

(1) Background

POL and other types of groundwater contamination were discovered at
a number of sites at an Air Force base in an arid region. The Base is
located over a major aquifer system which has complex hydrologic
conditions, including fluctuating water levels and continually changing
aquifer conditions. Extensive groundwater pumping in the region has a
significant effect on the pattern of groundwater flow.

Three sites located at the Base are of particular interest. Site A
is the location of an underground POL storage tank which leaked
unspecified amounts of POL to groundwater. Site B is the location of a
jet fuel pipeline break which occurred in the late 1960s. Approximately
250,000 gallons of jet fuel leaked to the ground surface. 100,000
gallons of this may have been recovered, leaving 150,000 gallons
available for seepage to groundwater. There is, however, no permanent
water table below the site. Site C is the location of a leaky hydrant
at the end of a fuel line. An estimated 50,000 gallons of jet fuel
leaked to the ground surface and possibly to groundwater in the
mid-1970s. These sites represent the major sources of groundwater
contamination at the base.

(2) Problem Identification

The Base is located in an area of significant groundwater
resources. It was, therefore, important to identify potential sources
of groundwater contamination due to spills (such as Sites B and C) and

disposal practices (Site A). These investigations were the result of
Air Force concern about known sources of potential contamination of a
sensitive groundwater area.

(3) Problem Assessment

The first stage of problem assessment was to identify potential
contamination of groundwater and soil at these sites. This involved
installing three monitoring wells at Sites A and C, and two soil borings
at Site B downgradient of the sites. Drilling locations were based on
local topography, geology, available well logs, and USGS water level
data.

Based on the results of analyses of samples from wells and soil
borings, additional wells or borings may be required before a remedial
action decision can be made. For example, if wells at Site A indicate
POL contamination, more wells will be required to delineate the areal
extent of contamination. Additional soil borings would also be required
at Site B to determine the extent of soil contamination.

(4) Remedial Action

Two basic options exist for remedial action at these sites. If the
site is found to have POL or fuel floating on the groundwater surface,
these materials could be removed by means of a double pumping system.
One pump in the double pumping well lowers the water table around the
well while the other, a skimming pump floating on the water surface,
removes the floating contaminants.

On the other hand, if the site is contaminated with organic vapors,
a "no action" plan may be used. Over time, natural diffusion processes
should release these vapors to the atmosphere. The time involved,
however, may be extensive, and the vapors may continue to represent a
potential source of groundwater contamination. Although expensive,
three-dimensional computer simulations of the diffusive processes, based
on detailed geologic data, can be developed to estimate the amount of
time required. The "no action" option could also be used if the
contamination is immobile or if the potential for further contamination
is not too severe. Periodic monitoring is required for this option.
Specific remedial action options proposed for these three sites include:

(a) Site A

- seal abandoned wells which may cause
  contamination of deeper aquifers used for base
  water supply;
- empty the remaining underground storage tanks
  to avoid the potential of additional leakage;
- option of either removing the POL by a double
  pumping system or performing no remedial action
  with monitoring. Choice depends on the

characteristics of the contamination, as discussed above.

(b) Site B

• **Alternative 1**: If the fuel is confined by geologic conditions (a bedrock "knob") injection of water could cause the fuel to rise to the top of the water where it could be recovered by a double-pumping system.

• **Alternative 2**: Accelerate the vaporization of the fuel by moving air through the soil (air sweeping) and collecting the vapors by means of a vent system.

• **Alternative 3**: No remedial action with monitoring.

(c) Site C

Either remove the fuel by air surface of water bodies or moist soil, or in precipitation, acquire enough energy through solar radiation to escape the liquid phase and pass into the gaseous state. Sublimation differs from this only in that the water molecules are converted from the solid phase (snow or ice) directly to vapor, without passing through the liquid form. Transpiration is the process by which water absorbed by vegetation is evaporated into the atmosphere from plants' surfaces. When measuring the amounts of water being circulated into the atmosphere, it is usually very difficult to distinguish how much is contributed solely by evaporation and how much is contributed solely by transpiration. The two processes, therefore, are often considered together as evapotranspiration.

4. REGULATORY SUMMARY

a. Introduction

The guidelines for responding to groundwater pollution are based on the legislation which gives the federal government (primarily through the EPA) and the states authority and responsibility to control groundwater pollution. The primary federal legislation which pertains to groundwater is summarized in Table 5 and described briefly below.*

   b.   Federal Statutes

      (1)   Resource Conservation and Recovery Act of 1976
            (RCRA)

RCRA amends the 1965 Solid Waste Disposal Act and outlines the federal government's program to manage solid and hazardous wastes and establish standards for treatment, storage, and disposal of hazardous wastes. The Act's guidelines are directed toward the protection of water users rather than groundwater, but the federal government's role in controlling the sources of groundwater contamination (land disposal of municipal waste and all aspects of hazardous waste) was increased. RCRA provides for the permitting of disposal facilities and a "cradle to grave" manifest system for hazardous wastes.

In addition, the new regulations for land disposal facilities (40 CFR 264) set groundwater protection standards for new and existing facilities. This standard has four parts (Inside EPA, 1982):

   ● Hazardous chemicals are to be monitored and removed if necessary.

   ● Maximum concentration limits established in the Safe Drinking Water Act will be used as groundwater standards where possible. Otherwise, there is to be "no increase over background levels."

   ● Standards must be met at the edge of the waste management area (the compliance point).

   ● If standards are exceeded, a corrective action program must be submitted and implemented until the standard has not been exceeded for a period of 3 years (the compliance period).

      (2)   Comprehensive Emergency Response, Compensation and
            Liability Act of 1980 (CERCLA, or Superfund)

CERCLA authorizes federal and state governments to remove hazardous substances and wastes and perform remedial actions at sites that are a danger to public health and welfare or the environment. Section 101(3) defines the environment as including " . . . groundwater, drinking water

---

*These summaries are based primarily on "The Federal Response to Ground Water Protection" by Kevin McCray, Waterwell Journal, Volume 36, Number 6, pp. 42-3; and memoranda between the EPA Administrator and the Ground Water Policy Group, published in the Environmental Reporter, The Bureau of National Affairs, Inc., 6/25/82, pp. 292-3.

TABLE 5. SUMMARY OF FEDERAL LEGISLATION PERTAINING TO GROUNDWATER

Act	Emphasis	Applicability to Groundwater Response
RCRA	Sources of contamination	Monitoring and cleanup requirements at disposal facilities
	Controlled sites	Response to contamination at RCRA-permitted sites
CERCLA	Remedial Action	Response to contamination at sites not permitted under RCRA
	Uncontrolled sites	
Clean Water Act	Protection of surface waters	Notification requirements for spills, any discharge of hazardous wastes or waste constituents in reportable quantities
		Establishes reportable quantities for hazardous materials
Safe Drinking Water Act	Sets maximum concentration levels	Regulates use of injection wells
		Protection of sole-source aquifers
Surface Mining Control and Reclamation Act	Protection from adverse effects of mining operations	Hydrologic studies required
		Provides for alternate water supply when mining disrupts groundwater supply of an adjacent landowner

TABLE 5. (CONCLUDED)

Act	Emphasis	Applicability to Groundwater Response
TSCA	Manufacture, distribution and use of hazardous materials	Gives EPA regulatory authority over hazardous materials which may affect the environment
Uranium Mill Tailings Radiation Control Act	Active and inactive uranium mill tailings site	Establishes standards for all environmental media
FIFRA	Pesticide control	Gives EPA responsibility to control pesticides

supply, land surface or subsurface strata, or ambient air within the United States . . ."

(3) Safe Drinking Water Act of 1974

This Act authorizes EPA to set maximum contaminant levels and monitoring requirements for public water systems. It also regulates the uses of underground injection wells to protect drinking water aquifers and provides for the protection of sole-source aquifers. The EPA Administrator may designate an aquifer as the sole or principal drinking water source if contamination "would create a significant hazard to public health." According to the Act, no federally funded projects may be constructed which would lead to the contamination of a designated sole-source aquifer.

(4) Clean Water Act

This statute is actually a series of amendments made in 1972 and 1977 to the Federal Water Pollution Control Act of 1948. The stated purpose of the Act is to "restore and maintain the chemical, physical and biological integrity of the nation's waters." Its emphasis, however, is on surface ("navigable") waters. The Act seeks to eliminate the discharge of pollutants into navigable waters by 1983, and establish national policies to prohibit the discharge of toxic pollutants in toxic amounts and to develop area-wide waste treatment management planning. Discharge to groundwater must be considered in the Comprehensive Programs for Water Pollution Control (Section 102) and the Area-wide Waste Treatment Management Plans (Section 208). Section 402 establishes a requirement that the states control the discharge of pollutants into wells, and Section 303 requires the states to establish groundwater quality standards where it is shown that groundwater has a "clear hydrologic nexus" with surface water. The Act also establishes notification requirements for hazardous chemical spills.

(5) Surface Mining Control and Reclamation Act of 1977

This Act gives the Department of Interior authority to protect the public and the environment from the adverse effects potentially caused by surface and underground mining operations. Hydrogeologic studies are required prior to the covering or burial of hazardous materials as well as when mines are to be used for the disposal of any type of waste material. In addition, if mining activities seriously disrupt the groundwater or surface water supply of an adjacent landowner, an alternative water supply must be provided.

(6) Toxic Substances Control Act (TSCA)

This Act authorizes the EPA to restrict or prohibit the manufacture, distribution, and use of products which may adversely affect health and the environment. While groundwater is not specifically mentioned, it is assumed that "the environment," which is defined in Section 6(e), includes groundwater.

(7) Uranium Mill Tailings Radiation Control Act

This Act establishes health and environmental standards for active and inactive uranium mill tailings sites. The standards protect "public health, safety and the environment" and apply to all media, both above and below ground, including groundwater.

(8) The Federal Insecticide, Fungicide, and Rodenticide Act (FIFRA)

This Act gives EPA responsibility for the control of pesticide use. Environmental impacts of pesticide use must be considered in the registration process, including effects on groundwater quality.

c. Notification Requirements

Response to groundwater pollution problems depends primarily on the types of facilities involved. Three possibilities are likely:

(1) Discovery of Groundwater Contamination While Monitoring an RCRA-Permitted Facility

This case falls under the RCRA regulations found in 40 CFR 265.93-4.

- ·If contamination is in wells upgradient of the site (i.e., background concentrations have changed) this should be reported in the annual report required under 40 CFR 265.75.

- If contamination is found and confirmed in downgradient wells, the EPA Regional Administrator must be given written notice, within 7 days of the confirmation, "that the facility may be affecting groundwater quality." In addition, within 15 days after the notification, a plan for a groundwater quality assessment program which has been certified by a qualified geologist or geotechnical engineer must be submitted to the Regional Administrator. The plan must specify:

  - the number, location, and depth of wells to be used;

  - sampling and analytical methods for those hazardous wastes or hazardous waste constituents in the facility;

- evaluation procedures, including use of previously gathered groundwater quality information; and

- a schedule of implementation.

The plan must then be implemented and used to determine:

- the rate and extent of migration of the waste or constituents in the groundwater; and

- the concentrations of waste or constituent in the groundwater.

Within 15 days after the determination is made, a written report must be submitted to the Regional Administrator containing an assessment of the groundwater quality. If no waste or constituents are determined to have entered the groundwater, this is indicated in the above report and the assessment program is discontinued. If wastes or constituents have entered the groundwater, then groundwater quality assessments and reports to the Regional Administrator must be continued on a quarterly basis until final closure of the facility.

(2) Discovery of Groundwater Contamination Related to a Facility not Permitted Under RCRA

This case falls under 103a of CERCLA. If one pound or more of hazardous waste or a reportable quantity of wastes specified under 40 CFR 117.3 is determined to have been released in a 24-hour period, the National Response Center should be notified "by means of rapid communication." If notification in this manner is impractical, alternate officials have been designated. Officials to notify, therefore, in order of priority are:

1. Duty Officer, National Response Center, U.S. Coast Guard, 400 7th Street, S.W., Washington, D.C. 20590. 800-424-8802.

2. The On-Scene Coordinator as specified in the Regional Contingency Plan for the geographic area in which the discharge occurs.

3. Commanding Officer Office-in-Charge of any Coast Guard unit in vicinity of the discharge.

4. Commander of the Coast Guard district in which the discharge occurs.

This notification is also appropriate for hazardous material spills or any reportable hazardous materials discharge.

(3)   Discovery of Groundwater Contamination not Related
      to Any Known Source

In this case, neither RCRA nor CERCLA directly apply. However, it is recommended that the procedure in Case 2 be followed so that the National Response Center can determine which government agencies to notify.

In addition to notification requirements under federal legislation, states may also have notification requirements. The state Board of Health or the Public Health Department should be contacted to find out the specific requirements of the state in which the facility is located.

d.   Developing Issues

Two policies which are now being developed by the EPA may also affect the regulations of groundwater and the response to groundwater pollution. The first is the EPA's Ground Water Protection Policy, expected to be released by September 30, 1982. This policy is expected to emphasize that "states should have the lead role in groundwater protection." Each state will be asked to develop a groundwater protection strategy "commensurate with each state's own needs" by the end of FY 1984 (Environmental Reporter, 6/25/82, pp. 290-293). These strategies may make notification of groundwater pollution more specific and more stringent at the state level.

The second policy is being developed by the EPA in conjunction with the Department of Defense. The goal of the policy is to protect the environment while taking into account "the important national security ramifications" of environmental issues. This may make response requirements for defense facilities less stringent in certain circumstances than for other facilities. The EPA and key Congressional leaders, however, are opposed to exempting the Defense Department from complying with environmental laws. The Department of Defense currently has a special status regarding environmental response. It has primary authority for responding to releases of hazardous substances for its own facilities and cannot use Superfund monies for long term remedial action activities. It can, however, use Superfund for emergency response (Environmental Reporter, 5/28/82, p. 91). Response procedures to groundwater pollution could eventually be affected by this policy change.

5.   CHARACTERIZATION OF POLLUTED GROUNDWATER

Following the identification of a groundwater contamination incident, characterization of the polluted groundwater is desired in one or more ways, including:

(1)   Extent and degree of contamination (i.e., how much groundwater has been polluted and how badly);

(2)   Altered potential for groundwater use (e.g., is groundwater fit for drinking, for irrigation, or for process water);

(3)   Treatability (i.e., to what extent might the groundwater be treated by various methods).

In all cases, groundwater sampling and analysis will be required to assess the degree of contamination. However, characterization under Items (b) and (c) (following paragraphs) will require substantially more chemical-specific analyses than for Item (a).

a.   Extent and Degree of Contamination

Samples collected from various locations (e.g., wells, springs), at various depths, and at various times can be used, in conjunction with relatively rapid and inexpensive tests, to generally assess the extent and degree of contamination.

Portable instrumentation, some with small probes that can be lowered into wells, is available to measure such parameters as pH, specific conductance, dissolved oxygen, total organic carbon, and individual anions (e.g., Cl^-, NO_3^-, $SO_4^=$) or cations (Na^+, K^+, Ca^{++}) that may show altered or elevated concentrations due to contamination. Specific conductance is an especially good indicator for many (but not all) classes of soluble pollutants since it is rapid (small probes are available for in situ measurements) and will reflect the presence of inorganic ions and highly soluble organics which, because of their high mobility, will be at the forefront of the plume of contaminated groundwater.

Other simple indicators of groundwater contamination include: (1) color, odor, and taste, (2) organic vapor concentration, which can frequently be monitored with small, semiconductor-type probes; and (3) measurements of the amounts of (organic) material that can be extracted with such solvents as methylene chloride or chloroform.

In all cases, it is necessary to obtain data over a wide enough area around the site of suspected contamination to clearly establish background levels of the selected parameters. With a good data base, it will frequently be possible to establish not only the extent and degree of contamination, but the flow direction of the plume of contaminated groundwater.

b.   Altered Potential for Groundwater Use

To assess to what degree a contaminated aquifer may still be useful, it will be necessary to obtain chemical-specific analyses of the water to be used. In general, although there are exceptions, the highest standards (and thus the lowest allowable concentrations or criteria) will apply to drinking water, with lower standards applying to livestock water, irrigation water and (industrial) process water, in turn.

A characterization or classification with regard to the potential for use will seldom be a simple task since water quality criteria and standards have been set for only a limited number of chemicals, and the focus has been on the protection of drinking water and aquatic life; much less attention has been given to criteria and standards for irrigation, livestock, and process water.

Some of the documents which may be of help in a use-classification include the following:

- EPA's 1980 Water Quality Criteria Documents (USEPA, 1980a) which "contains recommended maximum permissible pollutant concentrations consistent with the protection of aquatic organisms, human health, and some recreational activities." The documents cover 64 of the 65 pollutants designated as toxic under Section 307(a)(1) of the Clean Water Act (i.e., the priority pollutants).

- EPA's Quality Criteria for Water (the "Red Book") (USEPA, 1976a). The Red Book provides recommended criteria levels focusing on the protection of aquatic life and domestic water supplies. In a few cases, separate criteria are listed for agricultural or industrial uses. The report focuses on inorganics (mostly metals) and pesticides. This book should be used only if the chemicals are not covered by the 1980 criteria (EPA, 1980).

- Water Quality Criteria, 1972 (the "Blue Book") (NAS, 1972). This major study by the National Academy of Sciences (NAS) for the EPA provides major sections on criteria for the protection of: (1) recreation and aesthetics; (2) public water supplies; (3) fresh water aquatic life and wildlife; (4) marine aquatic life and wildlife; (5) agricultural uses of water (livestock and irrigation); and (6) industrial water supplies. The coverage of organics, except for some pesticides, is small.

- Drinking Water and Health (NAS, 1977). This is an important review of the problems associated with chemicals in drinking water with important conclusions (similar to criteria) for many classes of pollutants. For organics, the report covers 74 nonpesticides (chosen from a listing of over 300 that have been found in drinking water) plus 55 pesticides.

- EPA's National Interim Primary Drinking Water Regulations (USEPA, 1976b). These primary standards cover only a few pesticides within the general class of organic compounds.

- EPA's National Secondary Drinking Water Regulations (USEPA, 1979a). These regulations cover properties or pollutants that may adversely affect the aesthetic quality of drinking water, such as taste, odor, color, and appearance. The proposed secondary levels represent federal goals, but are not federally enforceable.

It should be noted that the criteria published by the EPA are not legally enforceable standards. In general, only states can set water quality standards; however, they must be at least as stringent as the EPA criteria. The state should always be contacted for the latest standards applicable to the site of a groundwater contamination incident. In addition, both the EPA and the Food and Drug Administration have standards for (pesticide) residues in food which should be consulted if irrigation uses are contemplated.

The Clean Water Act requires the states to review and revise their water quality standards at least once every 3 years. While all states have a set of water quality standards, they are often highly variable in nature, and their coverage of toxic organics at present (except for some pesticides) is minimal.

In the absence of federal or state criteria and standards for specific chemicals, it may be necessary for special literature searches and/or laboratory tests to be conducted. If sufficient data exist, the calculation of a preliminary pollutant limit value (PPLV) may be carried out as described by Docre et al.. 1980. The PPLV value is a temporary, nonregulatory value that is based on information available in the literature and which relates primarily to human health effects.

c.   Treatability

Characterization of polluted groundwater with regard to treatability is discussed in Section VI ("Groundwater Treatment Methods").

2. Factors Affecting Contaminant Transport

1. HYDROLOGIC CYCLE

Water is present on the earth in the oceans, in the atmosphere, on the surface, and in the ground. The complex system whereby water moves these environments is called the hydrologic cycle. A general understanding of all aspects of the hydrologic cycle is essential to understanding its subsurface.

The earth's hydrologic cycle is driven by the heat of the sun and the pull of gravity. For example, at the surface of the oceans, water is heated by the sun, vaporizes, and escapes to the atmosphere. Conversely, water vapor in the atmosphere can be cooled and condensed as water droplets, which fall to the earth's surface. Water's circulation system is a complex loop, with many interconnections. Figure 8 illustrates this system. The following is an overview of the various means of transfer of water through the hydrologic cycle.

Water is released to the atmosphere through the combined actions of evaporation, transpiration, and sublimation. These are three variations of the same process driven by heat energy from the sun. According to Davis and DeWiest (1966), evaporation, or vaporization, is "the process by which molecules of water at the surface of water bodies or moist soil, or in precipitation, acquire enough energy through solar radiation to escape the liquid phase and pass into the gaseous state. Sublimation differs from this only in that the water molecules are converted from the solid phase (snow or ice) directly to vapor, without passing through the liquid form." Transpiration is the process by which water absorbed by vegetation is evaporated into the atmosphere from plants surfaces. Commonly, when measuring the amounts of water being circulated into the atmosphere, it is very difficult to distinguish how much is contributed solely by evaporation and how much is contributed solely by transpiration. The two processes, therefore, are often considered together as evapotranspiration.

The amount of water vapor which the atmosphere is capable of holding at any given time is a function of air temperature. As air temperature increases, the atmosphere's vapor capacity increases. As air temperature decreases, however, the atmosphere's vapor capacity is lowered and the excess water vapor condenses on small particles of dust or of salt that are also in the atmosphere. The principal way of decreasing air temperature is to lift the air higher in the atmosphere, and there are a number of ways to accomplish this. For example, winds blowing clouds toward mountain ranges carry the clouds upward over the mountainous obstacle. Similarly, when a warm, light air mass meets a cooler, heavy air mass in the atmosphere, the lighter one must rise over the obstacle of the heavier one. Finally, the earth's radiational cooling heats the air near the surface of the earth, causing it to rise.

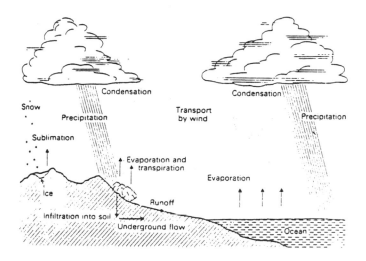

Figure 8.   Schematic Representation of the Hydrologic Cycle

Source: Domenico, P.A., Concepts and Models in Groundwater
        Hydrology, McGraw-Hill Book Company,© 1972.

When enough water condenses in the atmosphere to form clouds, the water will fall to the earth as precipitation. Precipitation is the part of the hydrologic cycle which moves water from the atmosphere to the earth's surface. Generally, water stays in the atmosphere for 10 to 14 days before it falls to the earth's surface. (Table 6). Less than 0.01 percent of the earth's water is in the air at any given time.

Water on the surface of the earth is found in the oceans and on the land. Water in the oceans will remain there until it is circulated back to the atmosphere through vaporization. The principal way in which water on the land surface is moved is via river flow. The term runoff applies to all water which is contributing to stream channel flow. There are two sources for runoff: overland flow and groundwater. Figure 9 (Davis and DeWiest, 1966) illustrates runoff.

Typically, as rain falls to the land surface, some is evaporated directly back to the atmosphere. Some of the precipitation will be intercepted by vegetation and will never reach the land surface. That which does reach the land will begin to infiltrate the soil and to be stored in surface depressions. When the soil's storage capacity has been reached and the surface depressions have been filled, the remaining volume of precipitation will begin to flow across the land in sheets. This is overland flow. The amount of overland flow which results from precipitation is a function of precipitation intensity, permeability of the ground surface, duration of precipitation, type of vegetation, area of drainage basin, distribution of precipitation, stream channel geometry, depth to water table, and slope of the surface. The overland flow which reaches river channels contributes to runoff. That which flows over the surface, but infiltrates the soil, or is stored in surface depressions before reaching a river, is not part of runoff.

Some of the water which exists below ground also contributes to surface runoff. Part of the infiltrate from precipitation will flow laterally at shallow depths to reach stream channels. Part will remain in the unsaturated soil. But some will also percolate down to the groundwater system and contribute to stream base flow.

The groundwater system is the subsurface component of the hydrologic cycle. Underground water moves downward due to gravity and flows laterally in response to potential gradients. It percolates through the pores of rock or soil, and through the cracks and joints of rocks which have very little porosity.

The groundwater system is typically broken down into components, as illustrated in Figure 10. The two major components to the system are the unsaturated vadose zone and the saturated phreatic zone. The vadose zone is further divided into the zone of soil moisture and the intermediate zone. The zone of soil water is the shallowest component of the groundwater system. It is distinguished from deeper unsaturated zones in that its water content is subject to large fluctuations due to evapotranspiration. The zone of gravitational water lies beneath the zone of soil water and is also unsaturated. This zone may be totally

TABLE 6. ESTIMATE OF THE WATER BALANCE OF THE WORLD

Parameter	Surface area $(km^2) \times 10^6$	Volume $(km^3) \times 10^6$	Volume (%)	Equivalent depth (m)*	Residence time
Oceans and seas	361	1370	94	2500	~4000 years
Lakes and reservoirs	1.55	0.13	<0.01	0.25	~10 years
Swamps	<0.1	<0.01	<0.01	0.007	1-10 years
River channels	<0.1	<0.01	<0.01	0.003	~2 weeks
Soil moisture	130'	0.07	<0.01	0.13	2 weeks-1 year
Groundwater.	130	60	4	120	2 weeks-10,000 years
Icecaps and glaciers	17.8	30	2	60	10-1000 years
Atmospheric water	504	0.01	<0.01	0.025	~10 days
Biospheric water	<0.1	<0.01	<0.01	0.001	~1 week

*Computed as though storage were uniformly distributed over the entire surface of the earth

Source: Freeze, R.A. and J.A. Cherry, Groundwater, Prentice-Hall, Inc., © 1979.

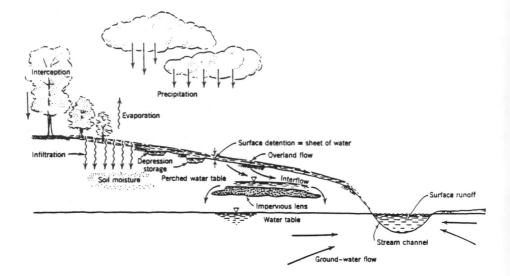

Figure 9. Simple Picture of a Runoff Cycle

Source: Davis, S.M. and R.J.M. DeWiest, Hydrogeology, John
         Wiley and Sons, Inc.,© 1966.

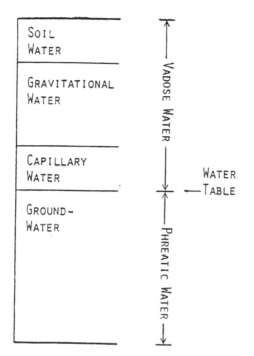

Figure 10.   Classification of Subsurface Water.

Source: LeRoy, L.W. et al., <u>Subsurface Geology -- Petroleum,
Mining, Construction</u>, Colorado School of Mines, 4th
Edition,© 1977.

absent in humid regions or thousands of feet thick in arid regions. Water moving through the vadose zone may be found in two forms: hydroscopic water - moisture held to the soil particles, and gravitational water - water moving downward under the force of gravity.

Lying below the unsaturated zone is the phreatic zone. Phreatic water is defined as water that will freely enter wells. Like the vadose zone, the phreatic zone is broken down into two components. The capillary zone is a transitional zone between the unsaturated and saturated groundwater zones. In this zone water is held to soil or rock particles by surface tension. The capillary zone is saturated at the bottom, becoming less so upward. The water table terminates the capillary zone from below. The water table is a theoretical surface at which hydrostatic pressure equals atmospheric pressure. All water below the water table is groundwater. Groundwater occurs in four modes: hydroscopically, as water free to move in response to potential gradients, in unconnected pores, and in chemical combination with rock.

Groundwater is discharged to the surface naturally as stream base flow and through springs, and artificially, through manmade wells. It may also flow directly to the ocean, or be trapped in pore spaces within rock. Soil water is usually circulated to the atmosphere by transpiration. Evaporation is not usually a significant means of groundwater discharge, unless the water table is very near the surface or the soil is nearly saturated.

2.   METEOROLOGICAL EFFECTS ON GROUNDWATER CONTAMINATION

Management of groundwater pollution incidents typically requires (1) definition of the extent of the contaminant zone; and (2) selection of a method to contain and/or treat the contaminant plume. The rate of future contaminant migration is among the governing factors in managing such incidents, and is a function of the recharge rate to the local groundwater system. Freeze and Cherry (1979) define groundwater recharge as "the entry into the saturated zone of water made available at the water table surface, together with the associated flow away from the water table within the saturated zone." The recharge rate to an aquifer depends on a number of factors including the available precipitation and temperature. These two parameters, precipitation and temperature, together with barometric pressure, wind velocity, humidity, and clouds define the condition of the earth's atmosphere, or weather. Meteorology is the study of the atmosphere, as it relates to predicting weather. This section addresses the effect that these meteorological factors, specifically precipitation and temperature, have on recharge and, hence, on the transport of groundwater contaminants.

Groundwater recharge rate is directly related to local precipitation. Precipitation and recharge will vary with overall climate, as well as seasonally. For example, in arid climates the depth to water is generally large. In more humid climates, the elevation of the water table tends to be shallower. As a result, pollutants in arid regions can accumulate in the unsaturated zone for a longer time, and be periodically released to the groundwater system. In humid areas, contaminants often enter the saturated zone immediately and provide a steady source to the groundwater system.

Seasonal meteorological fluctuations are generally reflected hydrologically as fluctuations in the elevation of the water table, or the thickness of the unsaturated zone. In general, the water table is at its highest in late spring, and at its lowest at the end of winter. This is because temperature also affects recharge. For example, precipitation in the form of snow is not available to recharge the groundwater system until it has melted. Also, ice, clogging pore spaces in frozen soil slows or prevents infiltration. In areas of substantial irrigation, the water table is typically at its lowest in late fall at the end of the irrigation cycle.

Seasonal fluctuations in the thickness of the unsaturated zone and in the elevation of the water table can affect contaminant movement in a number of ways. Flow in the unsaturated zone is driven by gravity, and is typically downward only. The thickness of the unsaturated zone is a principal factor required for a contaminant to reach the water table and be entrained in lateral groundwater flow. This is an important factor in gauging contaminant migration. In the case of a spill of a soluble material, fluctuations in the water table could affect the prediction of the amount of time required for a contaminant to be entrained in the groundwater flow, and begin lateral movement. Insoluble contaminants which float on the top of the water table are particularly sensitive to fluctuations in water table elevation. According to the American Petroleum Institute (1972), "If the water table drops, oil will follow and some of it will be absorbed by the soil it passes through. When the water table rises, oil previously absorbed by the soil will be picked up and then continue to move laterally with the groundwater."

The continuity of aquifer recharge is another important consideration in characterizing the extent of contamination. Many localities receive regular rainfall resulting in fairly continuous aquifer recharge. However, at others, precipitation may be more variable, producing bursts of recharge instead. This pattern should not be neglected when estimating the direction of contamination movement, because plumes will reflect such variations. Figure 11 illustrates a hypothetical hydrological setting receiving periodic recharge. In a hydrologic investigation there, water level readings at P and Q indicated a head gradient from left to right. Water quality samples taken at P and Q indicated contaminant gradient from right to left due to the concentration gradient. This set of contradictory information might be resolved with the consideration that an opposite head gradient could be possible within the error range for the water level measurement. However, this would be an inaccurate assessment. Even regular water quality sampling for changes in concentration would not aid in planning further investigation because groundwater flows so slowly that changes would be discounted by the error range for the sampling apparatus. In this situation, historical precipitation would

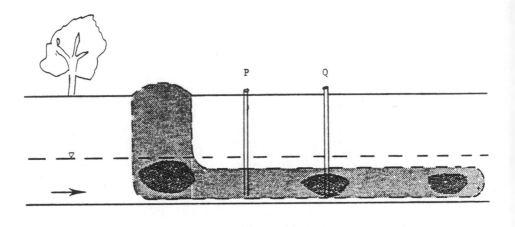

Figure 11. Effects of Periodic Recharge on Chemical
            Concentrations (Shading Darkness Reflects
            Concentrations)

be a key factor in characterizing the direction of contaminant migration.

Meteorological factors also affect the extent of aquifer contamination. Consider an example in which identical sampling programs are applied at two sites, one receiving periodic precipitation and one regular precipitation (Figure 12). Head gradient is the same at both sites. Wells A and B produce the same water quality data at both sites. In this example, an understanding of the recharge pattern is the key to accurately defining the extent of pollution at each site. Without this understanding, the extent of contamination could be underestimated, resulting in a misdirected control program.

Finally, recharge patterns are also important in analyzing the terminus of a contaminant plume. Historical precipitation data can indicate whether a higher quality sample at a contaminated site is the front of a plume or merely a hiatus in recharge. Similarly, higher water quality measured on the source side of the plume could be caused by diminished contaminant source or diminished recharge to transport the contaminant.

3.   GEOLOGICAL EFFECTS ON GROUNDWATER CONTAMINATION

The movement of contaminated groundwater is controlled by physical and geochemical properties of (1) the contaminant; (2) the groundwater; and (3) the geologic system through which the contaminated groundwater is flowing. The lithology, stratigraphy, and structure of a region control the distribution of aquifers and confining beds, and affect the direction and rate of groundwater and contaminant migration. This section addresses the ways in which geologic factors affect the movement of contaminated groundwater. It begins with a discussion of the porosity and permeability of different rock types, and is followed by a discussion of how those hydrologic properties govern the groundwater pathway within a local geologic system.

The properties which control groundwater flow are porosity and permeability. Flint and Skinner (1974) describe porosity and permeability as they relate to groundwater movement:

> "The limiting amount of water that can be
> contained within a given volume of rock
> material depends on the porosity of the
> material; that is, the proportion (in per
> cent) of the total volume of a given body of
> bedrock or regolith that consists of pore
> spaces (i.e., open spaces). So a very porous
> rock is a rock containing a comparatively
> large proportion of open space, regardless of
> the size of the spaces. Sediment is

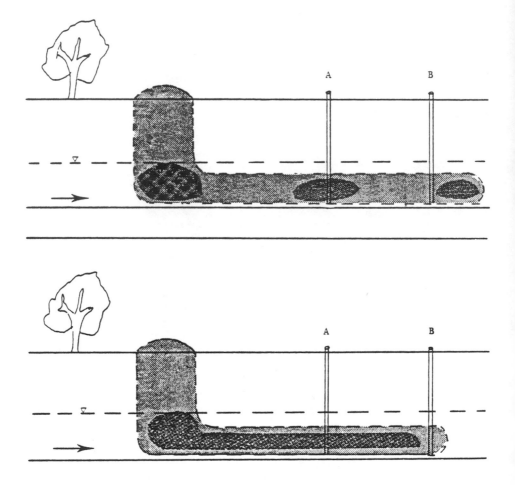

Figure 12.   Periodic Versus Continuous Recharge
              (Shading Darkness Reflects Concentration)

ordinarily very porous, ranging from 20 per
cent or so in some sands and gravels to as
much as 50 per cent in some clays. The sizes
and shapes of the constituent particles and
the compactness of their arrangement affect
porosity, as does the degree, in a
sedimentary rock, to which pores have become
filled with cementing substances."

The porosity which results directly from the soil or rock matrix is
termed primary porosity. The primary porosity of a rock unit may be
enhanced by structurally controlled fracturing or dissolution, and this
additional porosity is referred to as secondary porosity. Secondary
porosity is usually the principal source of porosity in crystalline
rocks. Flint and Skinner note that "igneous and metamorphic rocks
generally have low porosity, except where joints and cracks have
developed in them." Table 7 provides some representative ranges of
porosity values typically exhibited by various geologic media.

High porosity in a rock unit does not necessarily result in
groundwater movement. Only a combination of favorable porosity and
favorable permeability will permit groundwater flow.

"Permeability is capacity for transmitting fluids.
A rock of very low porosity is likely also to have
low permeability. However, high porosity values
do not necessarily mean high permeability values,
because size and continuity of the openings
influence permeability in an important way. The
relationship between size of openings and the
molecular attraction of rock surfaces plays a
large part. Molecular attraction is the force
that makes a thin film of water adhere to a rock
surface despite the force of gravity; an example
is the wet film on a pebble that has been dipped
in water. If the open space between two adjacent
particles in a rock is small enough, the films of
water that adhere to the particles will come into
contact. This means the force of molecular
attraction is extending right across the open
space, as shown on the left side of Figure 3-6. At
ordinary pressure, therefore, the water is held
firmly in place and so permeability is low. This
is what happens in a wet sponge before it is
squeezed. The same thing happens in clay, whose
particles are so tiny their diameters are less
than 0.005 mm.

By contrast, in a sediment with grains at least as
large as sand grains (0.06mm to 2mm) the open

TABLE 7. RANGE OF VALUES OF POROSITY (N)

	$n(\%)$
Unconsolidated deposits	
Gravel	25–40
Sand	25–50
Silt	35–50
Clay	40–70
Rocks	
Fractured basalt	5–50
Karst limestone	5–50
Sandstone	5–30
Limestone, dolomite	0–20
Shale	0–10
Fractured crystalline rock	0–10
Dense crystalline rock	0–5

Source: Freeze, R.A. and J.A. Cherry, Groundwater, Prentice-Hall, Inc. © 1979.

spaces are wider than the films of water adhering
to the grains. Therefore the force of molecular
attraction does not· extend across them
effectively, and the water in the centers of the
openings is free to move in response to gravity or
other forces, as shown in Figure 13. This
particular sediment is therefore permeable. As
the diameters of the openings increase,
permeability increases. With its very large
openings, gravel is more permeable than sand and
yields large volumes of water to wells." Flint &
Skinner (1974, p. 155).

The permeability ranges for various sediment and rock types are listed
in Table 8.

The porosity and permeability of a rock unit is controlled by its
lithology, geometry, and spatial variability. Individual rock types
have characteristic lithologic properties, and, depending on
depositional environment, characteristic structure and stratigraphy,
which govern groundwater flow within them (Figure 14). For example,
sandstones tend to have hydrologic characteristics which are good for
transmitting groundwater. Porosity in sandstone depends directly upon
the amount of cementation and degree of compaction. Unlithified sands
have porosities of 30 - 50 percent. Because the sediments become more
compacted with burial, porosity in sandstones will decrease with
increasing depth. Sandstone permeability follows well-defined trends in
relation to porosity (Figure 15). As porosity increases so does
permeability. However, the distribution of permeability within a unit
may vary, both laterally and vertically. These variations depend on
bedding, depositional environment, and stratigraphy. A zone which may
appear homogeneous upon visual inspection, may actually vary in
permeability by one to two orders of magnitude locally. This
anisotropic variability typically favors lateral groundwater flow
parallel to bedding planes. However, with increasing cementation and
compaction, secondary porosity in the form of fractures may play a major
role. In this case, the trend for increased permeability along bedding
planes changes to higher fracture permeability in the vertical
direction.

Shales often form extensive confining beds, but frequently they
occur as discontinuous lenses. Primary porosities in shales range from
0 - 10 percent, and, as a result, permeability is usually very low.
Permeability values can be on the order of 10^{-12} cm^2 to 10^{-15} cm^2
reducing groundwater flow to centimeters per century. Secondary
porosity in the form of hairline fractures can increase shale porosity
significantly to produce permeabilities on the order of 10^{-7} - 10^{-8} cm^2.

Like sandstone, hydrologic properties of limestone vary with depth,
due to compaction effects. Young limestone porosity values range from
20 - 50 percent, whereas older ones may have little, if any. Generally
such unaltered limestone deposits are not major sources of groundwater.

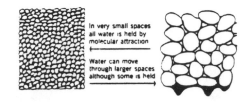

Figure 13.   Effect of Molecular Attraction in the
Intergranular Spaces in Fine Sediment
(Left) and in Coarser Sediment (Right).
Scale is Much larger than Natural Size.

Source: Flint, R.F. and B.J. Skinner, <u>Physical Geology</u>, John
Wiley and Sons, Inc.© 1974.

TABLE 8. RANGE OF VALUES OF HYDRAULIC CONDUCTIVITY AND PERMEABILITY

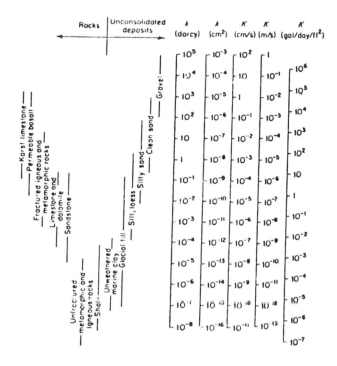

Source: Freeze, R.A. and J.A. Cherry, Groundwater, Prentice-Hall, Inc., © 1979.

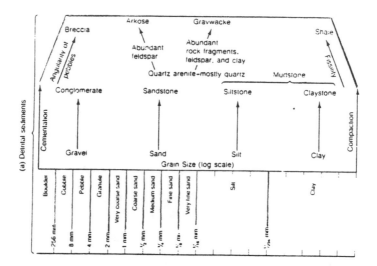

Figure 14.   Sedimentary Rock Classification.
(a) Detrital Sediments. (b) Chemical
Sediments.

Source: Press, F., and R. Siever, Earth, 2nd Edition, W.A.
Freeman and Company, © 1978.

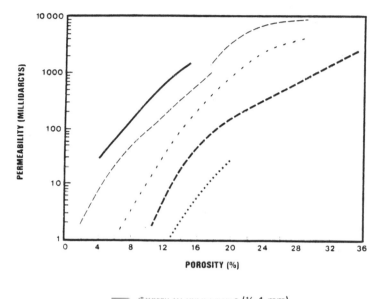

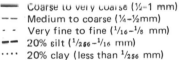

Coarse to very coarse (½–1 mm)
Medium to coarse (¼–½mm)
Very fine to fine ($^1/_{16}$–$^1/_8$ mm)
20% silt ($^1/_{256}$–$^1/_{16}$ mm)
20% clay (less than $^1/_{256}$ mm)

Figure 15.   Relationship Between Porosity, Permeability, and Grain-Size
            Distribution of Sands and Sandstones.

Source: Adapted from Chilingar, Proc. Intern. Sedimentol. Congr.,
          Amsterdam, Antwerp, 1963.

However, dolomitization, folding, and dissolution along fractures and in openings along bedding planes contribute to significant amounts of secondary porosity and permeability in these units.

Glacial drift is a special type of sedimentary deposit formed from continental glacial environments, and is the most abundant material that was deposited on the land surface during the Pleistocene. The lithology of a deposit can vary greatly. It can be sandy, with variable amounts of silt and minimal clay. This type of unit can form local aquifers. Others have high silt and clay content, with very little sand, resulting in a low permeability. Deposits of this type act as confining layers.

Plutonic igneous and unfractured metamorphic rock units have average primary porosities on the order of 2 percent. They can, however, develop significant secondary porosity resulting from fractures and dissolution. Fracture orientation may be vertical due to tectonic and thermal stresses or they may be horizontal, from overburden unloading. However, it is characteristic of crystalline rock for fracture permeability to decrease with depth. Extrusive igneous rocks, such as basalt can be more permeable from bubbles from entrapped gases, rubble zones from flow and differential solidification, interbeds of soils and stream channels, and columnar jointing from cooling. Extrusive igneous rocks tend to have dominantly lateral flow on a regional scale.

The course by which groundwater, and any accompanying contaminants travel to a discharge point is controlled by the rock units present in the hydrogeologic setting with water following the path of least resistance. Figure 16 from Freeze and Cherry (1979) illustrates some regional flow regimes which may result from various hypothetical hydrogeologic settings. Elements a through f are vertical cross-sections of identical dimensions. All cases represent a major valley running perpendicular to the page on the far left side of the system, with an upland valley to the right. Figure 16 illustrates a homogeneous system of a single rock type in which flow is effectively horizontal. In Figure 16, a higher permeability unit has been introduced below the original surface layer. This new layer exhibits essentially horizontal flow and is being recharged from above.

> "If the hydraulic conductivity contrast is increased (Figure 16), the vertical gradients in the overlying aquitard are increased and the horizontal gradients in the aquifer are decreased. The quantity of flow is increased. One result of the increased flow is a larger discharge area, made necessary by the need for the large flows in the aquifer to escape to the surface as the influence of the left-hand boundary is felt.
>
> In hummocky terrain (Figure 16) the presence of a basal aquifer creates a highway for flow

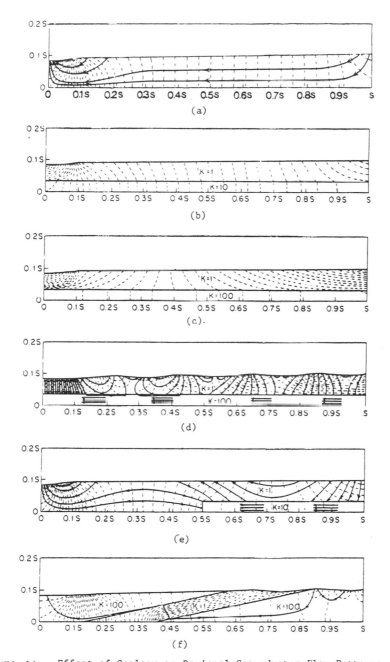

Figure 16   Effect of Geology on Regional Groundwater Flow Patterns.

Source: Freeze, R.A. and J.A. Cherry, Groundwater, Prentice-Hall,
        Inc., © 1979.

that passes under the overlying local
systems. The existence of a high
permeability conduit thus promotes the
possibility of regional systems even in areas
of pronounced local relief.

There is a particular importance to the
position within the basin of buried
lenticular bodies of high conductivity. The
presence of a partial basal aquifer in the
upstream half of the basin (Figure 16)
results in a discharge area that occurs in
the middle of the uniform upland slope above
the stratigraphic pinchout. Such a discharge
area cannot occur under purely topographic
control. If the partial basal aquifer occurs
in the downstream half of the system, the
central discharge area will not exist; in
fact, recharge in that area will be
concentrated.

In the complex topographic and geologic
system shown in Figure 16, the two flowlines
illustrate how the difference of just a few
meters in the point of recharge can make the
difference between recharge water entering a
minor local system or a major regional
system. Such situations have disturbing
implications for the siting of waste disposal
projects that may introduce contaminants into
the subsurface flow regime." Freeze &
Cherry (1979, p. 197).

The actual positioning of rock units in a region is determined from
depositional environment, stratigraphy, and structural history. For
example, glaciers advance and retreat leaving outwash plains of till and
stream-deposited sand. Lakes dry up and lacustrine clay lenses are
deposited. Marine shorelines transgress and regress leaving sands,
shales and limestones that interfinger, pinch out, grade in and out of
each other. Faulting disconnects continuous layers, positioning sand
units abutting shales. All of those processes come together in a region
to control a groundwater pathway. The variety of settings is almost
infinite, and no two sites are precisely the same.

Figures 17 to 32 illustrate some hypothetical hydrogeologic
landfill settings. They are only schematic, with simplified geology and
hydrology, and they illustrate leachate flow principles in a general way
only. They have been included as additional examples of how geology can
affect contaminant migration (USEPA, 1980b).

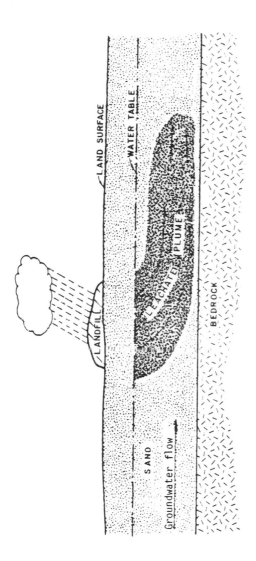

Figure 17.   Single Aquifer With A Deep Water Table: Leachate percolates downward from the landfill to the underlying aquifer and then moves downgradient as a bulb or plume in the direction of groundwater flow. The mass of leachate may: 1) Sink to the bottom of the aquifer if of a heavier specific gravity, or 2) float at or near the top of the water-bearing unit if the leachate is predominately hydrocarbon in nature.

Source: EPA, 1980b.

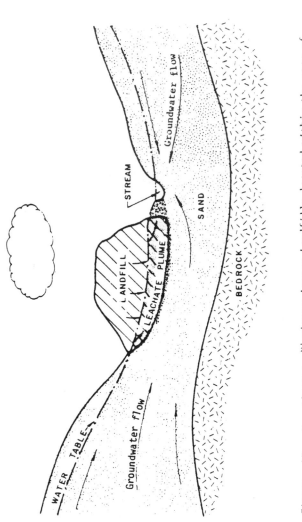

Figure 18.   Groundwater Discharge Areas: Landfills located within the zone of saturation area always in contact with ground water moving from topographically higher recharge areas to a stream discharge point. In such cases, leachate is transported with the ground water to the stream where it becomes diluted by surface water.

Source: EPA, 1980b.

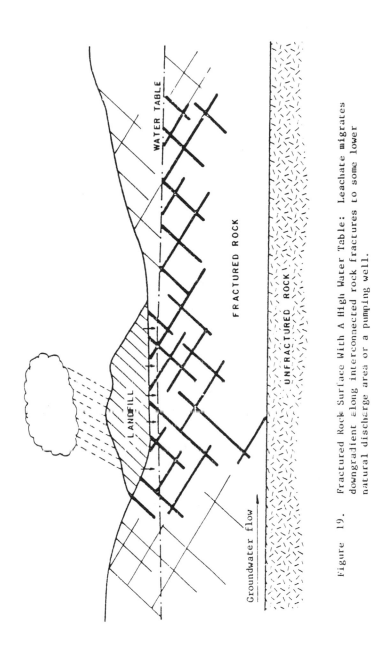

Figure 19.   Fractured Rock Surface With A High Water Table: Leachate migrates downgradient along interconnected rock fractures to some lower natural discharge area or a pumping well.

Source: EPA, 1980b.

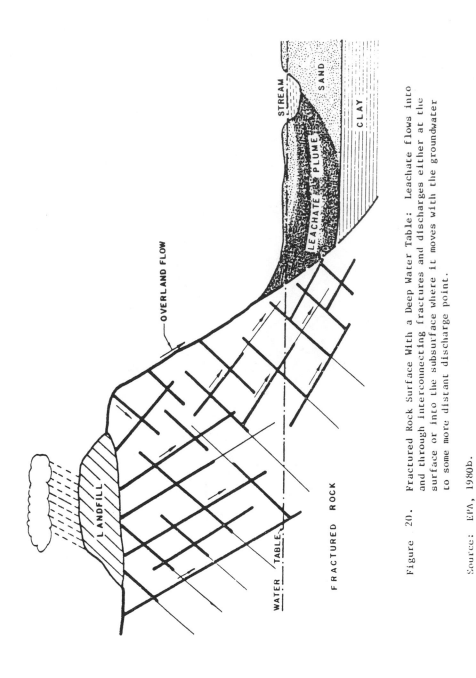

Figure 20. Fractured Rock Surface With a Deep Water Table: Leachate flows into and through interconnecting fractures and discharges either at the surface or into the subsurface where it moves with the groundwater to some more distant discharge point.

Source: EPA, 1980b.

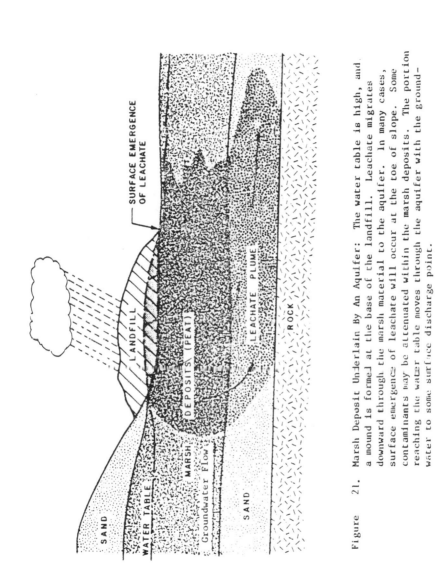

Figure   21.   Marsh Deposit Underlain By An Aquifer: The water table is high, and
a mound is formed at the base of the landfill. Leachate migrates
downward through the marsh material to the aquifer. In many cases,
surface emergence of leachate will occur at the toe of slope. Some
contaminants may be attenuated within the marsh deposits. The portion
reaching the water table moves through the aquifer with the groundwater to some surface discharge point.

Source: EPA, 1980b.

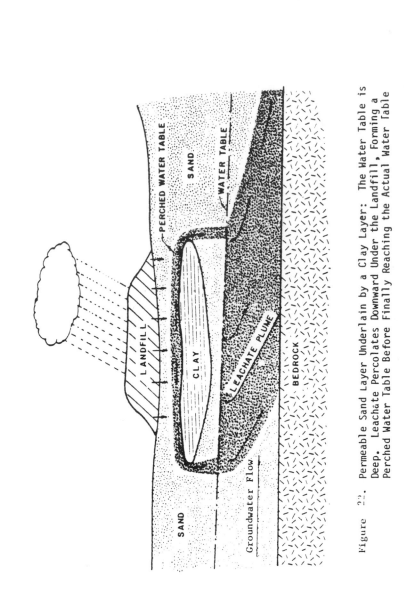

Figure 22.   Permeable Sand Layer Underlain by a Clay Layer: The Water Table is
Deep. Leachate Percolates Downward Under the Landfill, Forming a
Perched Water Table Before Finally Reaching the Actual Water Table

Source: EPA, 1980b.

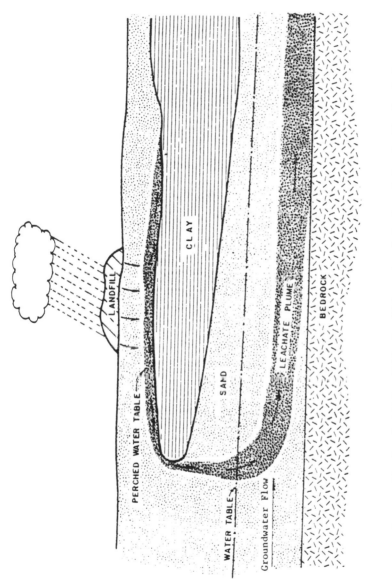

Figure 23. Perched Water Table Condition: Leachate Percolates to the Perched Water Table and Flows Downgradient to the end of the Confining Layer Where it Moves Downward to the Actual Water Table

Source: EPA, 1980b.

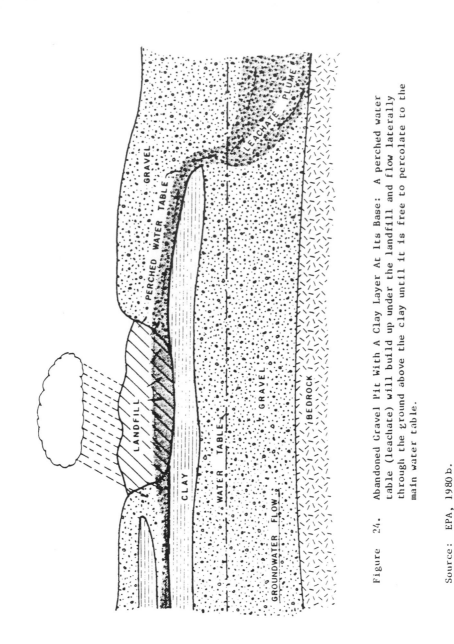

Figure 24.   Abandoned Gravel Pit With A Clay Layer At Its Base: A perched water
              table (leachate) will build up under the landfill and flow laterally
              through the ground above the clay until it is free to percolate to the
              main water table.

Source: EPA, 1980 b.

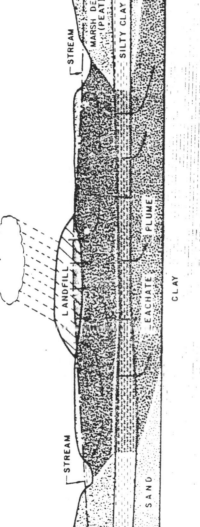

Figure   25.   Marsh Deposits Bounded On Either Side By Streams and Underlain By a
Shallow Aquifer: Leachate from the landfill may move horizontally
through the marsh materials to the stream, or vertically downward as
groundwater recharge to the aquifer.

Source: EPA, 1980b.

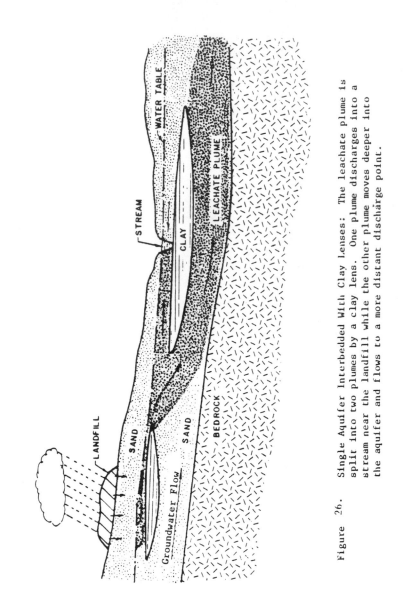

Figure 26.   Single Aquifer Interbedded With Clay Lenses: The leachate plume is split into two plumes by a clay lens. One plume discharges into a stream near the landfill while the other plume moves deeper into the aquifer and flows to a more distant discharge point.

Source: EPA, 1980 b.

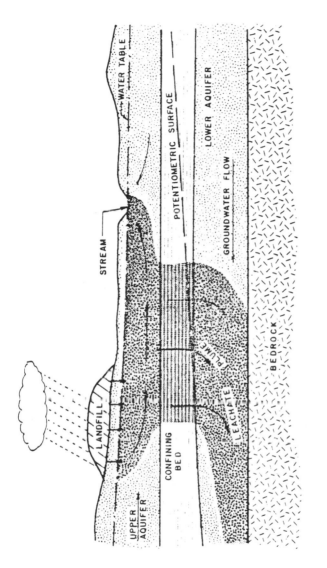

Figure 27.   Two-Aquifer System With Opposite Flow Directions: Leachate first moves into and flows with the groundwater in the upper aquifer. Some of the leachate eventually moves through the confining bed into the lower aquifer where it flows back beneath the landfill and away in the other direction.

Source: EPA, 1980b.

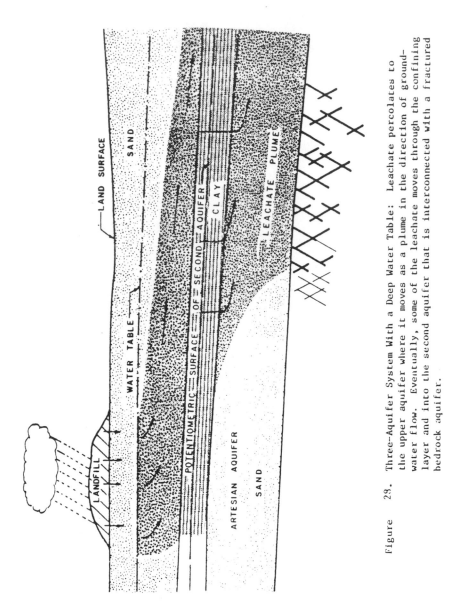

Figure   29.   Three-Aquifer System With a Deep Water Table: Leachate percolates to the upper aquifer where it moves as a plume in the direction of ground-water flow. Eventually, some of the leachate moves through the confining layer and into the second aquifer that is interconnected with a fractured bedrock aquifer.

Source: EPA, 1980 b.

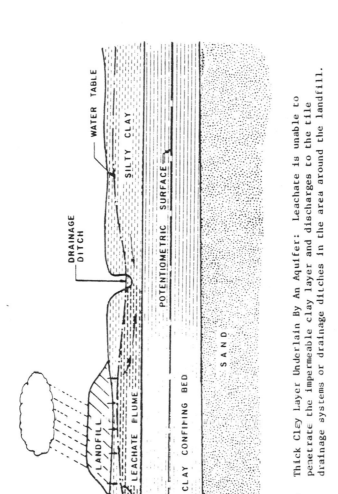

Figure 29.   Thick Clay Layer Underlain By An Aquifer: Leachate is unable to
penetrate the impermeable clay layer and discharges to the tile
drainage systems or drainage ditches in the area around the landfill.

Source: EPA, 1980 b.

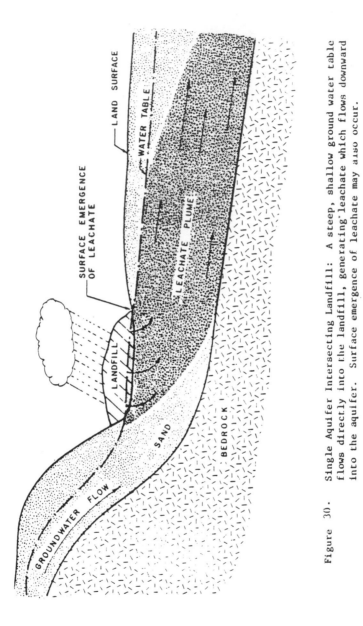

Figure 30.   Single Aquifer Intersecting Landfill: A steep, shallow ground water table flows directly into the landfill, generating leachate which flows downward into the aquifer. Surface emergence of leachate may also occur.

Source: EPA, 1980 b.

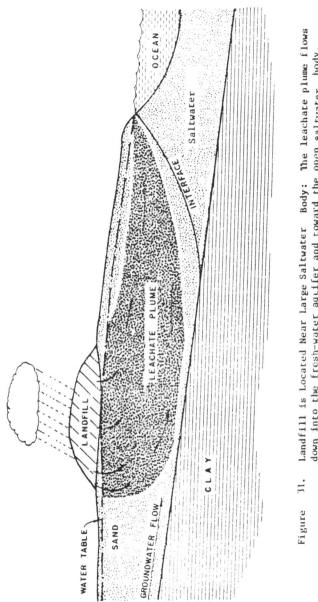

Figure 31.   Landfill is Located Near Large Saltwater Body: The leachate plume flows
down into the fresh-water aquifer and toward the open saltwater body.
As the leachate plume reaches the fresh-salt interface, it is forced upward
along the interface to discharge at or near the edge of the saltwater body.

Source: EPA, 1980b.

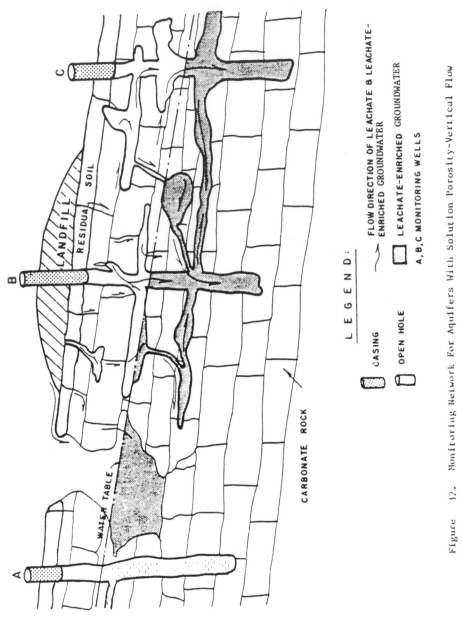

Figure 32. Monitoring Network For Aquifers With Solution Porosity-Vertical Flow

Source: EPA, 1980b.

4.   EFFECTS OF PUMPING ON CONTAMINANT PLUMES

Under natural conditions groundwater flow systems are in a state of
approximate dynamic equilibrium. Recharge to a system and discharges
from it are in approximate balance when considered over a season or a
climatic cycle. Human activities, including pumping water supply wells,
recharge wells, or landfills, can affect a stable groundwater system.
Such activities constitute stresses imposed on the system and must be
balanced by changes in the pattern or amount of recharge to the system
and discharge from the system by changes in the amount of water in
storage, or by a combination of these. Such stresses also have impacts
on the pathways and rates of migration of groundwater contaminants,
since such stresses cause distortions of the equilibrium positions of
equipotentials and streamlines.

In this section we are particularly concerned with the effects of
pumping on the shape, location and migration of contaminant plumes.
These pumping effects are of interest for two important reasons:(1) the
potential for contamination of municipal or domestic water supply wells,
and (2) the opportunities for manipulation of hydraulic gradients as a
means of controlling migration of contaminants.

For the purposes of this discussion, and to clarify the important
underlying principles, we will focus on a simplified physical situation
consisting of a confined aquifer of infinite areal extent with uniform
hydraulic properties and thickness.

Consider as a first case a flow regime consisting of a static
horizontal potentiometric head profile (no gradients). On this regime,
we superimpose the effects of a single pumping well. Figure 33 shows
the streamlines and equipotentials which result. The configuration of
streamlines (defining regions of equal flow) and equipotentials (contour
lines of equal hydraulic potential) is referred to as a flow net. For
this situation, the flow net shows flow occurring radially towards the
well. Equipotentials consist of concentric circles of decreasing head
as we move toward the pumping well. Darcy's law characterizes flow in a
porous medium as the product of a proportionality constant, known as the
hydraulic conductivity, and a gradient. By lowering the potential at a
well, a potential difference, or gradient, is created between the well
and adjacent materials, thus inducing flow towards the well. As a
result the pumping stress leads to decreases in potential at points in
the vicinity of the well. At a given point, the difference between the
prepumping potential and the new lower potential resulting from pumping
is called the drawdown. The new potentiometric surface is often
referred to as the cone of depression, deriving its name from the shape
of the surface of reduced potential observed around the well.

The shape of the cone of depression depends on the properties of
the aquifer and the pumping rate. At any point at a given time drawdown
is directly proportional to the pumping rate and inversely proportional
to aquifer hydraulic properties. The aquifer properties of greatest
importance are transmissivity, or the ability of an aquifer to transmit

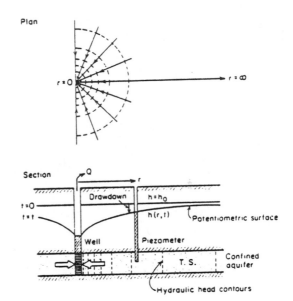

Figure 33.   Radial Flow to a Well in a Horizontal
               Confined Aquifer

Source: Freeze, R.A. and J.A. Cherry, <u>Groundwater</u>, Prentice-Hall,
         Inc.,© 1979.

water, and storativity, or the ability of an aquifer to store water.
Aquifers of high transmissivity develop shallow cones of depression of
wide areal extent. Aquifers of low transmissivity develop deep cones of
limited extent. Low storativity produces deeper drawdown than high
storativity. It should be noted that for a given aquifer, the cone of
depression increases in depth and extent with increasing time.

The discussion above is also valid for unconfined aquifers,
although the physical situation is more complex. The storativity of an
unconfined aquifer is referred to as the specific yield. Specific
yields of unconfined aquifers are generally much higher than
storativities of confined aquifers. Thus, cones of depression will be
less extensive and shallower than for a confined aquifer with a similar
transmissivity. In addition, pumping from an unconfined aquifer leads
to actual dewatering of the aquifer, which is not the case for a
confined aquifer. As a result, the thickness of the aquifer at any
given point changes with time. This in turn affects the transmissivity
of the aquifer. As Figure 34 indicates, flow lines toward a well in an
unconfined aquifer are not horizontal, but have a vertical component,
due to the variation in aquifer thickness induced by pumping. In a
confined aquifer, thickness does not change and flow lines remain
horizontal after imposition of a pumping stress. The practical result
of these differences is that exact mathematical treatments of flow to a
well in an unconfined aquifer are more complicated than treatments in a
confined aquifer.

The preceding discussion assumed a static head profile as an
initial condition in the aquifer. Water in an aquifer is actually
moving in response to a gradient in potential. In other words, there is
a slope to the potentiometric surface in a confined aquifer, and a slope
to the water table in an unconfined aquifer. Furthermore, a velocity
field characterizes the natural, prepumping conditions in an aquifer, on
which the effects of a pumping well are superimposed.

Figure 35 shows the flow net and cone of depression for a well
withdrawing water from an unconfined aquifer with an initially sloping
water table. The flow is not radial towards the well, nor are the
equipotentials concentric. This is a direct result of the previously
existing flow conditions. We observe for this cone that there is a
definite region of the aquifer upgradient from the well from which the
well captures water. Outside this region, the streamlines do not
terminate at the well, and water will flow past the well. This is in
contrast to the static case where all streamlines terminate at the well.

The significance of this difference relates to the existence of an
area of capture, from which the well draws. If a landfill, waste dump
or other source of contamination exists within this area upgradient from
a well, pollutants from the source can be expected to appear in the
water derived from the well. Conversely, if a plume of contaminated
water exists in an aquifer, it should be possible to install wells
downgradient so as to intercept the plume.

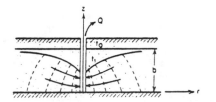

Figure   34.   Radial Flow to a Well in an
Unconfined Aquifer.

Source: Freeze, R.A. and J.A. Cherry, <u>Groundwater</u>, Prentice-Hall,
Inc.,© 1979.

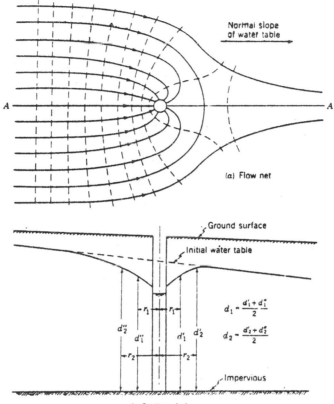

(a) Flow net

(b) Section A·A

Figure 35.   Flow to a Well in an Unconfined Aquifer
with an Initially Sloping Water Table.

Source: Linsley, R.K. and J.B. Franzini, <u>Water Resources
<u>Engineering</u>, McGraw-Hill, Inc., New York,© 1972.

Frequently, more than one well may be operating in a particular area. If this is the case, pumping from one well may interfere with pumping from an adjacent well. The observed effect will be increased drawdown at points in the vicinity of the two wells. As shown in Figure 36 the drawdown at any point caused by the interference of several wells will be the sum of the drawdowns caused by each individual well. The increased drawdowns imply increased gradients and, consequently, increased velocities toward the wells.

Many confined aquifers are not perfectly bounded above or below. There may be several aquifers in a system, each pair separated by a confining bed (layer of less permeable material). For example, as shown schematically in Figure 37, an unconfined aquifer near ground surface may be underlain by a confining bed (e.g., clay layer or glacial till), which may in turn overlay a confined aquifer. Under natural conditions, a head difference will frequently exist between the upper and lower aquifers, such that flow occurs through the aquitard between the two aquifers. A situation such as this is termed a leaky aquifer system. The significance of pumping in such a system is in its potential for changing hydraulic gradients between the two connected aquifers. Depending on the original direction of flow and the location of the pumping well may increase flow in the original direction, decrease the flow or possibly reverse the direction of flow. As an example, consider the two-aquifer system mentioned above, and assume a higher potential in the confined aquifer than in the overlying unconfined aquifer, with a source of contamination existing in the unconfined aquifer. Development of wells in the confined aquifer will cause drawdown in potential. If the potential is reduced to values below those of the unconfined aquifer, flow will be induced into the confined aquifer, with the possible consequence of migration of contaminants between the two aquifers and degradation of water quality in the lower aquifer.

The preceding discussions of pumping effects all assume idealized representations of actual aquifer configurations. To quote Freeze and Cherry (1979):

> In the real world, aquifers are heterogeneous
> and anisotropic; they usually vary in
> thickness; and they certainly do not extend
> to infinity. Where they are bounded, it is
> not by straight-line boundaries that provide
> perfect confinement. In the real world,
> aquifers are created by complex geologic
> processes that lead to irregular
> stratigraphy, interfingering of strata, and
> pinchouts and trendouts of both aquifers and
> aquitards.

As a result these discussions should be taken as indicative of the principles involved in understanding the effects of pumping on flow regimes rates, and the potential consequences with respect to pollutant migration. Application of these principles requires a detailed

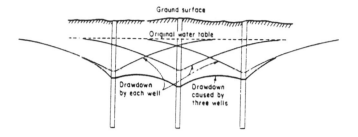

Figure   36.   Effect of Interference Between Wells.

Source: Linsley, R.K., Jr., et al., Hydrology for Engineers, McGraw-Hill, Inc., New York,© 1975.

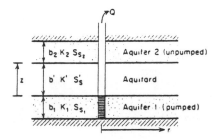

Figure   37.   Schematic Diagram of a
                Two-Aquifer Leaky System.

Source: Freeze, R.A. and J.A. Cherry, Groundwater,
          Prentice-Hall, Inc.,© 1979.

understanding of the particular geology, stratigraphy, and flow regime
of a particular region or site.

5.   EFFECT OF SOLUBILITY AND DENSITY OF ORGANIC CONTAMINANTS

The rate and direction of movement of contaminants which have
entered a groundwater system are functions of the local geology,
groundwater flow regime, and the chemical and physical properties of the
contaminant. This section qualitatively introduces the effect that
variations in the solubility and the density of organic contaminants
have on contaminant plume migration (the quantitative approach will be
discussed in a later section). The first classification of contaminant
is by solubility. Soluble contaminants dissolve into the groundwater
and their subsequent movement is governed by Darcy's law, combined with
hydrodynamic dispersion. Solids with low solubility are less likely to
be transported in the groundwater. The movement of fluids that do not
mix with water (immiscible fluids, such as oil) can be predicted using a
more complex form of the general flow law that includes a term to
describe the interactions between the different fluids and between each
fluid and the solids matrix.

a. Soluble Contaminants

In saturated flow through a porous medium a portion of the flow
domain is assumed to contain a certain mass of solute known as a tracer.
As flow takes place, the tracer gradually spreads out and occupies more
of the flow domain, beyond the region predicted by the average water
flow alone. This spreading phenomenon is called hydrodynamic dispersion
(also dispersion, immiscible displacement) in a porous medium (Bear,
1979). Figure 38 illustrates the difference between concentration
levels with dispersion and without dispersion.

The actual mixing of the tracer with the uncontaminated water is
caused by two microscopic processes: mechanical dispersion and molecular
diffusion. Mechanical dispersion is the result of tracer velocity
variation in direction and magnitude within a single pore space and
between pore spaces of different sizes (Figure 39). This causes the
individual stream lines to fluctuate with respect to the average flow
(Figure 39). These phenomena cause the spreading of initially close
groups of tracer particles until they occupy a larger and larger portion
of the flow domain. Simultaneously, molecular diffusion produces an
additional flux of tracer particles from regions of higher concentration
to those of lower concentrations. This causes continual equalizing of
tracer concentrations, first within a single streamline and then between
two streamlines (Figure 39).

An "ideal tracer" is a solute that is inert with respect to its
liquid and solid surroundings and does not affect the liquid's
properties (Bear, 1972). Movement of this kind of solute can be
predicted using the overall Darcy flow with a component for dispersion.
However, solutes with large density contrasts with respect to water have
a larger vertical component of dispersion. Figure 40 illustrates how a

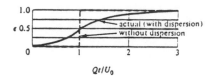

Figure   38.   Transition Curve with Dispersion Versus
Abrupt Change with No Dispersion

Source: Bear, J., Hydraulics of Groundwater, McGraw-Hill Book Company,
New York,© 1979.

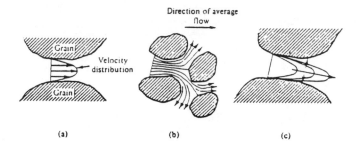

Figure   39.   Hydrodynamic Dispersion Caused by
Mechanical Dispersion (a,b) and Molecular
Diffusion (c)

Source: Bear, 1979.

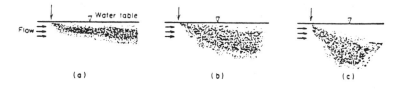

Figure   40.   Effect of Density on Migration of Contaminant
Solution in Uniform Flow Field, (a) Slightly
More Dense than Groundwater; (b) and (c) Larger
Density Contrasts

Source: Freeze, R.A. and J.A. Cherry, Groundwater, Prentice-Hall,
Inc.,© 1979.

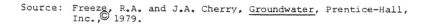

larger density contrast increases the importance of the dispersion component in accurately predicting tracer migration (Freeze and Cherry, 1979). The equations used to model solute transport by dispersion will be discussed in a later section.

b. Immiscible Flow

Fluids that do not readily mix with water (immiscible fluids, such as oil or nonaqueous phase liquid waste) do not flow according to the laws of hydrodynamic dispersion. The presence of these fluids causes water to adhere to the surface of the grains of the porous medium, a force known as surface tension. The immiscible fluid then flows through larger pore spaces, rarely displacing water from the smaller spaces. Capillary pressure is defined as the difference in pressure across the interface between the immiscible fluid and water (Davis and DeWiest, 1966). Capillary pressure only exists when two immiscible fluids are present, thus causing surface tension. The magnitude of capillary pressure strongly depends on the grain size of the porous material and on wetting properties of the fluids. This pressure term is an additional factor in the overall flow system. The existence of capillary pressure in a two-phase flow system means that the migration of an immiscible fluid is not solely dependent on the flow of groundwater. In fact, an immiscible fluid can migrate in a direction in complete opposition to the dominant flow system.

An example of contamination at the site of an oil spill is shown in Figure 41. The oil flows through the unsaturated zone, leaving residual oil absorbed by the soil particles until it intersects the water table. Once at the water table, oil floats on the surface of the water and is moved along the general flow of water. However, the oil will tend to be absorbed by soil particles until the volume of oil being transported is minimal (API, 1972). The added force of the capillary pressure can allow some of the insoluble oil to move upstream of the dominant flow (shown by the oil to the left of the original spill in Figure 41). Some of the components of oil are soluble and enter the groundwater itself (shown above the dotted line on Figure 41). These soluble components move by hydrodynamic dispersion causing a larger volume of groundwater to be contaminated downstream of the spill (Williams and Wilder, 1972). In the case of petroleum products, the more dense products (crude oil) have fewer soluble components than the less dense products (gas). One of the biggest problems with spills of light hydrocarbons is that the relative solubility increases the volume of groundwater that is contaminated (Matis, 1972).

An immiscible fluid that is more dense than water will also move according to the combined effects of the density difference, and the fluid-fluid and fluid-solid interfacial pressures. Because of the density contrast, the fluid will, in general, sink within the groundwater. A thorough understanding of the basic flow system is still required to predict the migration, and knowing all of the interactions between the fluids and the fluids and solids is fundamental to the prediction of fluid flow.

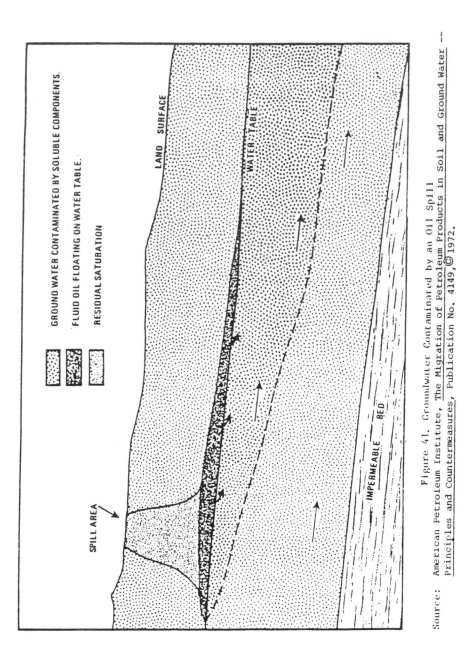

Figure 41. Groundwater Contaminated by an Oil Spill

Source: American Petroleum Institute, The Migration of Petroleum Products in Soil and Ground Water --
Principles and Countermeasures, Publication No. 4149,© 1972.

A major problem in utilizing field measurements in two-phase flow cases is caused by a residual saturation of the immiscible liquid on soil particles that it has flowed past. Figure 41 shows a residual saturation in the zone above the water table. Fluctuation in the water table can increase the zone of residual saturation (Figure 42). A prolonged dry spell can cause a lowering of the water table that the petroleum will follow, thus deepening the zone of residual saturation. When the water table rises, at least a portion of the residual oil will be displaced by the water and will move vertically with the groundwater (API, 1972). Field measurements under such conditions must be carefully analyzed to discern the continuity of the phase as well as the concentration levels.

Immiscible fluids will tend to travel at velocities lower than the associated groundwater and will persist longer in a given area. Soluble contaminants (and soluble components of immiscible fluids) will tend to contaminate a larger volume of groundwater and at a rate faster than the dominant water flow.

6.   FATE OF ORGANIC GROUNDWATER POLLUTANTS

    a. Overview

In assessing the probable transport and fate of organic groundwater pollutants, the key processes to be considered may be listed in three groups:

    (1) Equilibrium Partitioning of the Chemical:

        ● soil ↔ soil water (adsorption of solutes)
        ● water ↔ soil air (volatilization from solution)
        ● soil ↔ soil air (adsorption of vapors)

    (2) Degradation of the Chemical:

        ● biodegradation
        ● hydrolysis (or elimination)
        ● oxidation or reduction (low importance)

    (3) Transport of the Chemical:

        ● leaching (through unsaturated zone)
        ● transport with (and dilution in) groundwater
        ● volatilization and transport to the atmosphere
        ● erosion or entrainment of surface soils

These processes are considered in conjunction with the major phases of the soil/groundwater system: soil, soil water and (in the unsaturated zone) soil air. Figure 43 provides a schematic diagram of the system and processes.

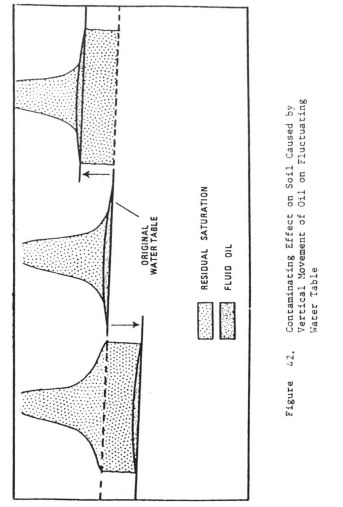

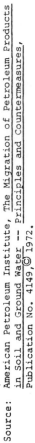

Figure 42.   Contaminating Effect on Soil Caused by
              Vertical Movement of Oil on Fluctuating
              Water Table

Source: American Petroleum Institute, The Migration of Petroleum Products
          in Soil and Ground Water -- Principles and Countermeasures,
          Publication No. 4149, © 1972.

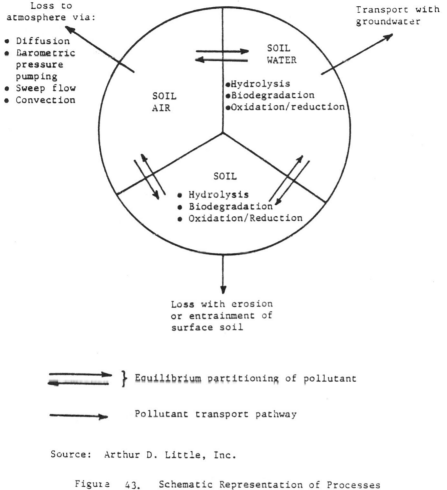

Source: Arthur D. Little, Inc.

Figure 43.   Schematic Representation of Processes
Affecting Organic Pollutants in the
Soil/Groundwater System

The list of processes given above (and the following discussion in this section) presumes that no second (or immiscible) phase of organic material is present. A discussion of important processes associated with the fate of second-phase organics in the soil/groundwater system is provided in Section V.

An assessment of the relative importance of each of these transport and fate processes can be a difficult task because of the numerous chemical-specific and environment-specific properties affecting them (see Section V), and the likelihood that many of these properties will not be known.

An example of the variable behavior that may be seen for different groundwater pollutants is shown in Figure 44. The curves shown in this figure represent different concentration-time profiles that might be seen in an aquifer observation well following the continuous introduction of pollutants into the aquifer feeding the well. Some highly mobile and conservative (i.e., nondegradable) species, e.g., the chloride ion (Cl^-), will be affected principally by dispersion, and their observation-well concentrations will rise rapidly. Chemicals subject primarily to dispersion and adsorption, but not degradation, will also eventually reach high concentrations in the observation wells. Chemicals subject to degradation (e.g., hydrolysis or biodegradation), as well as adsorption and dispersion, will show a slower rise in concentration, may show a maximum, and will level off at some concentration (C) less than the injection concentration (C_o).

b. Equilibrium Partitioning

For phases in close physical contact, it will usually be reasonable to assume that the concentration of pollutants in the different phases is an equilibrium distribution, i.e., the distribution that would result - after a reasonable time period - when the rate of transfer from phase A → B is the same as from B → A. The assumption of equilibrium may be poor in some cases such as for: (1) rapid movement of groundwater through fissures and porous soils; (2) cold weather, especially when the ground is frozen; and (3) chemicals with very low solubility (< 10 ug/L) or high soil adsorption coefficient ($K_{oc} > 10^5$).

(1) Soil ↔ Soil Water[*]

Adsorption of organic solutes by soils will generally involve one or more of the following:

> (a) Adsorption on (or absorption in) the organic fraction of the soil;
>
> (b) Adsorption onto the surface of the inorganic soil minerals (especially clays); and

---

[*] Much of the information in this section is from Lyman (1982).

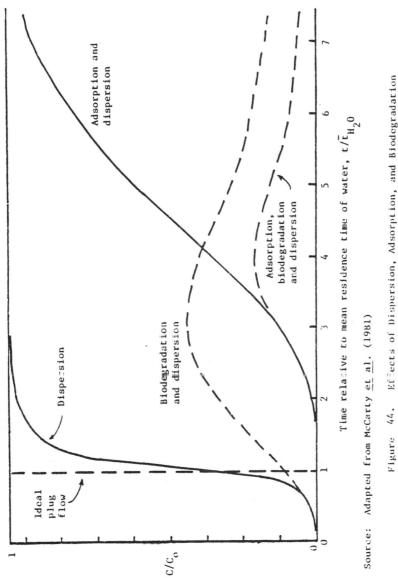

C_o = Injection concentration, C = observed concentration

Expected responses to a step change in concentration

Time relative to mean residence time of water, $t/\bar{t}_{H_2O}$

Source: Adapted from McCarty et al. (1981)

Figure 44.   Effects of Dispersion, Adsorption, and Biodegradation
on the Time Change in Concentration of an Organic
Compound at an Aquifer Observation Well

(c)   For chemicals which act as bases (i.e., tend to lose a hydrogen ion, H^+), complexation, cation exchange, or other forms of interaction with electron-negative sites in the soil minerals.

Processes (a) and (b) are most important for neutral organic chemicals (e.g., hydrocarbon fuels, chlorinated solvents), and the relative importance of these two depends primarily on the organic carbon content of the soil. (The nature of the chemical and the surface area of the soil particles are also important factors.) The organic carbon content of soils may range from a high of about 40 percent [by weight] (peat, forest top soils) to 0.1 percent or less (sand, clay). Agricultural soils will usually have from 1 - 10 percent organic carbon. In all soils, the organic carbon content can be expected to decrease with soil depth, and may reach negligible amounts (< 0.1 percent) below one or two meters depth. In these deeper regions, the relative importance of adsorption processes (a) and (b) may be very uncertain and can only be resolved by laboratory experiments with the chemicals and soils in question.

The extent of adsorption is frequently described with the Freundlich equation:

$$x/m = KC^{1/n} \qquad (2)$$

where   x = amount adsorbed on soil (μg)

m = mass of soil (g)

K = adsorption coefficient

C = concentration in water (μg/mL)

n = parameter, usually in range of 0.7 to 1.1

Table 9 provides, for example, the Freundlich constants K and 1/n for Trichloroethylene (TCE) on a variety of soils; for comparison, the values for activated carbon adsorption of TCE are also shown. The last column in the table shows the amount that would be adsorbed on the soil (μg per g of soil) if the equilibrium concentration in the groundwater (C) was 100 μg/L.

For soils with appreciable amounts of organic matter (> 0.1 percent for some chemicals, > 1 percent for others), it has been found that the amount of a chemical that is adsorbed - and, thus, the adsorption coefficient, K - is directly proportional to the organic carbon content of the soil. To represent this, a new adsorption coefficient, K_{oc}, is defined as K/f_{oc} where f_{oc} is the weight fraction of organic carbon in the soil ($0 < f_{oc} < 1$).

TABLE 9. SOIL ADSORPTION PARAMETERS FOR TRICHLOROETHYLENE

Material	K	1/n	x/m (µg at C = 100 µg/L)
Iron oxyhydroxide (goethite)	–	–	< 0.1
Amorphous manganese oxyhydroxide	0.0085	0.938	0.639
Kaolinite (a two-layer clay)	–	–	< 0.2
Montmorillomite (a three-layer clay)	0.0970	0.705	2.49
Soil from Wurtsmith AFB, Mich.	0.0759	0.324	0.337
Peat	0.148	0.737	4.41
Activated Carbon	377.0	0.623	6640.0

Source: Richter, 1981.

Numerous studies have shown that values of K_{oc} for a chemical are relatively constant (over a wide range of low concentrations) and reasonably independent of the soil used. Values of K_{oc} for organic chemicals range over 7 orders of magnitude (1 to 10,000,000) and, thus, extreme differences may be seen in the degree of adsorption for different chemicals. Table 9 provides K_{oc} values for a range of compounds. Reasonable estimates of K_{oc} may be obtained from regression equations which relate K_{oc} to other properties such as octanol-water partition coefficient, water solubility, bioconcentration factor in fish, and parachor (Lyman, 1982).

A final example will demonstrate the high variability that may be seen in soil adsorption of different organics. Table 10 provides information for benzene, trichloroethylene (TCE), and DDT, with assumed equilibrium concentrations in groundwater of 100 µg/L, 50 µg/L, and 1 µg/L, respectively. In these calculations, it is assumed that all adsorption is on (in) the organic carbon fraction of the soil, that f_{oc} = 0.001 (= 0.1 weight percent), that the soil has a bulk density of 2.5 g/cm^3 and a void volume of 50 percent, and that the Freundlich parameter $1/n$ has a value of 0.9. Values of K_{oc} are from Lyman (1982). The final column of Table 10 indicates that about 1/4 of the benzene and 1/2 of the trichloroethylene present in the soil/groundwater system are associated with the soil; the rest is in solution. For DDT, however, only 0.08 percent (1/1210) is in solution, while 99.92 percent is adsorbed on the soil.

The sample calculations above are representative of the type of analysis that may be carried out to assess the degree of adsorption for organics. Numerical predictions of this sort, however, should probably be considered to have an uncertainty of at least a factor of two (even when experimental chemical-specific and soil-specific data are used) when they are being applied to a specific site. This is due to a number of variables (see Section V) and the general inability to take all into account.

The effect of adsorption on the mobility of organic chemicals in groundwater is discussed later on in this section.

(2). Water -- Soil Air[*]

Partitioning of organic chemicals between groundwater and soil air (which may account for up to 50 percent of the volume of unsaturated zone soils) is important, not so much because of the mass distribution, but because transport of vapors may be significant. The important groundwaters here are clearly limited to the soil water in the unsaturated zone and the top layer of the groundwater in the saturated zone.

---

[*] Background information obtained from Thomas (1982a, 1982b).

TABLE 10. EXAMPLE ADSORPTION CALCULATIONS FOR BENZENE, TRICHLOROETHLYLENE, AND DDT[*]

Chemical	Water Concentration ($\mu g/L$)	K_{oc}	K	Amount Adsorbed $\mu g/g$	Amount Adsorbed $\mu g/cm^3$	Amount in Soil/Amount in Water
Benzene	100	83	0.083	0.0104	0.026	0.26
TCE	50	160	0.16	0.0108	0.027	0.54
DDT	1	243,000	243.0	0.485	1.21	1210.0

[*]See text for conditions of calculations.

Source: Arthur D. Little, Inc.

At low solute concentrations, the concentration of the chemical in the vapor phase will be directly proportional to its concentration in water:

$$C_{air} = H \cdot C_{water} \qquad (3)$$

The constant of proportionality, H, is called Henry's law constant. H is temperature-dependent, and, like K_{oc}, may range over 7 orders of magnitude for different chemicals.

The units used for H are variable, depending upon user preference. If both C_{air} and C_{water} are expressed in similar units (e.g., g/m^3), then a 'nondimensional' H is obtained. Such units are common in the older literature. More recently, H is reported with units of atm $\cdot$ m^3/mol or - most recently - in the SI units of $kP_a \cdot m^3/mol$. Table 11 lists values of H for several chemicals. A critical review of values of H for several organic chemicals was recently prepared by Mackay et al. (1982).

For chemicals of limited solubility (< 0.1 mole fraction), Henry's law constant may be estimated from the ratio of the chemicals vapor pressure (P_{vp}) to water solubility (S):

$$H = P_{vp}/S \qquad (4)$$

Since, for most chemicals, P_{vp} is a much stronger function of temperature than S, values of H will increase with increasing temperature in rough proportion to the increase in P_{vp}.

An inspection of the values of H' (nondimensional H) in Table 11 shows that the mass distribution of a chemical between groundwater and soil air is weighted on the water side for all except the most volatile chemicals (e.g., ethyl bromide, vinyl chloride, etc.). Table 12 indicates, for example, how three chemicals would be partitioned between the soil water and soil air (in the unsaturated zone) if it were assumed the air and water were of equal volume and that soil adsorption was negligible.

(3) Soil -- Soil Air

Relatively few data are available which describe the nature and extent of organic vapor adsorption onto soils, and this process is frequently ignored in models simulating the transport of organic vapors through soil.

To a first approximation, the relative concentrations of a chemical in soil air and soil should be proportional to the chemical's vapor pressure and inversely proportional to the square root of solubility:

TABLE 11. VALUES OF HENRY'S LAW CONSTANT FOR SELECTED CHEMICALS

	Henry's Law Const.			Henry's Law Const.	
	H $\frac{atm \cdot m^3}{mol}$	H' (Non-dim.)		H $\frac{atm \cdot m^3}{mol}$	H' (Non-dim.)
Low Volatility ($H < 3 \times 10^{-7}$)			**High Volatility** ($H < 10^{-3}$)		
3-Bromo-1-propanol	1.1×10^{-7}	4.6×10^{-6}	Ethylene dichloride	1.1×10^{-3}	4×10^{-2}
Dieldrin	2×10^{-7}	8.9×10^{-6}	Naphthalene	1.15×10^{-3}	4.9×10^{-2}
			Biphenyl	1.5×10^{-3}	6.8×10^{-2}
Middle Range ($3 \times 10^{-7} < H < 10^{-3}$)			Aroclor 1254	2.7×10^{-3}	1.2×10^{-1}
			Methylene chloride	3×10^{-3}	1.3×10^{-1}
			Aroclor 1248	3.5×10^{-3}	1.6×10^{-1}
Lindane	4.8×10^{-7}	2.2×10^{-5}	Chlorobenzene	3.7×10^{-3}	1.65×10^{-1}
m-Bromonitrobenzene	1.6×10^{-6}	7.4×10^{-5}	Chloroform	4.7×10^{-3}	2.0×10^{-1}
Pentachlorophenol	3.4×10^{-6}	1.5×10^{-4}	o-Xylene	5.1×10^{-3}	2.2×10^{-1}
4-tert-Butylphenol	9.1×10^{-6}	3.8×10^{-4}	Benzene	5.5×10^{-3}	2.4×10^{-1}
Triethylamine	1.3×10^{-5}	5.4×10^{-4}	Toluene	6.6×10^{-3}	2.8×10^{-1}
Aldrin	1.4×10^{-5}	6.1×10^{-4}	Aroclor 1260	7.1×10^{-3}	3.0×10^{-1}
Nitrobenzene	2.2×10^{-5}	9.3×10^{-4}	Perchloroethylene	8.3×10^{-3}	3.4×10^{-1}
Epichlorohydrin	3.2×10^{-5}	1.3×10^{-3}	Ethyl benzene	8.7×10^{-3}	3.7×10^{-1}
DDT	3.8×10^{-5}	1.7×10^{-3}	Trichloroethylene	1×10^{-2}	4.2×10^{-1}
Phenanthrene	3.9×10^{-5}	1.7×10^{-3}	Mercury	1.1×10^{-2}	4.8×10^{-1}
Acenaphthene	1.5×10^{-4}	6.2×10^{-3}	Methyl bromide	1.3×10^{-2}	5.6×10^{-1}
Acetylene tetrabromide	2.1×10^{-4}	8.9×10^{-3}	Cumene (isopropyl benzene)	1.5×10^{-2}	6.2×10^{-1}
Aroclor 1242	5.6×10^{-4}	2.4×10^{-2}	1,1,1-Trichloroethane	1.8×10^{-2}	7.7×10^{-1}
Ethylene dibromide	6.6×10^{-4}	2.8×10^{-2}	Carbon tetrachloride	2.3×10^{-2}	9.7×10^{-1}
			Methyl chloride	2.4×10^{-2}	3.6×10^{-1}
			Ethyl bromide	7.3×10^{-2}	3.1
			Vinyl chloride	2.4	99
			2,2,4-Trimethyl pentane	3.1	129
			n-Octane	3.2	136
			Fluorotrichloromethane	5.0	–
			Ethylene	>8.6	~ 360

Source: Thomas, 1982b

TABLE 12. EXAMPLE AIR-WATER PARTITIONING CALCULATIONS*

Chemical	H' (Non-dim.)	Amount in Water/Amount in Air
Benzene	0.24	4.2
Trichloroethylene	0.42	2.4
DDT	0.0017	590.0

*See text for conditions of calculation.

Source: Arthur D. Little, Inc.; values of H' from Thomas (1982b).

$$\frac{C_{air}}{C_{soil}} = \frac{H}{K} = \frac{P_{vp}/S}{K} = k\,P_{vp}/S^{\frac{1}{2}} \qquad (5)$$

where:

H = Henry's law constant
K = Freundlich soil adsorption coefficient
    ($1/n$ assumed = 1)
P_{vp} = vapor pressure
S = water solubility
k = constant of proportionality

With P_{vp} in units of atm and S in units of mol/m^3, k would appear to be on the order of 100; this would provide a dimensionless ratio for C_{air}/C_{soil}. Sample calculations using equation 3-4 (k $P_{vp}/S^{\frac{1}{2}}$ with k = 100) indicate that the C_{air}/C_{soil} ratios for DDT, trichloroethylene and benzene are about 1 x 10^{-5}, 5, and 2, respectively.

c. Degradation

Only two processes, biodegradation and hydrolysis, appear to have the potential to effect any significant amount of degradation of organic chemicals in the soil/groundwater system. In many instances, the refractory nature of the chemicals and/or environmental variables (e.g., sterile soils, cold weather) may reduce even these processes to minimal importance.

The third type of reaction mentioned in the overview, oxidation/reduction, can only be important when potent oxidizing or reducing agents are present. Such agents do not exist in natural soil groundwater systems. In natural systems, near-surface soils (and soil waters) will contain oxygen, a very mild oxidizing agent; and in deeper soils such mild reducing agents as H_2 and CH_4 may be present in small amounts. Direct oxidation or reduction by such agents is unlikely to be significant except for the most reactive of compounds.

(1) Biodegradation[*]

Biodegradation is one of the most important environmental processes that cause the breakdown of organic compounds. This capability is put to use, for example, in biological wastewater treatment plants and in the disposal of some wastes on land ("land farming") in a manner that promotes biodegradation. Natural biodegradation of a wide variety of organic chemicals can take place in surface water and soil systems.

The most significant group of organisms involved in biodegradation are microorganisms, including mostly a large variety of bacteria, fungi, and protozoa. The ability of various microorganisms to degrade

---

[*] Background information for this section was obtained from Scow (1982).

certain chemicals varies widely, and the importance of environmental conditions (warm temperature; adequate food, water and nutrients; pH; etc.) cannot be overstated.

The quantity of such organisms available to degrade chemicals is also a key factor. Near-surface soils in fields and woods may contain 100-1000 or more kilograms (wet wt) per hectare of these living micro organisms representing up to 10 individual microorganisms   per gram of soil (Scow, 1982). The population density of soil microorganisms, however, can drop off rapidly with increasing soil depth. This may be caused by a combination of factors, including decreasing amounts of food (organic matter) and nutrients with depth, lesser amounts of oxygen for anaerobic microorganisms, and a filtering of the organisms by the soil. Only under landfills, or in other areas where organic matter exists in deep soils, can biodegradation be expected to be a significant degradation process below 1 or 2 meters of the soil surface.

When a soil is heavily contaminated, e.g., by contamination from a leaking chemical lagoon, the combination of quantity and toxicity (to microorganisms) may effectively block biodegradation. In less severe cases, biodegradation may follow a period of acclimation lasting days or weeks.

Except for a few limited cases, rates of biodegradation in soil cannot be predicted for specific chemicals, and laboratory tests must be carried out. Similarly, the reaction pathways (through intermediate chemicals) leading sometimes to ultimate degradation must be determined in the laboratory.

Understandably, most of the existing data on rates of biodegradation in soils are for pesticides. Tables 13 and 14 provide an assortment of such data which may be illustrative only of rates for pesticides (often difficult to degrade) in near-surface soils. All of the rate constants in these tables are for primary degradation (i.e., any alteration of the initial compound) and imply a first order reaction, i.e.:

$$\frac{-dc}{dt} = kC \tag{6}$$

where   c = chemical concentration
        t = time
        k = degradation rate constant

The use of Equation 6 is exemplified by assuming a soil with 1 mg/L of lindane. The value of -dc/dt is thus (1 mg/L) x (0.0026/day) = 0.0026 mg/L · day. With first order reactions, the half-life (time for 50 percent to disappear) is 0.693/k. For lindane, the half-life in soil (under the test conditions) would, thus, be 270 days.

TABLE 13.   BIODEGRADATION RATE CONSTANTS FOR
ORGANIC COMPOUNDS IN SOIL[a] (DAY^{-1})

Compound	Test Method	
	Die-Away	$^{14}CO_2$ Evolution
Aldrin, Dieldrin	0.013	
Atrazine	0.019	0.0001
Bromacil	0.0077	0.0024
Carbaryl	0.037	0.0063
Carbofuran	0.047	0.0013
Dalapon	0.047	
DDT	0.00013	
Diazinon	0.023	0.022
Dicamba	0.022	0.0022
Diphenamid		0.123[b]
Fonofos	0.012	
Glyphosate	0.1	0.0086
Heptachlor	0.011	
Lindane	0.0026	
Linuron	0.0096	
Malathion	1.4	
Methyl parathion	0.16	
Paraquat	0.0016	
Parathion	0.029	
Phorate	0.0084	
Picloram	0.0073	0.0008
Simazine	0.014	
TCA	0.059	
Terbacil	0.015	0.0045
Trifluralin	0.008	0.0013
2,4-D	0.066	0.051
2,4,5-T	0.035	0.029

a.   All constants are from soil incubation
studies.   Except where noted, source is Rao
and Davidson (1980), a compilation of first
order rate constants derived from data published from other studies.

b.   Optimum degradation rate, from Donigan et
al. (1977).   Test method not specified.

TABLE 13.   BIODEGRADATION RATE CONSTANTS FOR ORGANIC COMPOUNDS
IN ANAEROBIC SYSTEM[a]

$$(DAY^{-1})$$

| Compound | In Soil[a] | | In Sewage Sludge[b] |
	Die-Away	$^{14}CO_2$ Evolution	
Carbofuran	0.026		
DDT	0.0035		
Endrin	0.03		
Lindane		0.0046	
PCP		0.07	
Trifluralin	0.025		
Mirex			0.0192
Methoxychlor			9.6
2,3,5,6-Tetrachlorobenzene			12.72
Bifenox			6.27

a.   Flooded soil incubation studies as reported in Rao and
Davidson (1980), a compilation of first order rate constants
derived from data published from other sources.

b.   As reported by Geer (1978).   Test method not specified.

(2) Hydrolysis[*]

Hydrolysis is a chemical transformation process in which an organic molecule, RX, reacts with water, cleaving a carbon-X bond and (generally) forming a new carbon oxygen bond. The net reaction is most commonly a direct replacement of X by OH:

$$R-X \xrightarrow{\text{H}_2\text{O}} R-OH + X^- + H^+ \tag{7}$$

Hydrolysis is not just one reaction type (as the example above), but a family of reactions involving attack by water at the sites of various functional groups (e.g., alkyl halides, esters, expoxides, nitriles, carbamates, and organophosphates). The reaction mechanism and products may differ significantly from compound to compound. Other types of reactions (of organic chemicals with water) that will also have to be considered in some cases include acid:base, hydration, addition, and elimination:

Acid: Base

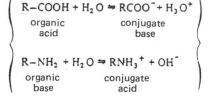

$$\left\{ \begin{array}{c} R-COOH + H_2O \rightleftharpoons RCOO^- + H_3O^+ \\ \text{organic} \qquad\qquad \text{conjugate} \\ \text{acid} \qquad\qquad\quad \text{base} \\[1em] R-NH_2 + H_2O \rightleftharpoons RNH_3^+ + OH^- \\ \text{organic} \qquad\qquad \text{conjugate} \\ \text{base} \qquad\qquad\quad \text{acid} \end{array} \right\} \tag{8}$$

Hydration

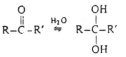

aldehyde/ketone   acetal/ketal

Addition

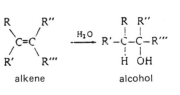

alkene   alcohol

Elimination

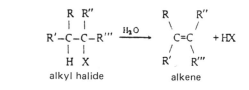

alkyl halide   alkene

---

[*]Background information obtained from Harris (1982).

Except for such reactions as acid:base, which are fast and reversible, most of the hydrolysis reactions of interest will involve highly variable rate constants, reaction pathways, and reaction products. Figure 45 shows, for example, that half-lives for degradation via hydrolysis range over at least 7 orders of magnitude under typical ambient conditions (pH 7, 25°C).

Temperature, pH and the presence of catalysts are very important variables determining the rate of hydrolysis. For temperature, if the energy of activation for hydrolysis is assumed to be 17-18 kcal/mol (for most chemicals, it is in the range of 12-25 kcal/mol), the following rules of thumb may be used for temperatures in the 0 - 50°C range:

- A 1°C increase (decrease) in temperature causes a 10 percent increase (decrease) in the rate constant;

- A 10°C increase (decrease) will increase (decrease) the rate constant by a factor of 2.5;

- A 25°C increase (decrease) will increase (decrease) the rate constant by a factor of 10.

Many hydrolysis reactions are catalyzed by the presence of OH^- (bases), H^+ (acids), or other constituents (e.g., certain heavy metals such as Cu^{++}), which may be present in the groundwater. Thus, tests with site-specific waters may be desirable in some cases. The extent of adsorption on soil is another factor which may affect the rate of hydrolysis. The available literature provides little guidance in this area.

It is generally observed that hydrolysis of organic chemicals in water is first order in the concentration of the organic species, [RX]; i.e.:

$$-d[RX]/dt = k_T [RX] \qquad (9)$$

where:

$[RX]$ = chemical concentration
t = time
k_T = total hydrolysis rate constant

To take acid and base catalysis into consideration, k_T is usually considered to consist of three terms representing acid-catalyzed, neutral, and base-catalyzed reactions:

$$k_T = k_H[H^+] + k_0 + k_{OH}[OH^-] \qquad (10)$$

Many organic functional groups (Table 15) are relatively or completely inert with respect to hydrolysis. Other functional groups which may hydrolyze under environmental conditions are listed in Table 16.

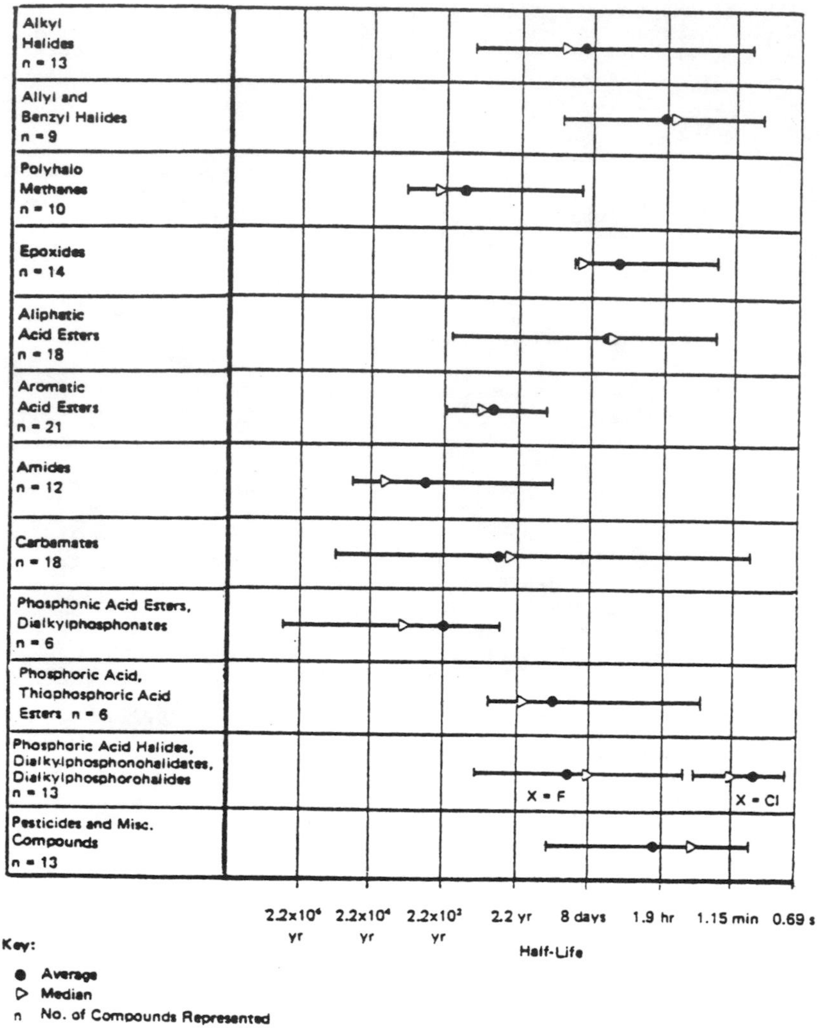

Key:
- ● Average
- ▷ Median
- n   No. of Compounds Represented

Source: Adapted by Fiksel and Segal (1980) from data of Maybey and Mill (1978)

Figure 45.   Examples of the Range of Hydrolysis Half-Lives for Various Types of Organic Compounds in Water at pH 7 and 25°C

TABLE 15. TYPES OF ORGANIC FUNCTIONAL GROUPS THAT
ARE GENERALLY RESISTANT TO HYDROLYSIS[a]

Alkanes	Aromatic nitro compounds
Alkenes	Aromatic amines
Alkynes	Alcohols
Benzenes/biphenyls	Phenols
Polycyclic aromatic hydrocarbons	Glycols
Heterocyclic polycyclic aromatic hydrocarbons	Ethers
Halogenated aromatics/PCBs	Aldehydes
Dieldrin/aldrin and related halogenated hydrocarbon pesticides	Ketones
	Carboxylic acids
	Sulfonic acids

a. Multifunctional organic compounds in these categories may, of course, be hydro-
lytically reactive if they contain a hydrolyzable functional group in addition to the
alcohol, acid, etc., functionality.

Source: Harris (1982)

TABLE 16. TYPES OF ORGANIC FUNCTIONAL GROUPS THAT
ARE POTENTIALLY SUSCEPTIBLE TO HYDROLYSIS

Alkyl halides	Nitriles
Amides	Phosphonic acid esters
Amines	Phosphoric acid esters
Carbamates	Sulfonic acid esters
Carboxylic acid esters	Sulfuric acid esters
Epoxides	

Source: Harris, 1982

In a few limited cases, it is possible to predict the rate of hydrolysis for organic compounds (Harris, 1982), but even here the uncertainties involved would suggest that estimated rate constants be considered order-of-magnitude estimates.

d. Transport

As was shown in Figure 43, the three transport processes that require consideration for organics in the soil/groundwater system are:

- transport with water - (deletes percolation through the unsaturated zone and movement with the groundwater);

- volatilization and transport of the vapors, through the unsaturated zone, to the atmosphere;

- erosion or entrainment of (contaminated) surface soils.

Only the first two processes are significant for pollutants already in groundwater, and some additional discussion on them is provided below.

(1) Transport With Water

Pollutants will be transported with groundwater, primarily in solution. The most important factor to understand is the chromatographic effect whereby chemicals with higher soil adsorption coefficients will be retarded with respect to those with lower soil adsorption coefficients.

A significant body of literature exists on leaching, including numerous soil column leaching tests, the measurement of retention factors (R_f), and the development of predictive mathematical models (See Hamaker. 1975; Tinsley, 1979; Letey and Farmer, 1974; Thibodeaux, 1979; Fried and Combarnous, 1971; Leistra, 1973; Letey and Oddson, 1972; and Section V of this report).

If simple equilibrium adsorption, molecular diffusion, and mixing were the only processes involved, then - in homogeneous soils - the concentration-depth (or distance) profile of a pollutant in response to a pulse of injected pollutant would include a symmetrical peak; this is shown by the dashed curve in Figure 46. In fact, nonsymmetrical peaks (involving a longer tail and sharper front) are often seen, especially at higher water flow rates. Figure 46 shows schematically the nature of such curves. The nonsymmetrical profiles may be due to nonequilibrium adsorption kinetics (i.e., rate of desorption < rate of adsorption), and/or to changes in the effective adsorption coefficient due to the pressence of previously adsorbed chemical.

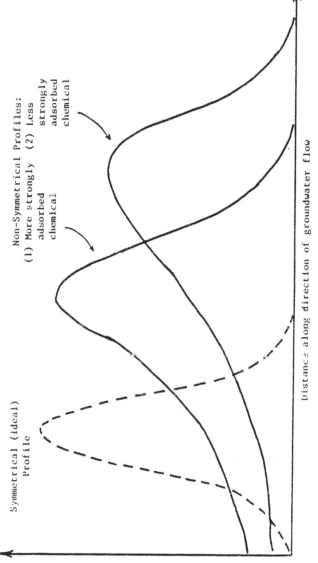

Source: Arthur D. Little, Inc.

Figure   46.   Schematic Representation of Concentration-Distance Profiles for
Pollutants Moving through Soil (All Profiles Assume a Discreet,
One-Time Injection of the Pollutant)

(2)   Volatilization and Vapor Transport

Four processes appear to be involved in chemical volatilization (from groundwaters) and the subsequent loss of vapors to the atmosphere:

(a)   Diffusion

- Driven by concentration gradient, in soil-air, of chemical between groundwater and soil surface.

- Surface winds deplete surface concentrations.

(b)   Sweep Flow

- Bulk transport of the pollutant vapors along with gases (e.g., methane, carbon dioxide) generated in the soil.

- Very important for landfills.

(c)   Barometric Pressure Pumping

- Continual compression and expansion of soil air follows changes in barometric pressure.

- The short term volatilization flux of pollutants is significantly affected.

- Most important where depth to contaminated groundwater or waste is small compared to depth of unsaturated zone.

(d)   Thermal Gradients

- Diurnal, short term and seasonal air temperature changes will lead to thermal gradients in the top few meters of the soils.

- In winter, the presence of cold (more dense) air at soil surface and warmer (less dense) air below could lead to convection currents.

Many factors are involved in these processes (Table 17). Models for the prediction of volatilization rates frequently ignore all processes except diffusion. The current knowledge on this volatilization process is rather limited, although a number of experimental (laboratory and field) and theoretical studies have been conducted (See Thomas, 1982a; Thibodeaux, 1979, 1981; Shen, 1981; and Farmer et al., 1980).

TABLE 17   FACTORS AFFECTING RATE OF VOLATILIZATION FROM SOIL

METEOROLOGICAL

- Changes in barometric pressure
- Air (and soil) temperature
- Rainfall, infiltration
- Snow cover
- Ground frost
- Wind speed
- Relative humidity

SITE AND SOIL

- Type and extent of soil coverage
- Soil porosity
- Moisture content of soil
- Lateral extent (reach) of site
- Protection from wind
- Adsorption capacity of soil (organic carbon content)

WASTE

- Form of waste (pure chemical, mixture, aqueous solution)
- Area covered by waste
- Physicochemical propoerties (D_S, H, γ, K_{OC}, P_{VP}, S, . . .)
  relating to volatilization, dissolution, adsorption,
  diffusion, partitioning, etc.

- Stability (resistance to hydrolysis, etc.)

Source: Arthur D. Little, Inc.

What is clear, however, is that the volatilization loss pathway can be significant, not only for chemicals with high concentrations in soil air, but also those (e.g., DDT, polychlorinated biphenyls, polynuclear aromatics) which have very high soil adsorption coefficients and are resistant to degradation. For this latter group of chemicals, downward migration through soils is almost negligible, and slow volatilization is the only transport pathway open.

3. Field and Laboratory Techniques for Assessing Groundwater Contamination

1. DRILLING METHODS

As part of the subsurface investigation of a site, wells may have to be bored or drilled in order to sample the sediment and rock, and/or to monitor groundwater flow. The method (of boring or drilling) employed to sink a well depends on the depth required, the use of the well, and on the material (sediment or rock) present at the site.

Most unconsolidated sediment (except sand and gravel) can easily be penetrated by auger boring, but only to limited depths. For drilling into rock, cable-tool percussion equipment works the best while rotary drilling is well suited for deep holes in unconsolidated sediments (especially sand and gravel). A brief description of each of these drilling methods follows, along with a summary of the advantages and disadvantages of each method.

a. Auger Boring

Figures 47 and 48 show examples of several augers. These tools are used to bore holes according to the procedure described in EPA (1980b):

> "In auger boring, the hole is advanced by rotating and pressing a soil auger into the soil and withdrawing and emptying the auger when it is full. Since water tends to prevent accumulation of soil in the auger, the borehole is kept dry as much as possible. Hand augering can be easy or difficult depending upon whether clay, sand, or gravel, respectively, is being removed. Small-diameter helical or posthole augers can be used to advance 5 to 30 cm (12-inch) diameter holes by hand to depths of 6 to 9 meters (20 to 30 feet) (Figure 47). If a tripod and pulley are set up to aid in pulling the auger from the hole, depths of 24 meters (80 feet) can be reached. If the hole can be kept open below the water table (usually only in cohesive material), screen and casing can be set, backfilled, and developed.
>
> The process becomes much simpler and less time consuming if power augers are used. Here, flights of spiral, hollow-stem augers are forced into the ground while being

Figure 47.   Small Helical Augers (left) and
a Posthole or Iwan Auger (right).

Source: EPA, 1980b.

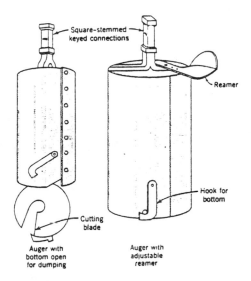

Figure 48.   Augers Used for Boring Wells.

Source: Todd, D.K,, <u>Groundwater Hydrology</u>, John Wiley and Sons,
         Inc., New York,© 1959.

rotated; the spiral action of the augers
conducts cuttings to the surface (Figure 47).
On completion of drilling, a small-diameter
casing and well point are pushed to the
desired depth. When bucket augers are used,
a large diameter barrel (up to 122 cm (48
inches)) fitted with cutting blades is
rotated into the ground until it is full
(Figure 48). The earth-laden bucket is then
brought to the surface, pulled to one side,
and dumped. This process is repeated to
completion depth."

The procedure for securing a collapsing bored hole (or drilled
hole) is through the use of metal, concrete or tile casings. Casings
are cylindrical tubes inserted to physically hold up the sides of the
hole for continued boring through the casing. Augers work best in
formations that do not collapse and are as effective as any other
penetrating device where cohesive clay formations are encountered (Todd,
1959).

The following list is a cumulative summary of the advantages and
disadvantages of auger boring equipment from EPA (1980b), Sowers (1970)
and Todd (1959).

Auger Boring

Advantages	Disadvantages
Inexpensive.	Limited penetration; normally 30 to 46 meters (100 - 150 feet) maximum (6 meters for hand augers).
Small, high-mobility rigs can reach most sites.	
Can be used to quickly construct shallow well clusters.	Vertical leakage through sediment left in borehole through which drive point is forced to completion depth. No method to isolate screened zones of aquifer.
If borehole prematurely reaches refusal depth, setup time is low and rig can be moved rapidly.	
No drilling fluids introduced into the borehole; no possibility of diluting formation water.	Careful attention during drilling is required to obtain correct log of formation materials penetrated.
	Unable to collect groundwater samples during drilling.

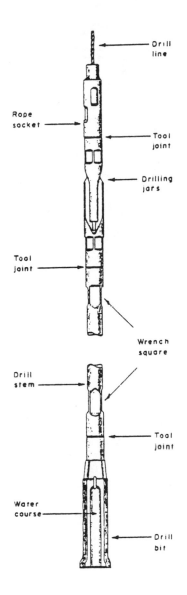

Figure   49. Four Components of the String
              of Drill Tools for Cable-tool
              Percussion Drilling.

Source: Johnson Division, Groundwater and Wells, UOP Inc., © 1982.

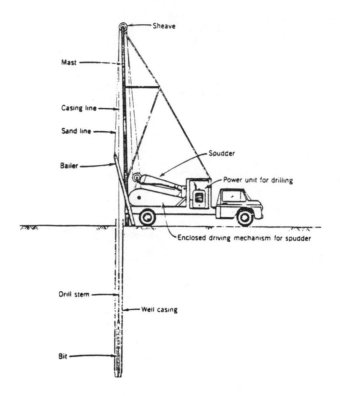

Figure 50. Truck-mounted Cable-tool Drilling Equipment.

Source: Davis, S.M. and R.J.M. DeWiest, _Hydrogeology_, John Wiley
and Sons, Inc.,© 1966.

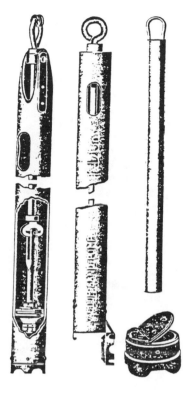

Figure 51.   Sand Pumps and Regular Bailer used
During Cable-tool Drilling.

Source: Johnson Division, Groundwater and Wells, UOP, Inc.,© 1982.

Advantages	Disadvantages
	Core sampling is possible only if hollow-stemmed auger flights are used.
	Can be used only in unconsolidated sediments or soft rock.
	Borehole will collapse in cohensionless sediment (sand and gravel.
	Often impossible to use in soils below the water table.

b.   Cable-Tool Percussion Drilling

Regular lifting and dropping (percussive action) of a heavy string of tools (cable-tools) deepens a hole being drilled with cable-tool equipment. A chisel-shaped bit on the end of the drill string breaks or crushes hard rock into small fragments or loosens unconsolidated sediments. The vertical motion of the drill string mixes the crushed or loosened particles with water to form a slurry. If no water is present in the formation being penetrated water must be added to form the slurry. When water-bearing formations are penetrated, they must be cased or grouted in order to deepen the hole (USEPA, 1980b).

Figure 49 shows all the components of a string of cable tools - a rope socket, a set of jars (to aid in loosening tools stuck in a hole), a drill stem (for weight and length), and a drilling bit (Walton, 1970). The drilling rig (Figure 50) for the cable-tool method consists of a mast, a multiline host, and an engine usually mounted on a truck (Davis and DeWiest, 1966).

After about 4 to 5 feet of drilling, the crushing action of the bit becomes impeded by the accumulation of the cuttings. At this point, the bit is removed from the hole and a bailer (Figure 51) is "allowed to fall to the bottom of the hole where it strikes the water, causing a rapid surge of water and cuttings upward within it" (Walton, 1970). The bailer is then withdrawn from the hole with the cuttings.

In unconsolidated formations, casings should be driven into the bottom of the hole to avoid caving (Todd, 1959). In formations prone to caving, unconsolidated sand and gravel, this method is least effective. This approach is best in consolidated rock such as limestone and sandstone. It is not capable of drilling as quickly or as deeply as rotary methods.

The cumulative summary of advantages and disadvantages of the cable-tool percussion drilling methods has been gathered from EPA, (1980b),Todd,(1959)Walton,(1973),and Campbell and Lehr,(1973).

Cable-Tool Drilling

Advantages	Disadvantages
Simple equipment and operation.	Slow.
Good seal between casing and formation if flush joint casing is used.	Use of water during drilling can dilute formation water.
Good disturbed soil samples. Known depth from which cuttings are bailed.	Potential difficulty in unconsolidated sand gravel.
Core samples can be collected.	No formation water samples can be taken during drilling unless open-ended casing is pumped, or a screen set.
If casing can be bailed dry without sand heaves, a formation-water sample can be collected.	Heavy steel drive pipe is used and could be subject to corrosion under adverse contaminant characteristics.
Can be used in unconsolidated sediments and consolidated rocks.	Cannot run a complete suite of geophysical well logs because of casing.
Only small amounts of water are required for drilling.	
Once water is encountered, changes in static or potentiometric levels are readily observable.	
Suitable for rugged terrain.	
Low initial investment in equipment.	

c.   Hydraulic Rotary Drilling

Figure 52 illustrates the equipment used for hydraulic rotary drilling consisting of a derrick, or mast, a rotating table, a pump for drilling mud, a hoist and the engine. The method is described simply in the Johnson Division Report (1972):

> "Hydraulic rotary drilling consists of cutting a borehole by means of a rotating bit and removing the cuttings by continuous circulation of a drilling fluid as the bit

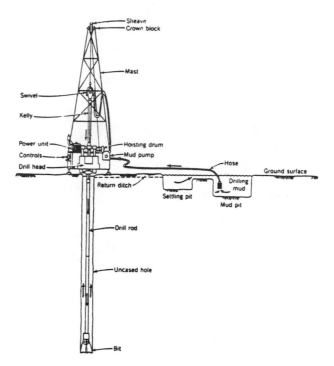

Figure   52. Major Components of Rotary Drilling
              Equipment. Arrows Indicate Direction
              of Mod Circulation.

Source: Davis, S.M. and R.J.M. DeWiest, Hydrogeology, John Wiley
         and Sons, Inc.,© 1966.

penetrates the formation materials. The bit
is attached to the lower end of a string of
drill pipe. In the conventional rotary
system, drilling fluid or drilling mud is
pumped down through the drill pipe and out
through nozzles in the bit. The mud fluid
then flows upward in the annular space around
the drill pipe to the surface, with the
cuttings carried in suspension. At the
surface, the fluid is channeled into a
settling pit and then into a storage pit. It
is again picked up by the pump after dropping
the bulk of its load of cuttings."

The direct rotary drilling method is heavily dependent on fluid
circulated through the hole during drilling (Figure 52). The fluid is
generally drilling mud (bentonitic clay mixed in water), and it
contributes to the hole stability by coating the hole (Freeze and
Cherry, 1979). When even heavy drilling mud does not stabilize the
hole, casing must be emplaced (as in the other drilling methods) for
drilling to continue. The most common drilling problems encountered
with this method are caving, lost circulation, and artesian water
flowing into the hole.

Overall, the various rotary rigs are the fastest and most
convenient means of drilling, especially in unconsolidated sediments.
The list of advantages and disadvantages has been compiled from EPA
(1980b), Johnson Division (1972), Freeze and Cherry (1979), and Todd
(1959).

Hydraulic Rotary Drilling

Advantages	Disadvantages
Fast.	Expensive.
Dilution of formation water is limited by formation of a filter cake on borehole walls.	Requires complex equipment and operation.
Formation water sample can be obtained with a special technique.	There is a potential for vertical movement in formation stabilizer material placed between casing and borehole wall after completion.
Good disturbed soil samples from known depths if travel time of borehole cuttings is taken into account, although sorting may occur.	

Advantages	Disadvantages

Flexibility in final well
construction.

Can run a complete suite of
geophysical well logs.

Core samples can be collected.

Can be used in unconsolidated
sediments and consolidated rocks.

    d.   Sampling

    Sediment samples can be obtained during any of these boring or
drilling procedures. The two types of samplers used most frequently are
the split-spoon samplers and coring samplers. At regularly spaced
intervals and at every change in sediment or rock type a sample is
collected by either method to document all subsurface lithologies.

    The split-spoon sampler (Figure 53) consists of a thick-walled
steel tube split lengthwise (Sovers, 1970). A cutting shoe is attached
to the end lowered into the hole first. This shoe penetrates the
sediment and prevents the sediment from falling back out of the tube.
Samples can be collected using this tool in most unconsolidated
sediments (except gravel) and even soft rock (Sowers, 1170).

    For collecting samples in consolidated formations a sampling method
known as coring is required. Coring is basically a method of drilling
during which a sample is obtained. Cores collected as samples are
between 4 and 30 inches in diameter and about 10 feet in length
(Campbell and Lehr, 1973).

    Double tube core barrels with an upper core-catcher assembly with
spring fingers and a lower catcher with a spring activated pivoted
attachment behind the drill bit are a common type of core barrel
(Campbell and Lehr, 1973). Figure 54 shows an example of a simple core
barrel.

    The sampling procedure takes place during drilling where the cure
barrel and bit rotate while water or thin drilling mud are forced down
the barrel and into the bit under high pressure. This pressure forces
the rock core upward into the barrel where the lower cone-catcher
assembly prevents it from falling back out (Sowers, 1970). This method
of sampling does not work well in unconsolidated sediments or in
decomposed rock. However, when used, it yields a continuous record of
subsurface formations.

2.   MONITORING WELL CONSTRUCTION

    In hydrogeologic investigations of groundwater contamination
incidents, monitoring wells are frequently used for sampling. A number

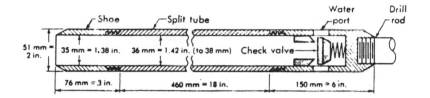

Figure 53.   Standard Split-Spoon Sampler.

Source: Sowers, G.F., Soil Mechanics and Foundations: Geotechnical
        Engineering, Macmillan Publishing Company, New York,© 1970.

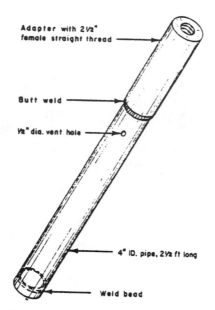

Adapter with 2½"
female straight thread

Butt weld

½" dia. vent hole

4" ID. pipe, 2½ ft long

Weld bead

Figure   54.   A Small Diameter Core Barrel
Suitable for Coring Inside
6-Inch Pipe

Source: Campbell, M.D. and J.H. Lehr, <u>Water Well Technology</u>,
McGraw-Hill Book Company, New York,© 1973.

of well configurations have been designed to meet the varied sampling needs and evaluate: the most effective monitoring configurations.

The sampling design used in a monitoring well depends upon the objective of the monitoring program. Groundwater quality monitoring is usually done to determine what has already happened at an existing site, to provide a warning of what is starting or may start to happen, or to provide a baseline description. The most important consideration is establishing the vertical interval to be sampled. Generally, monitoring wells can provide two types of samples: depth-integrated samples, and point-source samples. With the depth-integrated approach, samples are drawn from wells or piezometers with long screens. With the point-sample approach, water samples are drawn from discrete levels in the groundwater zone. Each has its relative merits depending on the monitoring objectives.

The most popular monitoring well designs used to accomplish groundwater sampling are: (1) a single well, screened or open over a single vertical interval; (2) well clusters; (3) a single well with multiple sampling points; and (4) sampling during drilling. A discussion of each of these methods follows.

According to the EPA (1980b), "wells screened over a single vertical section of an aquifer are the most common construction method used to obtain groundwater samples from unconsolidated sediments or semi-consolidated rocks." (Figure 55) This type of design involves depth-integrated sampling. In consolidated rocks, the same effect can be obtained by sampling from an uncased hole. When the sampling interval includes the entire thickness of the aquifer, this method can aid in evaluating the areal distribution of aquifer contamination. However, data accuracy would decrease significantly in a study made on a smaller interval, since contamination distribution may not be uniform within the aquifer. The entire plume might not be sampled. A limitation of this type of well is that the vertical distribution of contaminant cannot be studied using depth-integrated sampling. In this type of sampling, the connate water may mask contamination or dilute the concentration of the contaminant. A single screen well is effective in studying areal extent of contamination when the sampling interval includes the entire aquifer. The data can then be used to plan more sophisticated monitoring wells. This sampling can also be useful when taking point-source samples over a short interval. However, depth-integrated sampling over a short interval provides incomplete data.

Wells with multiple sampling points overcome the dilution and vertical sampling problems of single-screen wells. Figure 56 illustrates some of the alternatives available. One of these alternatives is to cluster a number of single-screen wells. Such clusters consist of closely spaced, small-diameter wells completed at different depths (Figure 57). According to the EPA (1980b) "well clusters are by far the most common and successful technique to date for delineating groundwater contamination." The limitation that remains however is that there will always be an unsampled interval using this

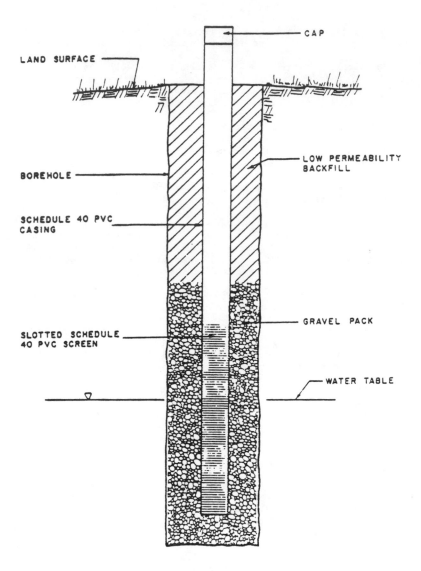

Figure   55. Typical Monitoring Well Screened Over a Single
              Vertical Interval.

Source: EPA, 1980b.

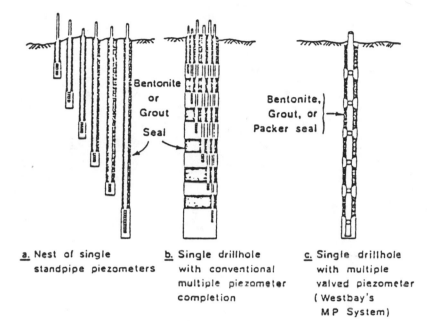

Figure   56.   Examples of Piezometer Configurations for Groundwater
               Monitoring at Numerous Depths at a Site.

Source: Cherry, 1981.

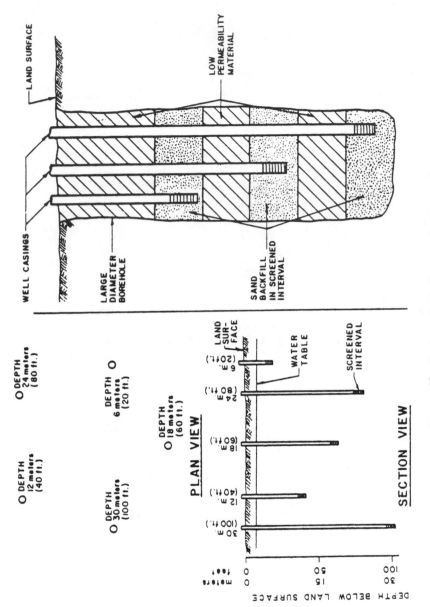

Figure 57. Typical Well Cluster Configurations

Source: EPA, 1980b

technique. For this reason, completion depths should be carefully planned to maximize exposure. The EPA (1980b) lists several approaches to selecting depths:

- "a pair of wells, one screened at the top and the other at the bottom of the aquifer;

- a three-well cluster with screens set on the top, middle and bottom of the aquifer under investigation,

- clusters in which the screened intervals are separated by preselected intervals, such as:

    - the 3-, 6-, 9-, 12-, and 18-meter (10-, 20-, 30-, 40-, and 60-foot) screen depths;

    - the 6-meter (20-foot) separation from 6 to 30 meters (20-100 feet);

    - terminating 2 to 3 wells at 3 to 4.5 meters (10 to 15 foot) intervals."

Finally, the EPA (1980b) provides this insight into the use of well clusters:

> "Some uncertainty will always exist as to the actual vertical distribution of the contaminant. Construction of more wells per cluster is not the answer; only a limited number of wells can be constructed close enough together to delineate vertical contaminant distribution at one particular point. Also, construction costs and the time required to complete the cluster would become prohibitive factors. The only way to obtain the most complete picture of leachate distribution is to collect groundwater samples during drilling."

Another method to obtain multiple sample points involves setting screens or casing perforations at intervals within a single borehole. Figure 58 illustrates a bundle piezometer and a multilevel point sampler which can be used in this type of operation. Vertical spacing of the sampling points depends on the data needs and the funds available for the particular investigation. In general, cost increases as the number of sampling points increases. When using this technology it is important to ensure the isolation of each sample point from the others. There can be no communication between intervals or unreliable samples will result. To accomplish this isolation, the sampling intervals are usually separated by packers of grout or bentonite. Besides careful packer placement, this configuration requires low pumping rates. This ensures that the samples are drawn only from the screened horizon, with

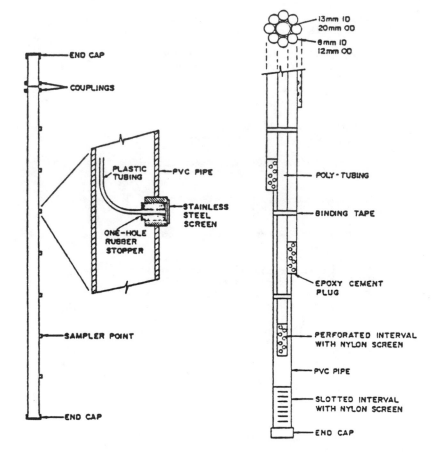

Figure   58.   Two Multilevel Sampling Devices for Monitoring of
Groundwater Quality in Cohesionless Deposits Below
the Water Table.

Source: Cherry, 1981.

no vertical warping. Otherwise, an unrepresentative sample would be taken. With this technique, as with well clusters, however, there will be an unsampled interval. This problem actually turns out to be a major drawback because stratification of contamination within an aquifer is a common phenomenon. If sampling depths are not chosen carefully to intersect these zones, the vertical distribution obtained will be erroneous. Sampling during drilling can help overcome this problem of fixed point sampling of the other multisample methods. Figure 59 illustrates the configuration for sampling while drilling. This has been a very effective technique as long as precautions are taken to prevent contaminating the sample with drilling fluids and mud which would ruin the accuracy of the data. According to the EPA (1980b), "the main advantage of this type of sampling is that the stratification of contaminated slug can be defined with reasonable accuracy prior to setting a permanent casing and screen. With this information, the well can be designed for the most advantageous sampling or withdrawal of contaminant at that point in the aquifer. Changes in the vertical distribution can then be monitored closely."

3.   GEOPHYSICAL TECHNIQUES APPLICABLE TO GROUNDWATER INVESTIGATIONS

   a. Introduction

   Geophysical techniques are directly applicable to groundwater movement and pollution investigations in several important respects:

- Surface geophysics are customarily used for an initial nondestructive, general site survey in order to describe the geologic framework and to help define the extent of a groundwater contamination problem in a cost-effective manner.

- Certain higher resolution surface geophysical methods (cf. metal detection) can be used to locate buried waste drums and, thus, pinpoint sources of contamination.

- Based on the interpretation of surface-geophysical data, the optimal location for drilling observation and monitoring wells can be determined.

- Borehole geophysical logging techniques should then be used to confirm, refine, and calibrate the interpretation of surface geophysical data. These techniques permit direct measurements of fluid flow and chemistry and, thus, an evaluation of the nature and extent of the pollution problem. In-situ sampling and monitoring programs can then be formulated.

- Finally, the geophysical data set provides spatially and temporally continuous information which

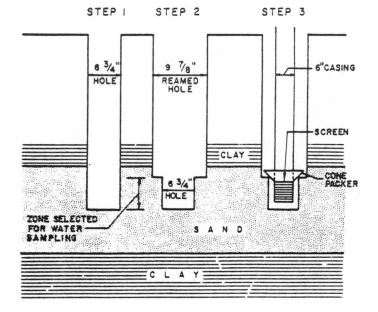

Figure   59.   Procedure for Water Sampling During Drilling.

Source: EPA, 1980b.

complements discrete sets of laboratory
measurements. Models of pollutant identity and
migration in groundwater are severely constrained by
geophysical and geochemical data derived from field
surface and borehole measurements. Hence,
confidence in model predictions is bolstered by
field geophysics and laboratory testing.

In addition, the geophysical field investigations can be used to support
test drilling:

- Borehole geophysical logging provides for lithologic
  sections and rock property analysis as well as for
  borehole cross-correlations between geologic strata.

- Surface geophysics can be used to supplement and
  extrapolate test drilling information after
  calibration against driller's logs.

The limitations inherent in the use of surface and borehole
geophysics are discussed here in general terms. Organic contaminants,
such as solvents, petroleum distillates, pesticides, and herbicides can
change the properties of the pore fluid and matrix geology in several
general ways:

- act as low-solubility, immiscible fluids, thus
  coating the grains and plugging the natural
  porosity;

- cause precipitation of certain complexing ions in
  formation waters, thus decreasing their activity in
  solution and the total dissolved solids (TDS)
  concentration;

- act as polar or ionic components, thus affecting the
  pH and Eh of groundwater and changing its electrical
  conductivity (resistivity).

These mechanisms must be identified, understood, and calibrated by
laboratory and field testing for suspected contaminants before
interpreting geophysical data. If a field resistivity anomaly of a
certain porous or waterbearing formation is to be ascribed with
confidence to organic pollution, calibration of in-situ natural
characteristics is necessary. Seasonal fluctuations of water table
levels at various locations and the degree of variation in porosity and
saturation of unpolluted formations must be assessed with geophysical
techniques before inferring the presence and effects of organic
pollutants on the natural system.

b.   Surface Geophysical Survey Methods

Surface geophysical methods are used widely for a nondestructive
and cost-effective initial survey and assessment of potentially
contaminated sites. The survey can be carried out by airborne
instrumentation, flown on gridded patterns, for gross resolution; or by
ground-based portable equipment, for finer-scale information. The
standard techniques fall within the following main categories:

- gravity surveys
- magnetic surveys
- electrical conduction
- electromagnetic induction
- elastic wave or acoustic sounding techniques such as
  seismic reflection and refraction.

Those geophysical methods based on potential theory (gravity, magnetic
or electrical) lack unique source solutions and require the combined use
and correlation of several sounding curves, and borehole calibrations
and sampling, to insure correct interpretation of results. The
principles of geophysical survey methods are discussed in detail in the
textbooks of Grant and West (1965) and Telford et al. (1976).

Several techniques applicable to the study of groundwater pollution
are briefly discussed and illustrated with examples. The relative
merits and disadvantages of the most widely used geophysical survey
methods in groundwater studies are compared in Table 18.

(1) Gravity Surveying

Gravity survey field work consists of measuring the gravitational
acceleration ($g = 980$ gals $= 980$ cm/sec^2) at grid stations covering the
area of interest. Corrections for topography, latitude, elevation and
earth tides must be applied, with respect to a local reference
equipotential surface. Lateral variations in density give rise to a
Bouguer gravity anomaly, which is the significant datum at each station.
However, an interpretation of Bouguer gravity profiles (in milligals) is
ambiguous in the sense that the shape, size and depth of the anomaly
source is nonunique. Sensitive gravimeters or gradiometers require
calibration and drift corrections to achieve practical sensitivities of
0.01 mgal. Since the gravitational acceleration varies inversely with
distance squared, finer grid spacings are needed to resolve shallow
anomalies. Ground surveys by a three- to four-person team are needed for
accurate contouring of anomalies which have a positive sign for excess
buried mass and a negative sign for low density, water-filled
sedimentary formations. The interpretation of gravity residuals from
regional gravity trends and contours requires use of graphical
techniques, analytical models and computer analysis. To insure a
correct interpretation, the true depth of strata and their respective
densities must be determined from test drilling and sampling. The

TABLE 18. COMPARISON OF REMOTE SENSING TECHNIQUES

Technique	Purpose	Advantages	Limitations
Electrical Resistivity Lateral profiling	Determine lateral extent of contaminated ground water Facilitate placement of monitoring wells and optimize their number Monitor changes in plume position and direction	Procedure less expensive than drilling Procedure more rapid than drilling Equipment lightweight, able to be hand carried Survey may be conducted in vegetated areas	Limited ability to detect nonconductive pollutants Technique unsuitable if no sharp contrast between contaminated and natural ground water Interpretation difficult if water table is deep Interpretation difficult if lateral variations in stratigraphy exist Interpretation difficult if radical changes in topography are not accounted for in choice of A-spacing Technique unsuitable in paved areas or areas of buried conductive objects
Depth Profiling	Indicate change in contamination with depth Establish vertical control in areas of complex stratigraphy	Same as above	Same as above
Seismic Refraction (nonexplosive method)	Determine depth and topography of bedrock Determine depth of trench containing buried drums	Procedure less expensive and safer than coring or excavation Procedure more rapid than coring or excavation	Technique unsuitable if no sharp velocity contrast between units of interest (e.g., trench containing buried drums and surrounding soil)

(continued)

Table 18. (continued)

Technique	Purpose	Advantages	Limitations
Seismic Refraction (continued)		Survey may be conducted in vegetated areas	Survey requires access road for vehicle. Depth of penetration varies with strength of energy source Low velocity unit obscured by overlying high velocity units Interpretation difficult in regions of complex stratigraphy
Metal Detection	Locate areas of high metal content (e.g., buried drums)	Procedure less expensive and safer than excavation or radar Procedure more rapid than excavation or radar Equipment lightweight, able to be hand carried Survey may be conducted in vegetated areas	Technique unsuitable for the detection of nonmetallic objects Technique unsuitable for objects below five feet Technique unsuitable for determination of number or arrangement of buried objects
Ground-Penetration Radar	Locate buried objects (e.g., buried drums) Provide qualitative information regarding drum density Detect interfaces between disturbed and undisturbed soil (e.g., bottom of trenches) Detect plumes of high chemical concentration	Procedure less expensive and safer than excavation Procedure more rapid than excavation Procedure deeper-penetrating than metal detection Procedure yields more information than metal detection Procedure may be used over paved areas	Technique unsuitable for vegetated areas Data requires sophisticated interpretation Underlying objects obscured by those above Survey requires access road for vehicle

Source: Kolmer, 1981

idealization required for modeling a complex field situation limits the utility of gravity surveys. However, as Figure 60 illustrates, the technique is useful in identifying porous, permeable strata such as buried stream channels and unconsolidated sedimentary formations, as well as permit estimates of depth to the water table.

(2) Magnetic Surveys

Magnetic surveys permit the detection and delineation of buried metallic objects, such as barrels (see (6) below) and of geological anomalies such as hydrothermal ore deposits, magnetic mineral formations and buried mafic plutons with sufficient magnetic contrast to surrounding rock. The magnetic signal intensity at the detector depends on the type and contents of magnetic minerals present, the depth of the anomalous body (since the dipole magnetic field strength decreases steeply as the inverse cube of the depth of the source) and its size and geometry. The total field intensity has contributions from the ambient earth's field, the induced moment of the anomaly and its intrinsic magnetic remanence.

Thus, in field measurements, the background earth's field (0.5 oersted = 50,000 γ) is subtracted from the total field strength, measured by portable instruments such as flux gate, nuclear precession or rubidium vapor (optical pump) magnetometers. Magnetic variometers or gradiometers achieve greater sensitivity (10 - 20 γ) and measure differential changes in the magnetic field, rather than its total intensity.

Airborne magnetic (aeromagnetic) mapping requires repeated measurements, towed or fixed sensitive instruments in gridded patterns. For interpretation of magnetic profiles, a number of corrections are necessary for anomalous signals weaker than approximately 500 γ; for nearby magnetic objects (railroad tracks); background drift due to diurnal field variations, and topography. Field anomalies are matched to simple geometrical shapes, at model depths and of model compositions. The same ambiguities in source characterization exist as in the case of gravity surveys. Ground-truth sampling and laboratory testing is needed to verify each interpretation. Figure 61 shows a way to combine gravity and magnetic measurements for more accurate interpretation.

(3) Electrical (Resistivity) Surveys

One of the most useful surface survey methods in groundwater studies is based on measuring the electrical resistivity (or its inverse, the conductivity) of ground layers. Each class of rocks has a characteristic range of values and within each class the degree of saturation with pore water and the pore fluid composition (cf. ionic concentration) lead to systematic variations in its conductivity. Resistivity surveys have been successful in locating the water table, mapping buried stream channels and detecting clay strata (Figures 62 and 63). Electrical surveys provide information on subsurface geologic

APPLICATION OF SURFACE GEOPHYSICS

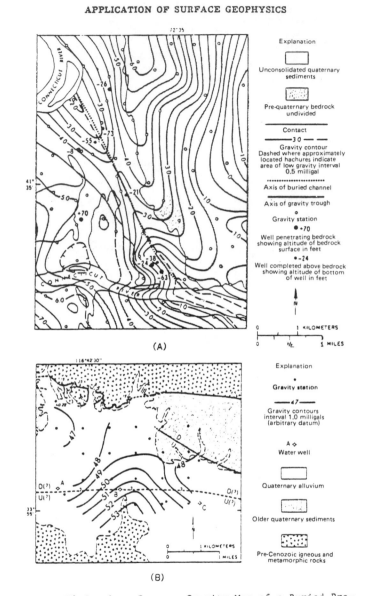

(A)

(B)

Figure 6C.   (A) Complete Bouguer-Gravity Map of a Buried Pre-
             Glacial Channel of the Connecticut River.(B) Complete
             Bouguer-Gravity Map of San Gorgonio Pass, California

Source: Zohdy, et al., 1974.

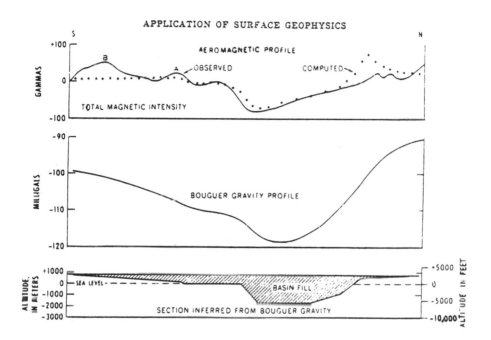

TECHNIQUES OF WATER-RESOURCES INVESTIGATIONS

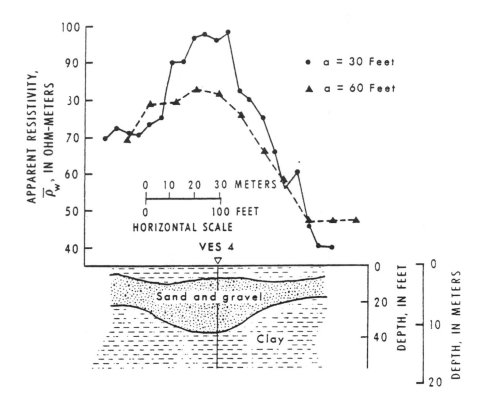

Figure   62.   Horizontal Profiles over a Buried Stream Channel
Using Two Electrode Spacings: a = 9.15 Meters (30
feet) and a = 18.3 Meters (60 Feet). VES 4 Marks
the Location of an Electrical Sounding Used to
Aid in the Interpretation of the Profiles

Source: Zohdy et al., 1974.

APPLICATION OF SURFACE GEOPHYSICS

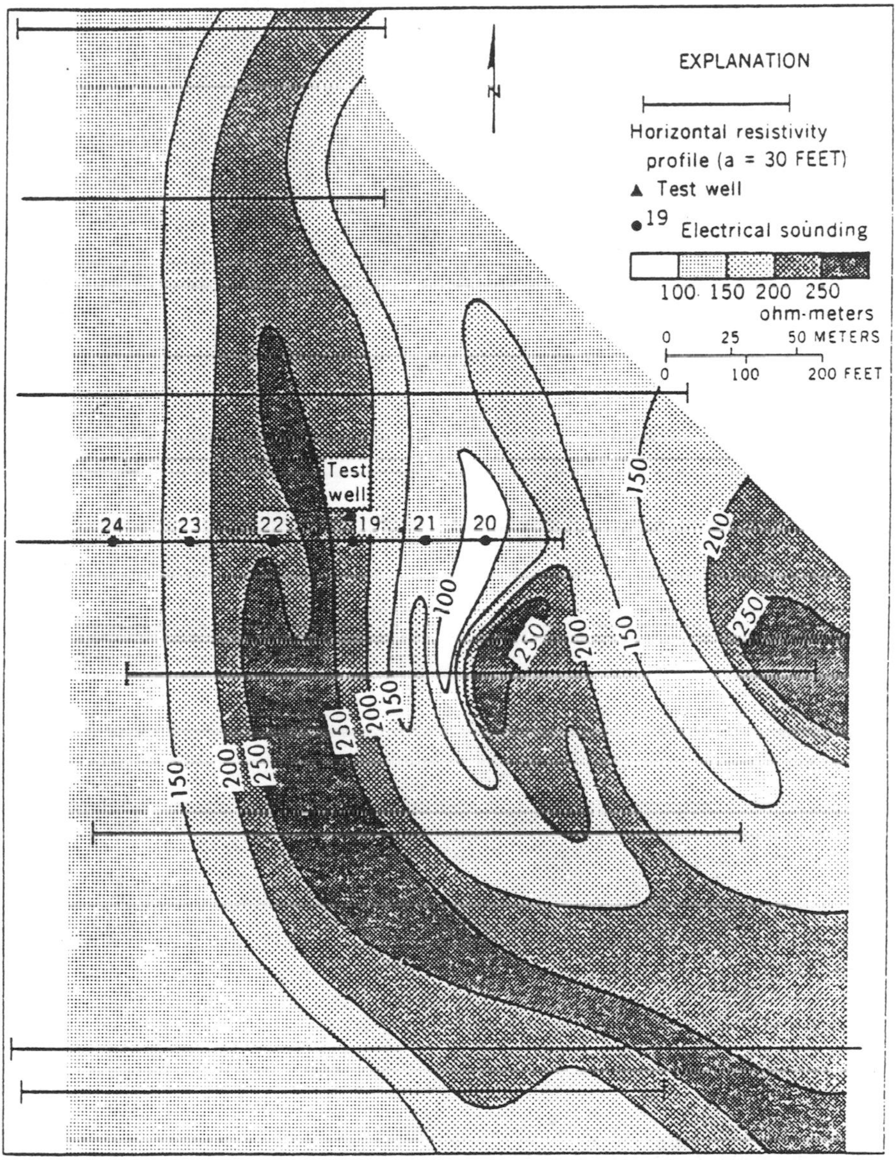

Figure 63.   Apparent-Resistivity Map near Campbell, Calif.
Unpublished Data Obtained by Zohdy (1964) Using
Wenner Array. Crosshatched Areas are Buried Stream
Channels Containing Thick Gravel Deposits. Stippled
Areas are Gravelly Clay Deposits

Source: Zohdy et al., 1974.

features (such as the depth of a fresh/saline water interface or depth
of a thick clay layer between two aquifers) which are not obtainable by
gravimetric, magnetic or seismic survey methods. For surface
resistivity measurements, metal electrodes are placed in the ground in
various configurations (e.g., as Wenner, Schlumberger and Dipole-Dipole
arrays) and the potential drop across two of them is measured directly.
The current flow paths can be characterized by isoresistivity contour
maps (Figures 62 and 63) obtained by moving the electrodes between
various stations, for lateral profiling at a given depth; or by numerous
measurements at a given station at various depths, for depth sounding.
Commonly, both horizontal profiling and depth electrical sounding
techniques are combined to determine the presence, location and extent
of permeable layers. However, interpretation of isoresistivity maps or
apparent resistivity profiles require either calibration and
confirmation by borehole geologic and geophysical measurements or
matching of modeling assumptions to computer predictions to optimize
accuracy.

(4) Electromagnetic (EM) Methods

These depth sounding methods are best suited to detection of good
electrical conductors at shallow depth. They encompass various methods
involving the propagation of time-varying (continuous or pulsed)
electromagnetic fields through the earth layers. An EM source
(transmitter) introduces energy into the ground, inducing secondary
currents and fields in the conducting body, which are detected by an
antenna receiver at the surface.

One field method makes use of atmospheric signals resulting from
worldwide thunderstorm activity (sferics) and detects induction
responses as audio-frequency magnetic fields (AFMAG).

Other methods make use of very low frequency (VLF) (5 - 25 kHz)
signals from fixed ground stations and yet others, of moving radio
frequency (RF) signals in the frequency range 10 kHz - 10 MHz, to
achieve a range of depth of penetration. The power frequency and
distance characteristics of available systems are selected as
appropriate to the field conditions and resolution desired. A great
variety of transmitter loop-receiver coils configurations are practiced
for ground and airborne surveys: long wire, two dipoles in phase or in
quadrature, variable dip-angle systems in parallel or broad orientation
along the traverse line, horizontal loop systems, rotary field systems
for two aircraft, etc. In order to isolate a small secondary response
from a larger primary signal, transient INPUT (Induced Pulse Transient)
systems have been introduced for time-domain, rather than
frequency-domain, operation. The INPUT airborne systems have greater
penetration depth and afford better characterization of the conducting
bodies. The height of the aircraft and transmitter-receiver separation
control the sensitivity for a given system. Data interpretation
requires theoretical modeling of conductor shape and sizes to match the
response of conducting sheets, spheres, cylinders, or lines in uniform
or dipole fields. For complex geometries and variable conductivity,

computer analysis of data is necessary. Table 19 summarizes the range of dielectrical constants and electrical conductivities for typical earth materials.

(5) Electrical Sounding Using Natural Sources

Several practical methods of deep electrical sounding take advantage of natural electric and electromagnetic phenomena and require only passive sensors and no power sources. These are:

- The Self-Potential (SP) Method which detects background potential drops due to fluid streaming through a formation, varying electrolyte concentration and changing electrochemical activity of groundwater. Regional gradients of 10 mV/1000 feet are typical of these sources (streaming potentials). Polarization potentials associated with mineralization zones give rise to SP anomalies of order $\leq$ 200 mV. pH variations above and within the water table also contribute to current and potential flowing around conducting zones. Detection of SP voltage drop requires simple field equipment, basically a potentiometer and two electrodes. However, the data interpretation may be difficult.

- Telluric and Magnetotelluric Methods detect large scale current and low frequency magnetic fields induced in surface ground layers by atmospheric (ionospheric) currents. These vary diurnally and with latitude and are subject to high frequency fluctuations due to electric and magnetic storms. These methods are still under active development and require sensitive equipment and careful interpretation of field data, after subtracting the undistorted background fields from distortions due to the local geology. The techniques have been applied to deep sounding and large scale interpretation of basement rock structures.

(6) Metal Detectors

Metal detectors and magnetic sensors offer reasonable sensitivity and reliability to depths of approximately 5 feet. Metallic targets at or near the surface can be reliably detected with magnetic induction detectors, while magnetometers (total field flux gates and magnetic gradiometers) can "see" up to 5 feet depths. However, radar-type techniques are needed to penetrate down to 20 feet or more of ground cover. The penetration effectiveness of electromagnetic radiation (pulsed or continuous wave CW) depends on the power of the emitter/receiver system, on its frequency domain and on the attenuation properties of the ground cover and its contrast to the target. Ground penetrating radar (GPR) and other pulsed radio frequency (RF) systems are becoming more widely applied for detecting shallowly buried waste barrels, whether made of metal or plastic. Metal detectors are based on

TABLE 19.   APPROXIMATE VHF ELECTROMAGNETIC PARAMETERS OF EARTH MATERIALS

Material	Approximate Conductivity (MHO/M)	Approximate Dielectric Constant	Depth of Penetration
			Max (km)
Air	0	1	
Limestone (Dry)	10^{-9}	7	
Granite (Dry)	10^{-8}	5	
Sand (Dry)	10^{-7} to 10^{-3}	4 to 6	
Bedded Salt	10^{-5} to 10^{-4}	3 to 6	
Fresh Water Ice	10^{-5} to 10^{-3}	4	
Permafrost	10^{-4} to 10^{-2}	4 to 8	
Sand, Saturated	10^{-4} to 10^{-2}	30	
Fresh Water	10^{-4} to 3×10^{-2}	81	
Silt, Saturated	10^{-3} to 10^{-2}	10	
Rich Agricultural Land	10^{-2}	15	
Clay, Saturated	10^{-1} to 1	8 to 12	
Sea Water	4	81	Min (cm)

Source: Arthur D. Little, 1980.

electromagnetic induction: AC currents in a surface coil give rise to a time-varying primary magnetic field at a metallic conductor and hence to induced eddy currents and a secondary magnetic field. This secondary magnetic field is then picked up at the surface with a coil connected to a sensitive electronic amplifier, meter or potentiometer bridge. The anomalous secondary field may be due to a metallic object or to a high-conductivity anomalous zone in the ground. Ordinarily, operating frequencies for ÉM induction prospecting instruments are $\leq$ 5 kHz, although they may reach 50 kHz.

(7) Ground Penetrating Radar

GPR systems, also more generally called pulsed RF, or impulse-radar systems, have been applied to detection of buried objects, rock cavities, faults and to locating the water table in different soil types. Pulsed RF systems have been operated in the frequency range of 1 to 6000 MHz, for maximum penetration depths of 225 to 3 meters, respectively.

The GPR uses pulsed transmission of electromagnetic signals and reflected target energy to detect the presence and depth of a target. The attenuation loss of the signal in the ground increases with ground conductivity and with frequency for a given material. A 10 dB loss indicates that only 1/10 of the energy survives passage through a given earth material. Depending on moisture content, attenuation to a given depth can vary from a fraction of dB/foot for dry soils up to 30 dB/foot in wet clay. The velocity (v) of propagation of radar in the ground is slower than in air (c), by the fraction

$$\frac{v}{c} = \frac{1}{\sqrt{\epsilon}}$$

where ϵ is the relative dielectric constant of the ground material (Table 19). For typical ϵ = 2-10, the radar wave ground velocity is 0.3 - 0.7 ft/nanosecond, so that signal round trip time delays are of order 10^{-8} seconds. The dielectric constant varies also with moisture content (percent saturation), as well as with mineralogical makeup (Table 19). Metal targets are near-perfect reflectors of radar energy and show a characteristic hyperbolic signature or profile radar charts.

A radar system developed by Geophysical Survey Systems, Inc. (GSSI) has been applied to profiling the ground to a maximum depth of 25 feet in sand and 5 feet in varved clays. This method is called Electromagnetic Subsurface Profiling (ESP) and has been used to detect buried pipes. The ESP system is a broad band videopulse radar, with variable gain, which generates graphic radar displays requiring careful interpretation in terms of frequency content and amplitude of the reflected signal, by a technique called Time-Domain Reflectometry (TDR). Interpretation of charts must be based on field calibrations and on laboratory and borehole information to minimize ambiguities. Several GSSI survey instruments are available (e.g., Models 3105 AP and 3102), operating at different center frequencies (e.g., 300 MHz and 600 MHz)

with different spatial resolution and penetration depth (which varies
roughly as the inverse square of the frequency).

(8) Seismic Profiling

Seismic profiling is a relatively deep acoustical (elastic wave)
sounding technique. Seismic energy from an explosive, vibrating
(vibroseis) or percussive (hammer) surface source is reflected from
geological interfaces and discontinuities and detected by arrays of
geophones. Continuous subsurface profiles can be obtained via a fixed
array of detectors (geophones) distributed on the surface, or by gridded
surface coverage from a moving vehicle. The geophones are transducers
which convert the reflected acoustic signal to an electrical signal,
which is subsequently amplified and recorded digitally on a magnetic
tape or graphically on a strip chart recorder. The principles of
geometrical ray optics are used for analyzing the propagation path of a
spherical seismic wave from the source to the receiver. Both the
frequency content and the amplitude of the reflected signal, as well as
its time of arrival at the detector carry information on the
homogeneity, depth, density contrast, porosity and fluid content of
earth layers. The presence of pore fluid changes the sonic velocity and
attenuation properties of a formation. The interpretation of reflected
seismic data is complicated by the presence of refracted waves and by
inelastic-attenuation losses (which increase with frequency), as well as
by ambiguities in the number and homogeneity of reflecting layers
(Figure 64).

The relative transmission and reflection coefficients and the
vertical and lateral homogeneity of formations, i.e., anisotropy and the
round-trip travel paths for the compressional (p-type or body) waves are
interpreted in terms of "earth models" (Figures 64).

Logitudinal shear (S-type) seismic waves are sometimes more useful
in detecting the depth of an aquifer or inferring pore water
saturations, since they are severely attenuated by fluids. Also,
seismic refraction methods are often useful in measuring the depth to
the water table, since the zone of saturation acts as a strong
refractor. There is usually sufficient velocity contrast between dry
and saturated sediments (0.1 - 1 km/sec), to insure good resolution of
wet zones from signal travel time differences. The acoustic velocity of
refracted seismic waves is directly related to the density, porosity and
saturation of rocks, in the geologic section.

(9) New Techniques and Instruments

Other new nondestructive geophysical techniques for applications to
environmental assessments of hazardous waste sites include:

- Time-domain reflectometry (TDR) which has been used
  to map soil moisture content along buried
  transmission lines and breaches in landfill liners.
  TDR is analogous to a 1-D radar along a transmission

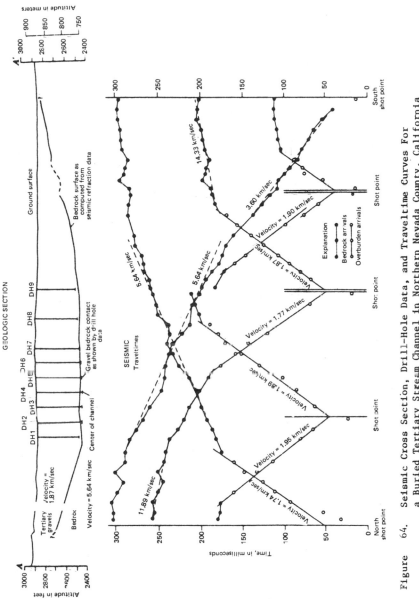

Figure   64.   Seismic Cross Section, Drill-Hole Data, and Traveltime Curves For
a Buried Tertiary Stream Channel in Northern Nevada County, California

Source: Zohdy et al, 1974.

line and detects discontinuities along it.   It
requires the installation of buried parallel cables
in the ground to monitor the wetness of a backfill
or landfill.

- Acoustic-emission (AE) can be used to detect
  resaturation and fluid flow in soils and rocks,
  assuming that laboratory calibration data on seepage
  flow and fracture flow AE exist, to aid in
  interpretation of field data. This monitoring
  method requires installation of borehole hydrophones
  and surface geophones to "listen" to stress changes
  in rock due to fluid flow.

A recent trend towards more systematic geophysical data acquisition
for less ambiguous interpretation has been to pair complementary
techniques for mapping subsurface geological and geochemical features.
For example, one can combine electrical resistivity (ER) with impulse
radar (GPR), and use formation resistivities to assess the depth of
penetration of radar pulses and use radar signatures for geological
interpretation of ER results.

Another trend has been to integrate remote sensing techniques with
direct environmental sampling programs, in order to reduce costs and
optimize locations for the latter and to improve the accuracy of the
former.

Recent advances in the remote sensing geophysical instrumentation
have focused on portability, sensitivity and expanded data storage and
processing capability:

- Electrical Methods: Both frequency-domain and
  time-domain electromagnetic (EM) systems were
  recently improved such as the airborne Pulse
  Electromagnetic (PEM) Crone System, the SIROTEM II
  System by GEOEX and the EM-37 System by Geonics.
  Improved ground systems include the Scintrex Genie
  time-domain EM system and new IPR-11 induced
  polarization spectral receiver; as well as the
  Phoenix   new   100   kW   1P/R   (induced
  polarization-resistivity) instrument and a real-time
  magnetotelluric (MT) device.

- Gravity and Magnetic Methods: airborne (helicopter)
  gravity surveys have recently achieved 0.5 milligal
  accuracies although on gridded flights the cost of
  surveying is high (approximately $200/km). Both
  GeoMetrics and EDA have recently introduced
  proton-precession field magnetometers for
  total-field intensity measurements, with data
  storage and processing capabilities.

● Seismic Methods: the recent improvements have focused on higher resolution, greater recorder capability and higher speed data transfer. For example McSEIS-1500 (an OYO Instruments system) has a 24-channel recorder with digitized output to a floppy disk.

c. Borehole Geophysics

Borehole geophysical techniques are necessary for calibrating and verifying information derived from and the interpretation of surface geophysical surveys. Borehole logging, coring, and sampling methods provide vital information on the geologic context and distribution of liquid water and waste vs. depth. Geophysical well logging provide data on the location, thickness, and lateral continuity of waste-storage zones and the confining beds based on estimates of porosity and permeability. Other physical properties, lithologic boundaries, and structural discontinuities affecting the potential pollutant migration can also be determined. For example, by cross-plotting acoustic-velocity and neutron or gamma-gamma logs, one can discriminate between fracture vs. intergranular porosity. Also, the distribution and orientation of fractures can be determined by acoustic televiewer logs. By measuring the conductivity, temperature, viscosity, and density of interstitial fluids, logs can be used to infer the chemical nature of native and pollutant fluids. Logs can serve to monitor in-situ changes in the groundwater system, such as porosity changes due to plugging by immiscible organics, or precipitation of solutes after chemical treatment. Thus, a continuous borehole logging program can guide remedial action at a polluted site.

Geophysical well logging is necessary to measure and monitor fluid chemistry, density, and viscosity changes due to waste migration, to map the distribution of groundwater types, and to establish flow patterns and the distribution of pollutants relative to recharge area. The time-lapse technique compares well logs run at selected time intervals to detect the nature and extent of fluid changes.

Geophysical borehole logs are based mostly on in-hole, wire-line measuring techniques (Figure 65). These yield a set of continuous analog or digital records of physical or chemical parameters (Figure 66), which require interpretation in terms of lithology, geometry, resistivity, and formation factors; bulk density and porosity, permeability, and moisture content and composition, as well as yield of water-bearing formation. These parameters are often interrelated so that a redundant set of different measurement techniques is required to determine their value (Figure 66). The groundwater velocity is proportional to the product of the intrinsic permeability and hydraulic gradient and inversely related to the effective porosity. The porosity may be inferred from at least two types of neutron, gamma-gamma, resistivity, or acoustic velocity logs. The velocity of groundwater movement between two wells can be derived from radioactive tracer

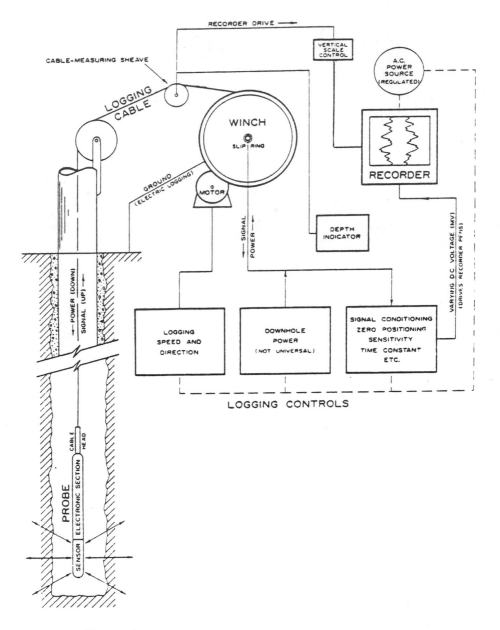

Figure 65.   Schematic Block Diagram of Geophysical Well-Logging
Equipment

Source: Keys and MacCary, 1971.

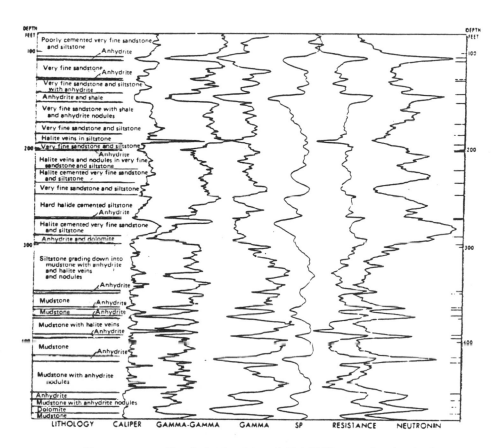

Figure   66.   The Relationship of Six Different Geophysical
Logs to Lithology, Upper River Basin, Texas.

Source: Keys and MacCary, 1971.

injection and migration tests detected by gamma emission. Hence, the intrinsic formation permeability can be estimated.

The basic well logs most widely correlated, based on calibrated crossplots, to permit more accurate and less ambiguous interpretation for surface mapping are listed in Table 20 and are briefly discussed in the text. Computer processing of log overlays or log combinations is often used for more complete interpretation.

The SARABAND technique is used for shaly-sands and water-bearing simple lithologies to determine water saturation and hydrocarbon weight and volume, together with bulk volume, porosity and permeability index analysis of the formation. SP, gamma-ray, and caliper logs are run in conjunction with one of the following log types: resistivity (Induction Log, Dual Induction Log, Laterolog, or Dual Laterolog); density (FDC - Formation Density Compensated Log); neutron (SNP - Sidewall Neutron Porosity or CNL - Dual Spacing Neutron logs); sonic (BHC - Borehole Compensated Log); microresistivity (proximity log or microlaterolog).

For complex lithologies, CORIBAND is used as a general interpretation code used in the oil-gas and exploration industries. Three to four types of logs are run to insure a minimum of ambiguity in output. Reference to standard crossplots is made in data reduction and statistical averaging of parameters is used.

The most widely used logging techniques and their areas of application are:

(1) Natural gamma: measures natural radioactivity of borehole walls and detects changes in lithology.

(2) Neutron-gamma: neutron bombardment coupled with gamma ray capture is used to detect changes in pore water concentration above the water table and, if calibrated, is a measure of porosity.

(3) Gamma-gamma: gamma ray from a source is absorbed by the rock walls in direct proportion to their density.

(4) Caliper: a record of the average diameter of a drill hole which helps to locate fractures, cavities, swelling of hydrated clays, dissolution of matrix by pore fluids, and unconsolidated wet sands.

(5) Spontaneous Potential (SP)/Resistivity: is measured by several variants (normal, single-point, and focused-beam) (Table 21). They provide direct information on fluid chemistry, including total dissolved solids concentration and salinity (Figure 67). Correlative information for interpretation of the downhole mineralogical and zonation based on other logs is also obtained from electric logs. The

TABLE 20. SUMMARY OF LOG APPLICATIONS

Required information on the properties of rocks, fluid, wells, or the ground-water system	Widely available logging techniques which might be utilized
Lithology and stratigraphic correlation of acquifers and associated rocks.	Electric, sonic, or caliper logs made in open holes. Nuclear logs made in open or cased holes.
Total porosity or bulk density........	Calibrated sonic logs in open holes calibrated neutron or gamma-gamma logs in open or cased holes.
Effective porosity or true resistivity..........................	Calibrated long-normal resistivity logs.
Clay or shale content................	Gamma logs.
Permeability..........................	No direct measurement by logging. May be related to porosity, injectivity, sonic amplitude
Secondary permeability - Fractures, solution openings.	Caliper, sonic, or borehole televiewer or television logs.
Specific yield of unconfined aquifers	Calibrated neutron logs
Grain size...........................	Possible relation to formation factor derived from electric logs.
Location of water level or saturated zones.	Electric, temperature or fluid conductivity in open hole or inside casing. Neutron or gamma-gamma logs in open hole or outside casing.
Moisture content.....................	Calibrated neutron logs.
Infiltration.........................	Time-interval neutron logs under special circumstances or radioactive tracers.
Direction, velocity, and path of groundwater flow.	Single-well tracer techniques - point dilution and single-well pulse. Multi-well tracer techniques.
Dispersion, dilution, and movement of waste.	Fluid conductivity and temperature logs for some radioactive wastes, fluid sampler.
Source and movement of water in a well.	Injectivity profile. Flowmeter or tracer logging during pumping or injection. Temperature logs.
Chemical and physical characteristics of water, including salinity, temperature, density, and viscosity.	Calibrated fluid conductivity and temperature in the well. Neutron chloride loggins outside casing. Multielectrode resistivity.
Determining construction of existing wells, diameter and position of casing, perforations, screens.	Gamma-gamma, caliper, collar, and perforation locator, borehole television.

Source: Keys and McCary, 1971.

TABLE 21. HYDROLOGIC APPLICABILITY OF ELECTRIC LOGS

Properties to be investigated	Type of electric log								
	Single point	Short normal	Long normal	Lateral device	Wall resistivity (nonfocused)	Focused guard and laterolog	Micro focused	Induction	SP
Lithologic correlation..............	×	×	......	......	......	×	......	×	×
Bed thickness..............	×	×	......	......	×	×	×	×	×
Formation resistivity (low R muds)..........	......	......	×	×	......	×	....	×	......
Formation resistivity (fresh mud)..............	......	....	×	×	......	......	......	×	......
Invaded zone resistivity..................	......	×	......	......	......	......	......	......	......
Flushed zone resistivity..................	......	......	......	....	×	......	×	......	......
Mud resistivity[1] (in place in hole)....	......	......	......	....	×	......	....	......	......
Formation water resistivity..................	......	....	×	×	......	×	....	......	......

[1] Use mud kit for pit samples.

Source: Keys and McCary, 1971.

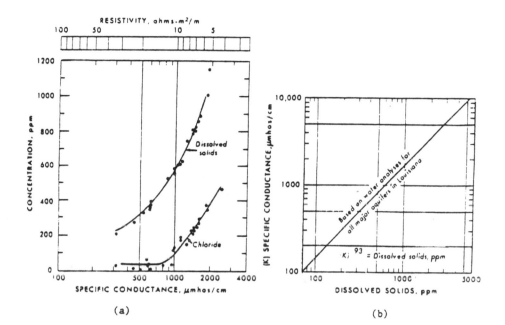

Figure 67. (a) Relation of Dissolved Solids and Chloride Concentration,
in ppm, to Resistivity and Specific Conductance.
(b) Relation between Specific Conductance and Dissolved
Solids.

Source: Keys and MacCary, 1971.

amplitude of SP deflections across beds is related
to electro-chemical and electrokinetic effects (ion
migration). Electric logs are widely used for
hydrologic applications (Table 21).

(6) Thermal logging: can be used to detect flowpaths
for groundwater and thermal anomalies due to
exothermic chemical reactions in solution.

(7) Tracer-tests: employ   radioactive   or
neutron-activated   chemical   tracers   to   chart
groundwater   flowpaths   and   plume   spreading
characteristics between wells.

(8) Sonic Televiewer (Teleseis): provides continuous
sound images of the borehole wall to detect or
verify discontinuities, porosity contrast, spacing
and orientation of fractures and help to identify
the depth intervals requiring in-situ testing.

Observation boreholes also permit integral sampling of cores
extracted and the installation of borehole monitoring instrumentation
(for seismic activity and water quality sampling, such as piezometers
and stream gauges). Less commonly used logging techniques are:

● The Mise-a-la-masse (MLM) logging technique, which
consists of simultaneous acquisition of both
resistivity and induced polarization, has recently
become widely used.

● The Thermal Decay Time (TDT) log records rate of
decay of thermal neutrons (produced by a neutron
source) in the borehole fluid and walls. It is a
good indicator of salinity of formation water, since
chlorine is a good neutron absorber. Porous
hydrocarbon or gas-bearing zones and shaly
formations can be resolved based on characteristic
neutron-capture capability.

Recent advances in borehole logging instrumentation consist of
complete logging units, which are light, portable, and controlled by
microprocessors. For example, the Mount Sopris (Series III) system is
compact, can be slung for helicopter transport, and records up to four
channels of simultaneous logging data on a magnetic tape. Various
sondes (spectral gamma-ray by Mt. Sopris, the magnetic susceptibility
Kappalog sonde by OYO Instruments) and various electromagnetic sounders
(Sirotem by Geoex and EM-37 by Geonics) have been recently refined for
borehole logging.

Geophysical field computers and calculator programs for geophysical
application (listed and compiled by the Society of Exploration

Geophysicists) are available for rapid reduction and interpretation of field logging data. The Schlumberger (1979) log interpretation charts, as well as companion reference volumes on the Principles and Applications of Logging Techniques (1972, 1974), offer a useful field reference for rapid interpretation of crossed log charts.

4.   MONITORING FIELD SAMPLING AND ANALYSIS METHODS OF GROUNDWATER
     INVESTIGATION

The following sections discuss various equipment and procedures that can be used to collect and analyze groundwater amples. This material is intended as a general overview of these topics and does not attempt to specifically define or recommend any procedure or technique. Generally, comprehensive sampling and analysis protocols can only be developed in response to a specific problem. Information presented in this section is intended to provide a background for decisions.

If specific information on sampling or analytical procedures is required, the reader is directed to the many references that exist in this area. A few specific references are presented throughout this section. Within the context of this discussion, the term "well" is meant to include both wells and piezometers.

   a.   Sampling Considerations

The objective of an environmental sampling operation is to collect a portion of some material (e.g., groundwater, wastewater, solid waste, etc.) and deliver it for analysis in a way that preserves the integrity of the sample. Preservation of integrity may require that the sample be unchanged (chemically, physically, or biologically) or it may allow for their controlled modification.

Obtaining a sample of groundwater from an aquifer or the vadose (unsaturated) zone above an aquifer being monitored may be done using a variety of sampling equipment and procedures. Different equipment/procedure combinations affect the sample. Thus, it is essential that all factors affecting the nature of the groundwater sample collected be considered before a sampling program is chosen.

Several types of sampling equipment are commonly used to collect samples from groundwater monitoring wells. These include devices that are as simple as a bucket and as complex as multistaged centrifugal pumps. Each of these have their own advantages and disadvantages (discussed later), and each may be used to fulfill the requirements of some particular program. Before any of these devices are applied to a particular program, however, their capabilities should be considered and reviewed with respect to the goals of the program and to factors that may influence how they should be used.

Several factors require consideration, including the depth of the water, the physical dimensions (e.g., overall well depth, length of the screened section, well casing diameter) and the rate at which water flows into the well after initial volumes are removed. In the selection of sampling equipment, the materials of construction should also be considered, as this may affect the integrity of the sample. The residence time of water in a well is important because collection of water that has become trapped in a well casing may produce samples that do not reflect the character of the groundwater that currently exists in the geological formations that are being monitored.

Several of these factors depend on the hydrology/geology of the study area and as such they cannot be changed. However, sampling procedures can be developed which partially control the way in which they occur, making it possible to collect samples with various types of equipment. A few of these procedures are described below.

The water above the open or screened section of a well is considered to be stagnant with respect to the water that is in the screened section. This is true because the water in the screened section is constantly being replaced with water from the aquifer and, therefore, has had more recent exposure to geological formations outside of the well casing. Water contained above the open area is trapped, and changes in its composition may have occurred due to extended periods of exposure to well casing materials and the atmosphere. Therefore, the stagnant water is not necessarily representative of the surrounding groundwater. To obtain a representative sample from the well, stagnant water must be excluded. The sample should be drawn from below the fresh/stagnant water interface in a way that keeps the stagnant water from the sample.

Two different procedures may be used to limit undesirable mixing. The first approach, frequently used, requires the removal of a large volume of water before the sample is collected. Generally, a volume equivalent to 3 to 5 times the amount of water originally contained in the entire well is recommended (USEPA, 1980b). This procedure is based on the assumption that as first-time water is removed, most of the stagnant water is displaced by fresh water from the aquifer. Removal of additional water is designed to flush the well casing, rinsing away any residual contaminants. Even if all of the stagnant water is not removed the first time the well is prebailed, the removal of five volumes of water will greatly reduce the amount that is contained in the sample. For example, if we assume that 100 percent of the well water is stagnant and that only 50 percent of the water originally contained in the well may be removed each time the well is bailed, then 5 prebailing repetitions will provide a final fresh/stagnant mixture that is roughly 97-percent fresh and 3-percent stagnant. If 33 percent of the initial well water is removed per volume bailed, then by the fifth prebailing repetition the fresh/stagnant mixture remaining is 87-percent fresh and 13-percent stagnant. Additional prebailings are recommended to further reduce the stagnant water content of the well.

A second approach is to selectively remove water from different levels of the well. Using this approach, one well volume of water is removed from the well from a location as high above the fresh/stagnant

water interface as is possible. The fresh/stagnant water interface is
assumed to coincide with the upper elevation of the open (screened)
section. The point from which the initial well volume is removed is
determined by the individual characteristics of each well; this includes
factors such as well depth, groundwater depth, and the rate of well
recharge. If the well does not recharge as quickly as water is being
removed, it may be appropriate to evacuate the stagnant water by
locating the sampling point at or just below the fresh/stagnant water
interface. Once the stagnant water is removed, the sampling location
would be dropped to a point near the bottom of the screened section
prior to sample collection.

The application of this procedure either completely removes the
stagnant water or keeps it at a point high enough above the final
sampling point to prevent undesirable mixing. This multilevel pumping
approach also requires the removal of considerably less total water
volume, thereby allowing each well to be sampled in less time than the
approach that requires the removal of 3 - 5 well volumes. If properly
performed, it will deliver a sample representative of the surrounding
aquifer.

Another factor to be considered before a well is sampled relates to
whether there are advantages to measuring some chemical parameters in
the field as the water is removed. Determinations of chemical
parameters, such as pH, conductivity, dissolved oxygen (DO) content,
oxidation-reduction potential, water temperature, and a few specific ion
(Cl^-, CN^-) concentrations, can easily be made in the field while the
well is being prepared for sampling and after collection of the sample.
Parameter measurement is not always appropriate during sample collection
because some of these techniques (pH, DO, etc.) use electrodes that
release small quantities of chemicals into the water. However,
determination of these parameters immediately before and just after the
sample is collected does provide some information on the homogeneity of
the water being removed from the well. For example, if two readings of
parameters that are separated by some predetermined volume of water
(e.g., one screen volume) prove to be equivalent, the water may have
reached some momentary level of equilibrium and a sample may then be
taken to represent that condition. If a set of readings taken
subsequent to sample collection shows equivalent values, then it may be
inferred that the water obtained is representative of the volume that
has been bracketed. If the readings are different, then provisions can
be made immediately to establish a new level of equilibrium within the
well and to collect a new sample.

Groundwater parameters can be measured regardless of the type of
sampling equipment used. If a noncontinuous supply of water is
available, parameters may be measured by placing aliquots of the
recovered well water into beakers, followed by the determination
desired. In systems where continuous sample streams are available, the
electrode or probe assemblies of many meters can be mounted directly
into a flow through cell and the parameters continuously monitored.

In programs concerned with organic species pollution, field extraction of organics is a convenient method of preparing the sample for analysis, while decreasing the possibility of sample degradation during shipment. Different systems can be set up for inline extraction of organics with equipment that provides constant flows of water from the well (Figure 68). With sampling equipment such as bailers, an additional pump must be used to channel the sampled water through the sorbent cartridges.

There are five general categories of groundwater monitoring equipment and one type of equipment normally used for vadose sampling. These include:

- Manual collection equipment
- Vacuum extraction equipment
- Pneumatic or pressurized collection equipment
- Mechanical collection equipment
- Gas entrainment equipment
- Lysimeters (vadose zone)

Each of these types of equipment will be more completely discussed in the following sections. During the review of each of these sections, the reader should keep in mind that it is essential to be able to control where and how the water is drawn from a well to insure that it is as unaffected by the combined sampling process (well installation and sample collection) as possible.

b.   Manual Sampling Equipment

Manual groundwater sampling equipment includes all devices that require or involve the use of repetitive manual manipulation. Generally, these devices may be grouped into two classes, one including bailers, and the other devices such as hand or foot pressure/vacuum pumps. Examples of these types of equipment are shown in Figures 69 to 72.

Both of these alternatives involve simple equipment, and are generally easy and inexpensive to implement. A major disadvantage in their use is that their operation is tedious and time consuming. If a specific monitoring location is expected to be sampled frequently, alternative sampling procedures and permanent installations may be more suitable.

Manual bailing is straightforward and basically equivalent for any of the bailers shown in Figures 69 to 71. Procedurally, the bailer is repeatedly lowered into the well, allowed to fill with water, raised to grade surface, and emptied until the required sample volume has been removed. Bailers that load from the top (Figure 69) or from the bottom (Figure 69) are common.

One disadvantage of top-loading bailers is that they cannot be used to evacuate a well to dryness. Another potential drawback of top-

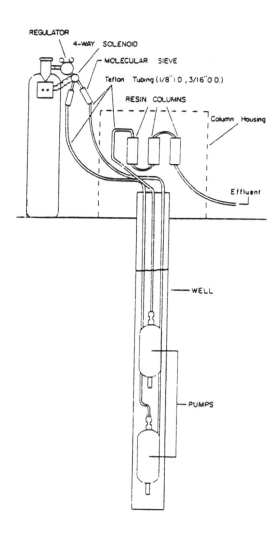

Figure 68.   Example of an Organic Extraction System

Source: Pettyjohn, W.A. et al., "Sampling Groundwater for Organic
           Contaminants," Ground Water, Volume 19, Number 2, pp. 180-189,©
           March-April 1981.

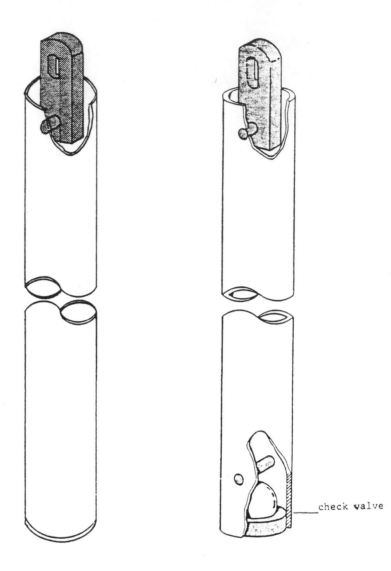

check valve

Figure 69. Top-loading and Bottom-Loading Bailer

Source: Timco Mfg.

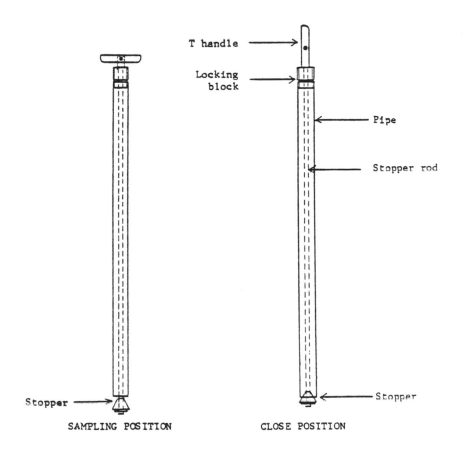

Figure 70, Coliwassa Sampler.

Source: EPA, 1980b.

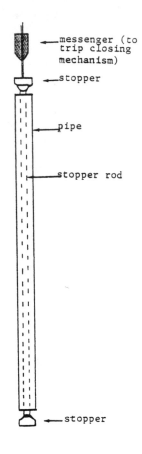

Figure   71.   Kemmerer sampler. (Flow through).

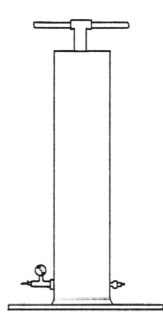

Figure 72.   Pressure-vacuum Hand Pump.

loading bailers is that they may be buoyant and difficult to fill if improperly constructed.

Bottom-loading bailers can recover more water from a well, and they will not be buoyant. Kemmerer and Coliwassa bailer designs, such as those shown in Figures 70 to 71, have the advantage in that they may obtain water from specified depths. When the Coliwassa bailer is used, the closed bailer is lowered to the desired horizon before it is opened. Once opened, water from that horizon enters and the bailer is then closed and raised. The Kemmerer bailer, on the other hand, is lowered into the well with both ends opened, allowing water to flow through it. Once the desired depth is reached, the sampler is closed and it is raised to the grade surface. One major disadvantage of all bailers is that the water column contained in the well is subjected to considerable mixing during sampling.. Stagnant and fresh water may combine; offgasing of volatile species may occur; or air may be introduced into the water. All of these disturbances may result in chemical changes in the water being sampled. Additionally, unless considerable care is exercised in handling the attached rope during sampling, large quantities of soil, pebbles, grass, etc., may be introduced into the well. Despite the many obvious drawbacks of the manual bailer, it is one of the most common methods of groundwater sampling.

Hand-pumping techniques are also time consuming and tedious. However, these techniques are better than manual bailer approaches because more control of water sample origin may be achieved by the use of various sample collection installations. One drawback of these devices is that suction lift and pressure capacities are somewhat limited. Generally, only 4.5 meters (15 feet) of suction lift is attainable, and pressure limits are on the order of 42 kg/cm^2 (60 lbs/in^2). Higher vacuum/pressure limits may be obtained, but this is usually done at the expense of sample volume throughput.

c.   Vacuum Collection Equipment

Vacuum pumps powered by gas or electric motors are a very convenient and efficient means of obtaining groundwater samples. The objective of this type of design is to produce a vacuum on some type of well installation (usually inert sampling lines), and draw water to the surface by means of a pressure differential. Peristaltic and diaphragm type pumps (supplied by Horizon/Ecology, Cole Parkman Instrument Corporation, or Fisher Scientific) can be hooked up to the inert sampling lines and can rapidly evacuate the desired volume of water. The major advantage of this type of well sampling is that water can be easily removed from different depths. Small volumes of water can be removed while still insuring that the well has been flushed properly, and reducing the sampling time. This equipment provides a constant supply of water from the well to the surface, helping in the logistics of the above ground sampling process.

A disadvantage of vacuum pump sampling collection is that there is a limit as to how far the pump is capable of lifting the water from

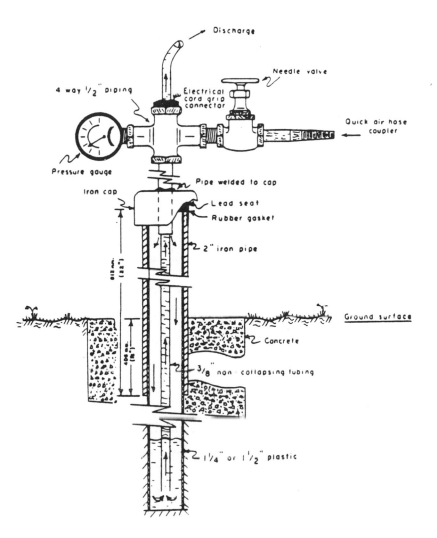

Figure   73.   Well Casing Used as Pneumatic Collection Vessel

Source: Sommerfeldt and Campbell, 1975.

the well (approximately 25 - 30 feet). For groundwater depths in excess
of this level, multistage vacuum systems would have to be used. These
systems are expensive and require extensive site preparation.

A second major disadvantage of vacuum sampling systems is that by
applying a vacuum, the volatile species dissolved in the water may be
driven out of solution, therefore, changing the chemical composition of
the water being sampled. This is of particular concern when sampling for
organic species.

    d.   Pneumatic or Pressurized Collection Equipment

Pneumatic pressure can be used very effectively to collect
groundwater samples. Pressure can be used to pump wells through many
various designs of equipment, but they all work on the same principle.
Water contained in some sealed area in the well is forced by gas
pressure up to the ground surface through a sampling line (Figure 73).
Two lines extend to the sampling device; one stops just below the seal
and the other extends to the bottom of the sealed area. Pressure is
exerted on the water from the shorter line, and the water is forced up
the longer line to the surface.

Pneumatic pressure can be applied by manual powered pumps, but as
has been previously discussed, these devices generally have pressure
limitations. Electric and gasoline-powered engines or bottled gas can
provide pressure up to 175 kg/cm^2 (250 lbs/in^2). Pressures of this
level are adequate for pumping wells of depths up to 100 meters (320
feet).

The simplest pneumatic device uses the well casing as the
collection vessel. A well cap is used to seal the well, and two lines
are introduced through the cap. One line extends to the point near the
bottom of the well, while the second stops just below the cap. An
example of this installation is shown in Figure 73. The shorter
sampling line is used to pressurize the well, while the longer channels
water to the surface. While this approach will deliver a water sample
to the surface, it is not recommended because it will also cause water
contained in the well to backflush into the aquifer, thereby minimizing
recovered volumes. Furthermore, if the well is improperly pressurized,
the compressed gas may be forced into the aquifer and be trapped outside
the well casing. This may cause changes in the flow patterns around the
well.

To prevent these problems, a separate collection vessel (i.e.,
pump) can be placed in the well and used to control movement of
pneumatic fluid and water. Many existing pump designs have been used
(Figures 73, 74, and 75). Although there are some differences in
configuration, they all operate on the same principle. These pumps have
many advantages; they can sample water from very deep wells; they can
pump water rapidly as compared to the previously discussed bailer and
vacuum methods; and they may provide a reasonably steady stream. It is
also possible to raise and lower the pump within the well during

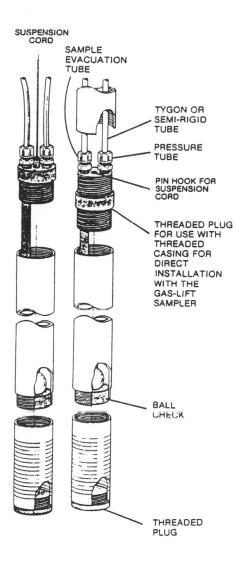

SUSPENSION
CORD

SAMPLE
EVACUATION
TUBE

TYGON OR
SEMI-RIGID
TUBE

PRESSURE
TUBE

PIN HOOK FOR
SUSPENSION
CORD

THREADED PLUG
FOR USE WITH
THREADED
CASING FOR
DIRECT
INSTALLATION
WITH THE
GAS-LIFT
SAMPLER

BALL
CHECK

THREADED
PLUG

Figure 74. Basic Pneumatic Pump

Source: Timco Mfg. Co., Inc.

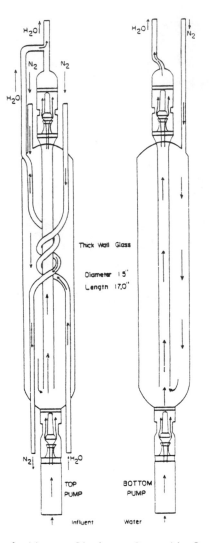

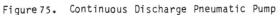

Figure 75.   Continuous Discharge Pneumatic Pump

Source: Pettyjohn, W.A. et al., "Sampling Groundwater for
        Organic Contaminants," Ground Water, Volume 19,
        Number 2, pp. 180-189,© March-April 1981.

sampling, thereby allowing for water to be obtained from specific levels within the well. This movement of the pump in the well induces a considerable amount of mixing of the water column; however, if the sampling procedure is designed properly, this should not present a problem as all the stagnant water originally contained in the well should be removed or positioned well above the sampling point before the sample is taken.

The major drawback to this type of sampling equipment is the constant exposure of the groundwater to the pneumatic fluid being used to drive the pump. Some designs eliminate the constant exposure of the water and the pneumatic fluid, such as piston type and bladder type (Figure 76) pumps. These designs are very expensive and difficult to construct. Many different sources of pressurized gas are in use: automobile engines (Trescott and Pinder, 1970), gas and electric compressors (Summerfeld and Rampbell, 1975), and bottled gas (Tomson et al., 1980). An understanding of the analysis to be done on the water samples is necessary to assess the benefits and disadvantages of each pneumatic fluid option. Using an inert gas such as purified nitrogen does not eliminate the possibility of chemical alteration of the sample, but reduces it to a minimum. Automobile engines and gas compressors do supply adequate pressures, however they may introduce impure and reactive gases which can affect the chemical composition of the water being sampled.

e.   Mechanical Pumping Systems

Groundwater samples may also be obtained by the use of any of a number of types of mechanical pumps. Within the context of this discussion, mechanical pumps include all submersible pumps that are driven by electric or gasoline-powered motors. Classic examples of this type of pump include reciprocating or piston pumps and centrifugal pumps. Reciprocating pumps move water by alternately compressing and decompressing a trapped volume of water confined in a fixed diameter barrel.

In their simplest form, reciprocating pumps consist of a single barrel or cylinder divided into two sections by the plunger. The lower section of this pump contains air that is isolated from both the well and the upper section of the barrel. The plunger is mechanically forced downwards, compressing the air and creating a partial vacuum in the upper chamber. Water from the well enters and fills this chamber through a check valve assembly. The direction of travel of the plunger is then reversed, causing the water to be compressed. Since the inlet check valve prohibits the water from flowing back into the well, it is forced upwards into the supplied discharge line. A check valve may be placed on the discharge line to keep discharged water from reentering the pump. By continually alternating the path of travel of the plunger, well water is repeatedly drawn into and expelled from the pump, forcing it to rise to ground surface. By including additional plumbing and modifying the design, these pumping systems can be made to move water on

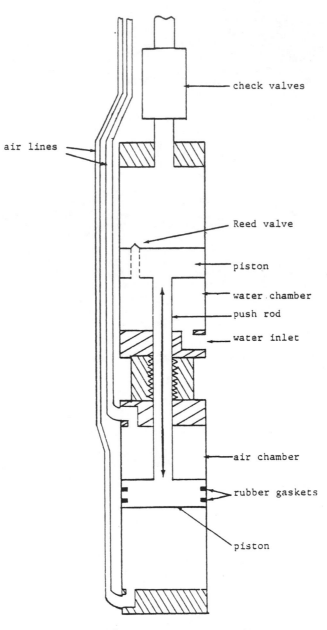

Figure 76.   Piston Pheumatic Pump. (Notice
Pneumatic Fluid does not Contact
Water During Operation).

Source: Hillerich, 1977.

both the up and downward stroke, providing a more continuous supply of water to the surface.

Centrifugal pumps force water out of a well by imparting a spinning motion to small, sequential volumes. The principal component of this type of pump is the pump head which contains an impeller. Water from the well enters the pump head through an inlet provided near the center of the pump head. When the impeller is set spinning, water contained in the pump head is thrown outward towards the outer edge of the pump head housing, causing a slight vacuum to be created in the center. This vacuum is immediately filled by water from the well. Water initially thrown outward is now at a slightly higher pressure than other water contained in the pump head, and it is forced through a discharge line to the ground surface. As the impeller continues to spin, more and more water is drawn into and spun out of the pump head, forcing a column of water to the ground surface. By mounting multiple-impeller assemblies atop each other, considerable water pressure can be exerted on the water, allowing for very deep wells to be sampled.

Designs of centrifugal and reciprocating pumps that use either compressed gas streams or manual manipulations to power the impeller or piston are available. These devices are considered in other sections of this document.

A major advantage of mechanical pumps is that they do not require much attention or manipulation after they are initially installed. Once in place, they can deliver large quantities of water at a steady pressure and a constant flow rate. Another advantage of this type of system is that the chemical, physical, and biological integrity of the sample is maintained because no external gases are introduced or mixed with the sampler. Furthermore, while high flow rates are obtainable, mixing and turbulence of the water stream is not violent.

A principal disadvantage of this sampling alternative is that pumps of this nature are generally large and usually cannot be used in wells smaller than 3-4 inches in diameter. A second disadvantage of these systems is that a power source may not be available at all sampling locations.

f.   Gas Entrainment Systems

Gas entrainment or air lifting represents another approach that may be used to obtain groundwater samples from a well. This technique is similar to the pneumatic systems previously described; except in this case, the gas is allowed to bubble through the groundwater rather than being controlled. An example of a typical gas entrainment sampling system is shown in Figure 77.

Gas entrainment systems require two sampling lines to be installed in a well. One of these lines is larger in diameter than the other, and both extend from the ground surface to a point that is below the groundwater surface. The larger of the two lines is used to channel the

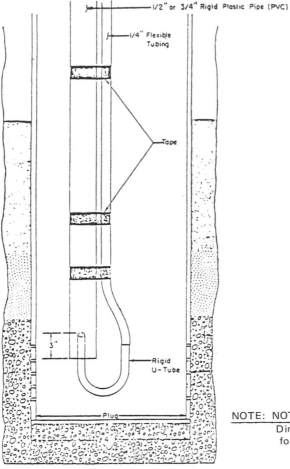

NOTE: NOT TO SCALE
Dimensions exaggerated
for clarity

Figure 77. Gas Entrainment Pumping System
Source: Ginilka and Harwood, 1979

sample water/gas mixture to the ground's surface, and it is generally open to the atmosphere. The smaller of the two lines is used to channel compressed gas down into the well to a point inside the larger pipe where it is released. Once released, the compressed gas expands, forming a bubble whose size is limited by the inner walls of the larger tube. The density difference that exists between the gas bubble and the surrounding water causes the gas to rise and forces the water trapped above it to rise at the same time.

At land surface, the compressed gas/water mixture is directed to a point away from the well where the gas is allowed to escape and the water is collected. Compressed gas requirements for this sampling approach may be fulfilled using either bottled gases or with gasoline or electric-powered compressors. Of these sources, bottled gases are generally preferred because a more inert gas can be supplied for pumping.

Gas entrainment sampling systems have numerous disadvantages and are generally not recommended for studies which require precision analyses. The principal disadvantage of these systems is that the violent or turbulent mixing of the compressed gas with the sample water may alter chemical, physical, and biological integrity of the sample.

Another disadvantage of these systems is that they are inefficient. Generally, less than half of the groundwater present in a well can be recovered, meaning that more time must be spent in developing the well prior to sample collection.

g. Vadose (Unsaturated) Zone Sampling Systems

The vadose (unsaturated) zone, is sampled using devices called lysimeters. There are two basic versions of this device, one that uses both pressure and vacuum to deliver a sample, and a second type which uses only vacuum. Examples of these two lysimeters are shown in Figures 78 and 79.

In operation, a vacuum is drawn on the porous cup assembly which induces water to flow into the cup. Water collects in the cup and is sampled in one or two ways. Using the pressure-vacuum lysimeter, the cup is pressurized, forcing the water up through a sample collection tube. This procedure is identical to that described in an earlier discussion on the pneumatic sampling system.

The vacuum system uses a vacuum source to draw the water contained in the cup up to the ground surface. This technique option suffers the same suction lift limitations as the vacuum pumping systems previously discussed. Both techniques may lose volatile organics during sample collection.

A specialized lysimeter design has been developed by personnel at the Robert S. Kerr Environmental Research Laboratory at the U.S.

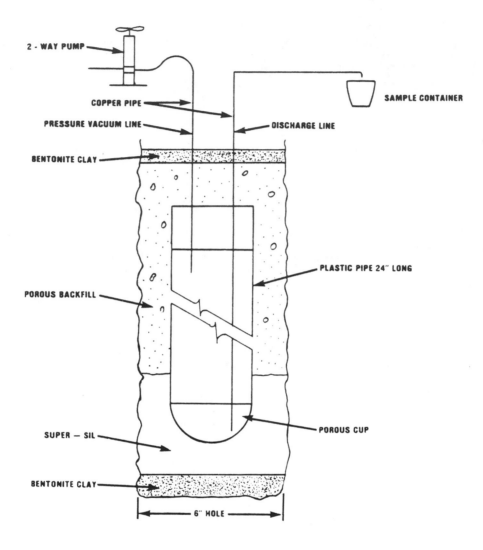

Figure 78. Example of a Pressure – Vacuum Lysimeter Installation

Adapted From: Parizek, R.R. and B. E. Lane, "Soil-Water Sampling Using Pan and Deep Pressure-Vacuum Lysimeters," _Journal of Hydrology_, Volume II, pp. 1-21, © 1970.

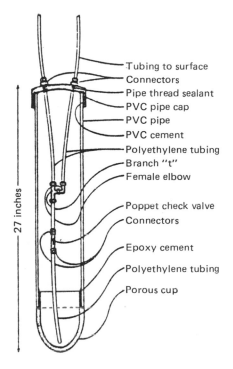

Figure 79.   Vacuum-Vacuum Lysimeter Installation

Source: Wood, 1973.

Environmental Protection Agency. This system combines a pressure vacuum lysimeter with a purge and trap apparatus and is intended to provide a means by which volatile organic species may be collected and concentrated in the field. A diagram of this unit is shown in Figure 80.

   h.   Chemical Analysis

   The analysis of groundwater samples, like the analysis of any material, may involve the utilization of equipment and techniques that span the entire spectrum of analytical chemistry. Particular problems encountered may be resolved simply, using nonsophisticated equipment and procedures such as pH meters or gravimetric determinations, while other problems encountered may require complex separation or fractionation techniques and sophisticated equipment such as gas chromatography/mass spectrometry. Thus, a comprehensive description of all procedures, protocols, and instrumentation that may be applied, is, by necessity, complex and beyond the scope of this manual. If specific information is sought, the reader is referred to one of the many references which describe analytical procedures.

   The following discussion sections address various important aspects of the analysis portion of a program, including: sample preparation considerations, general analytical techniques, qualitative techniques which may be utilized in exploratory or screening phases, and the quantitative analyses.

   i.   Sample Preparation

   The handling of a sample after collection and before analysis is, in most cases, defined by the specific analytical protocols being used. These are related to the nature of the sample being analyzed, as well as to the type of information needed. An example of a formal rationale for application of a set of analytical techniques is given in Figure 81. This protocol was designed specifically for characterization of water soluble, nonvolatile organic species in an aqueous stream.

   In general, such protocols should address four important aspects of sample integrity:

   • Protection of Sample Components - Losses of individual compounds of volatilization, adsorption or chemical reaction must be guarded against. Appropriate sampling containers, chemical preservation, and refrigeration will prevent the degradation/modification of the samples.

   • Extraction of Components - Most methods used for the analysis of complex mixtures require that the organic components be extracted into an organic solvent. Ultimate selection of an appropriate solvent should address solvent efficiency and

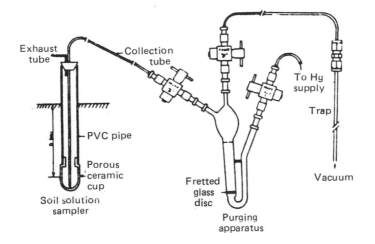

Figure 80. Lysimeter with Purge and Trap Setup

Source: Pettyjohn, et al., 1981.

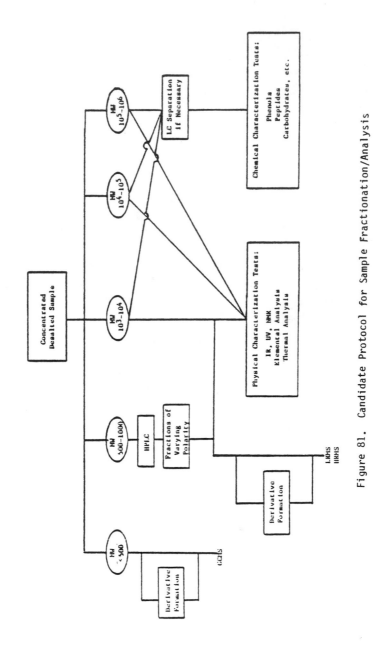

Figure 81. Candidate Protocol for Sample Fractionation/Analysis

boiling point, as well as sample component
volatility.

- Separation Schemes - Organic mixtures frequently
  contain such a large number of components that it is
  difficult to interpret the data obtained from the
  analysis of an undifferentiated sample. Various
  fractionation methods appropriate to the sample and
  the nature of the information desired must be
  considered.

- Derivative Formation - Before the analysis of some
  samples, derivation may be required to stabilize
  sample components or enhance the detection limit of
  an analytical method.

In characterizing complex environmental samples, it is frequently
necessary or desirable to perform some sample fractionation prior to
analysis. Many times a separation based on physical properties is
accomplished during sampling. For example, a sample of groundwater may
be filtered to remove suspended sediments and salts to allow for the
analysis of dissolved metals only.

Once the sample has been returned to the laboratory, a considerably
larger range of separation techniques is available. Liquid column
chromatography on media such as silica gel, alumina biobeads, or
Florisil® is often useful as a method for separating a sample into a
small number (e.g., 4-8) of fractions prior to bioassay or chemical
class characterization. This type of LC separation is also used to
remove interferences in the analysis of selected substances. Other
chromatographic procedures have been developed into high resolution
instrumental methods (HPLC, GC) that are discussed under
characterization. These high performance methods are frequently
preferred for applications involving detailed chemical analysis. Other
fractionation procedures based on physical or chemical properties of
sample components include: gel filtration chromatography, membrane
separation, chemical precipitations, sieving, and Bahco particle
separators.

j.   General Analytical Techniques

The modern analytical chemist has at his disposal a wide range of
techniques for the qualitative and quantitative chracterization of
environmental samples. In Table 22 many of the tools which have been
successfully used in these applications have been listed, along with an
indication of whether their primary utility is to provide elemental
composition or compound identification and whether they are primarily
useful for these purposes in a broad-screen survey or
specific-quantitative analysis mode. This listing distinguishes between
those methods which are primarily used for the analysis of organic
materials from those used principally for inorganics. These
distinctions are somewhat arbitrary; HV methods can be used for

TABLE 22. ANALYTICAL TECHNIQUES

Technique	Acronym	Analysis Category			
		Elements	Compounds	Screen	Quantitative
Primarily for Organic Analysis:					
1. Infrared Spectroscopy	IR		✓	✓	
2. Ultraviolet Spectroscopy	UV		✓		✓
3. Luminescence Spectrophotometry	LS		✓	✓	✓
4. Mass Spectrometry					
- Low Resolution	LRMS		✓	✓	
- High Resolution	HRMS		✓	✓	
- Automated GC/MS	GC/MS/DS		✓	✓	✓
5. Nuclear Magnetic Resonance Spectroscopy	NMR		✓	✓	✓
6. Chromatography					
- Thin Layer	TLC		✓	✓	✓
- Liquid	LC,HPLC		✓	✓	✓
- Gas	GC		✓	✓	✓
7. Combustion Analysis	C,H, etc.	✓			✓
8. Thermal Analysis	TA/TGA/DTA	✓		✓	✓
Primarily for Inorganic Analysis					
9. Spark Source Mass Spectroscopy	SSMS	✓		✓	
10. Optical Emission Spectroscopy	OES	✓		✓	
11. X-ray Fluorescence					
- Dispersive	XRF	✓		✓	
- Energy discriminating	EDXRF	✓		✓	
12. Neutron Activation Analysis	NAA	✓		✓	
13. Atomic Absorption Spectroscopy	AAS	✓			✓
14. Flame Emission Spectroscopy	FES	✓			✓
15. Anodic Stripping Voltammetry	ASV	✓			✓
16. Optical Microscope	OM	✓	✓	✓	✓
Scanning Electron Microscopy	SEM	✓	✓	✓	
Transmission Electron Microscopy	TEM		✓	✓	
17. Electron Diffraction	TEM/ED		✓	✓	
18. X-ray Diffraction	XRD		✓	✓	
19. Chemical Analysis					
- Anions	=		✓		✓
- Specific Analytes		✓	✓		✓

colorimetric assays of inorganics, as well as organics, and XRD can be applied to inorganic or organic crystalline materials. It is recognized that some methods provide adequate quantitative data in a survey mode, but the purpose of the table is to indicate the primary utility of a method.

At the onset of a task designed to characterize and treat environmental samples it is desirable to assess the applicability of each of the methods listed in Table 22. In this way, the most appropriate techniques can be utilized to ensure that all the necessary data are obtained for a given sample. The following discussion addresses both a general screening approach and the specific analytical techniques which will most likely be used in the conduct of this work.

k.   General Approach to Analytical Methods

Judicious selection among the techniques available for the analysis of complex samples requires a rationale that considers both the nature of the sample and the nature of the information being sought. Different selection rationales need to be developed to meet different types of environmental objectives.

For example, the Process Measurements Branch of EPA's IERL-RTP has promulgated a set of procedures for use in Level 1 Environmental Assessment sampling and analysis (USEPA 1978). These procedures are aimed at quantifying comprehensive multimedia environmental loadings of pollutants to within a factor of two to three by using one consistent protocol. To ensure that no potentially harmful pollutant species are overlooked, the Level 1 analysis protocol places heavy emphasis on techniques that can be used in a survey mode. An overview of the Level 1 analysis scheme is given in Figure 82. The complete Level 1 evaluation, which includes biotesting as well as chemical analysis, is a tool for comprehensive screening of all significant sources. Level 1 may be supplemented or followed by Level 2 -- directed detailed analyses of specific identified pollutants, and Level 3 -- process monitoring on selected pollutants, as the need arises.

In other types of programs, particular published protocols have been specified for the determination of some contaminants, such as the EPA's list of priority pollutants. The protocols for analysis of the priority pollutants in support of the Effluent Guidelines Division of EPA have been successfully applied in our laboratories for volatile organics, base-neutral organics, acid-extractable organics, pesticides, and metals.

Other methods may also be appropriate for use in programs involving groundwater monitoring. Considerations which must always be addressed in the selection of appropriate analytical methods include:

- The Chemistry of the Analyte - The solubility of the analyte in various solvents, its boiling point, its

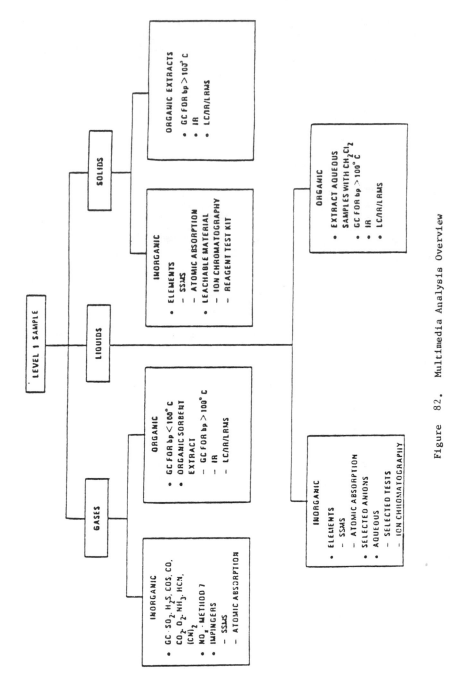

Figure   82.   Multimedia Analysis Overview

Source: EPA-600-7-78-201

UV spectrum (or UV maxima) and special precautions, such as decomposition conditions.

- The Method of Detection - Analyte—specific characteristics are a major consideration in selecting the method of detection. For example, the presence of electronegative groups in the chemical makeup of an analyte makes the use of electron capture detection possible when more than one type of group is present in the analyte; the advantages of specifically detecting each will be considered.

- Extraction Considerations - Initial considerations should be given to the solvent method of detection compatibility. The solvents used to extract the analyte from groundwater samples should be compatible with the developed analytical method. For example, samples should not be extracted with methylene chloride if electron capture detection will be used.

- Limits of Detection - Decisive efforts must be made to the lowest attainable levels of detection. Methodology such as derivation can enhance by orders of magnitude the detection of a compound, but the derivatization techniques must be reproducible.

- Statistical Requirements - Required Quality Assurance/Quality Control protocols must be followed in using or developing analytical methods.

The instrumentation necessary to determine chemical composition can be as simple as a density balance or a pH meter, or as advanced as a gas chromatograph or a mass spectrometer. Frequently, more than one method may be appropriate for the measurement at hand. One must carefully choose the equipment suitable for the task. Several considerations are important:

- Cost and availability
- Complexity of the groundwater sample
- Required sensitivity and accuracy
- Special considerations, such as trace impurities in the stream

It is often necessary to separate or resolve the component of interest from the process stream because of the lack of specificity of some instruments. In determining concentration, some property of a chemical which varies with composition is measured. For example, in measuring a solution pH (i.e., hydrogen ion concentration), the pH meter actually reads the EMF generated by the chemical potential of the hydrogen ions in the liquid. In other cases (e.g., mass spectrometery), the individual molecules themselves are detected electronically.

Because of the large number of instruments used in composition measurement, any listing is necessarily incomplete. The more important analytical tools which may be used are briefly described below.

1.   Gas Chromatography (GC)

Gas chromatography is used for separation of complex mixtures of organics with appreciable volatility. GC is also an exceedingly valuable technique for quantitative analysis of sample components. With the exception of Gas Chromatography/Mass Spectrometry (discussed below), qualitative analysis of GC is limited to inferences drawn from retention times of individual peaks and known detector selectivities.

A considerable variety of detection principles have been utilized for gas chromatography purposes. Of the detectors available, several are of interest in organic analysis.

The flame ionization detector (FID) is the most versatile general purpose device. The FID responds to any substance that will burn in the air/hydrogen flame to produce ions. Most organics, except for highly halogenated species, give strong FID responses. The lower limit of detection for organic species is on the order of one nanogram per microliter (ppm) of injected solution, and the dynamic range of the detector spans four or more orders of magnitude.

The electron capture detector (ECD) is specific for species that contain electronegative atoms or groups -- halogens, phosphorus, sulfur, and nitro-groups. For example, the high selectivity of the ECD has been used to good advantage in analyses of pesticides and polychlorinated biphenyls (PCBs). The lower limit of a detection is as much as 1000 times lower than that of the FID, but the dynamic range is generally smaller. The magnitude of the ECD response is its sensitivity to analyte structure, as well as to concentrations, since different species have different electron capture cross-sections. The detector response is also temperature-dependent, and an ECD cannot usefully be employed in temperature-programmed GC. All of these factors combined support the conclusion that GC/ECD is not a technique with wide applicability in organic analysis, but one which may be used for some specific categories of analytes. Octachloronaphthalene and ethylene glycol dinitrate are examples of compounds measured successfully using electron capture detection.

The flame photometric detector (FPD) is specific for species that contain sulfur or phosphorus. Its specificity, detection limits, and sensitivity are comparable to those of the ECD, although a completely different principle of detection is involved. When operated in the sulfur mode, the FPD response is logarithmic, rather than linear, with concentration. The FPD may be utilized in organic analysis of organophosphorus pesticides and some sulfur species.

The alkali flame ionization detection (AFID) is selective for nitrogen and phosphorous species. In the alkali flame ionization

detector, a cesium bromide or rubidium sulfate salt tip is incorporated
into a FID burner jet. When fuel/air flows are properly adjusted, the
detector provides enhanced selectivity for phosphorus or nitrogen
compounds compared to hydrocabons. Quantitative analysis procedures for
such species as nicotine and nitrosoamines have been developed using the
AFID. Although AFID can be used for nitro compounds, recent comparisons
of detection limits for a series of explosives have demonstrated that
the ECD is more sensitive for them by an order of magnitude.

The photoionization detector (PID) is specific to compounds with
ionization potential of < 10.5 eV, which are ionized after absorption of
UV radiation. The detector has a wide linear range and is highly
sensitive to species with high UV absorptivity, such as aromatics.   As
little as 2 nanograms/liter of tetramethyl lead has routinely been
detected using this technique.

m.   Combined Gas Chromatography - Mass Spectrometry (GC/MS)

Gas chromatographic separation with mass spectrometric detection
makes possible identification of GC peaks in the submicrogram sample
size range. For certain specific compounds, such as polycyclic aromatic
hydrocarbons (PAHs) detection limits using packed columns are in the
subnanogram range (on the order of 100s of picograms). The use of
capillary GC columns with their very high resolution allows separation
of a number of isometric species with detection limits 1-2 orders of
magnitude lower (1-10 picograms).

The marriage of a GC/MS to a dedicated minicomputer or to an
on-line data system is synonymous with the term GC/MS to many people.
Computer control of the mass spectrometer during data acquisition has
been enhanced by the availability of low cost microelectronic circuitry
for use in the control loop. Use of GC/MS with data systems is a
standard technique in both commercial and government laboratories (such
as the EPA laboratories) concerned with water quality analyses.
Analyses with capillary columns are only practical with automated GC/MS
instruments since the typical elution peak is only about 3 seconds wide.
To obtain more than a single mass spectrum in 3 seconds requires a rapid
scanning mass spectrometer so that the volume of hard copy generated
with a continuously operating oscillography would be unmanageable. A
further advantage of these systems is in their archival and retrieval
characteristics. The data recorded for any sample can be retrieved on
magnetic tape, for instance, and transported to other systems. Other
options, such as searching the data file of a sample for specific ions
to aid in the detection, identification, and quantification of specific
compounds, would be possible without a computer system for data
acquisition only by rerunning the sample. Thus, automated GC/MS is
rapidly developing into the method of choice for a majority of the
organic analyses required with environmental samples.

n.   High Performance Liquid Chromatography (HPLC)

High Performance Liquid Chromatography is an attractive alternative to more traditional methods, such as GC. The method's speed, sensitivity, and compatibility with aqueous systems make it an excellent candidate for use in anticipated in ground water monitoring programs.

HPLC is a separation technique with applications in quantification, isolation, and identification. It is differentiated from other LC methods by high speed, high sensitivity, and high resolution comparable to that of gas chromatography. These improvements have been achieved by columns using microparticulate packings with small diameter (5-50 micrometers) and high surface area (approximately 300 m^2/g particles). Separation may be achieved by differences in molecular size, number of types of functional groups, steric configuration, polarity, etc.

HPLC methods are likely to be used in one of two modes: normal or reverse phase. Reverse phase HPLC is based on separation by polarity differences in the mobile vs. stationary phase. The stationary phase is prepared by using long-chain alkyl silylating reagents to produce a hydrophobic layer on a silica solid substrate. Nonpolar solutes have a higher affinity for the stationary phase than do polar solutes. The mobile phase is generally programmed in a continuous gradient elution scheme from polar to nonpolar solvents. Water/methanol and water/acetonitrile are binary solvent systems frequently employed. In reverse-phase HPLC, polar sample components elute first. Normal HPLC, like reverse-phase HPLC, is based on separation by polarity. In normal phase separations, the stationary phase is more polar than the mobile phase. Gradient elution systems from nonpolar to polar solvents are used, and nonpolar solutes elute first. A guide to the selection of an analytical HPLC method is summarized in Figure 83.

The two most commonly used methods for detection of sample components in HPLC effluents are ultraviolet (UV) absorbance detectors and refractive index (RI) detectors. High sensitivity and specificity are achievable using a UV detector at fixed (e.g., 254 nm) or variable 200-800 nm wavelength. Lower limits of detection in the nanogram range have been reported for strongly absorbing sample species (i.e., molar absorbtivity $\sim$ 14,000).

The differential refractometer detector has lower sensitivity and less specificity than the UV detector. The RI detector responds to essentially all sample components and is a potential "universal" detector for HPLC, but lower limits of detection are in the microgram range. Furthermore, generality of the RI detector response requires matching of solvent system refractive indices during gradient elution HPLC; this is difficult to achieve in practice.

A third type of detection system is the spectrofluorometer. This fluorescence detector affords greater sensitivity and selectivity than UV does for those compounds which fluorescence. The specificity of this detector does limit its applicability in detecting a great number of

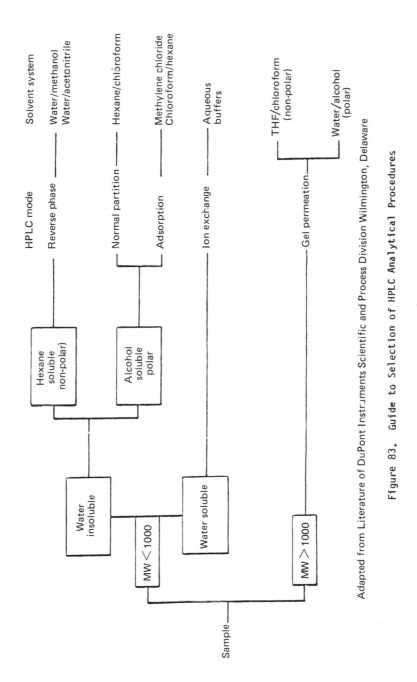

Figure 83. Guide to Selection of HPLC Analytical Procedures

Adapted from Literature of DuPont Instruments Scientific and Process Division Wilmington, Delaware

organic compounds. However, its use can be extended by the use of fluorescent reagents, such as Fluorescamine (4-phenylspiro[furan, 2H-1-phthalan]3'3-dione) and Fluoropa (o-phthaladehyde). These reagents react with primary amines to produce highly fluorescent derivatives.

Another type of detection system which may be useful is the electrochemical detector (EC). Electrochemical detection is a sensitive and selective method which will recognize a particular substance depending on the ability of the compound to be oxidized or reduced, and on the nature of the electrode and the mobile phase. Selectivity for one component of a mixture, as compared to another, may be achieved by adjusting the potential of the elctrode.

o.   Mass Spectrometry

- Low Resolution (LRMS) - Low resolution mass spectrometry is perhaps the single best technique to use to examine an unknown sample to determine whether or not organic compounds are present. In cases of simple mixtures, the LRMS data may be sufficient for compound identification, especially using modern file-searching techniques.

- High Resolution (HRMS) - High resolution and specific ion detection mass spectrometric techniques used in conjunction with data acquisition and processing have the ability to identify elements and compounds in the 10^{-9} - 10^{-12} g range with freedom from false positive results. Relatively uniform high sensitivity is obtained for all materials which are volatilizable at source temperatures and pressure. The technique is restricted to volatile material and usually excludes low polymers or salts, etc. A tremendous amount of data are present in the HRMS of an organic environmental mixture, and much of the value of the data lies in evolving a unique means of data reduction. A scheme based on an analysis of the chemical types present in a sample has been developed at Arthur D. Little, Inc., by Caragay et al. HRMS appears to have its greatest promise in determining the elemental composition of trace organic constituents in a sample and in veryifying the absence of certain sought species.

- Infrared Spectroscopy (IR) - IR is a general technique for group type characterization of samples. Virtually every organic and inorganic species of interest has at least one absorption frequency in the normal IR range. The technique is widely available in most laboratories and is capable of detecting less than 10 micrograms with modern attachments. The method is most attractive for its nondiscrimination and assurance that every major component of the sample will be represented in the spectrum. Species present at less than 5 - 10 percent in the mixture will generally not be detected. Combustion source samples which are black will suffer from low transmission and resolution because of black-body adsorption effects. However, the new Fourier ransform technique has helped overcome some of these problems.

- Nuclear Magnetic Resonance Spectroscopy (NMR) - NMR has not been extensively used in environmental analyses to date, primarily because of the high purity, liquid character, and large sample sizes which had been a requirement of older instruments. The introduction of Fourier Transform Nuclear Magnetic Resonance Spectrometers and the availability of Magic Angle Spinning probes has eliminated these earlier constraints. New environmental applications of NMR techniques are being developed and show particular promise for the characterization of fuels and solid effluent streams.

p.   Ion Chromatography (IC)

Ion chromatography (IC) is a separation and analysis technique which was recently developed for analysis of cations and anions. The technique involves use of a low capacity ion exchange resin column for separation, followed by another column designed to suppress the conductivity of the background electrode prior to detection of the ions of interest by means of a conductivity detector. The method of analysis is selective and sensitive, especially for many ionic species which cannot otherwise be detected by the usual ultraviolet or visible absorption photometric or refractive index liquid chomatographic detectors. Analysis of alkali and alkaline earth metals, aliphatic amines, ammonia, and many anions, including $P_2O_7^{=}$, NO_3^{-}, S^{2-}, $AsO_4^{=}$, carboxylates and sulfonates, has been achieved in complex aqueous mixtures. Detection limits have been around the high ppb region with a precision of less than 5 percent for most species. Incorporation of a preconcentrator column has increased sensitivity to less than 2 ppb for anions such as Cl^{-}, NO_3^{-}, SO_4^{-} and $PO_4^{=}$. This allows rapid and accurate analyses of many species which previously could only be performed using ion-specific electrodes and wet chemical methods on an individual ion basis.

q.   Atomic Emission Spectroscopy (AES)

Multielement analyses can be performed with the AES. Qualitative analysis for most elements can be performed by recording the entire emission spectrum of a given sample and inspecting for the presence or absence of emission lines corresponding to elements of interest. This may be especially useful, for example, for screening of samples for subsequent quantitative analysis.

Detection limits are, in general, comparable or superior to those obtained using conventional flame atomizers and atomic absorption spectroscopy. For certain elements (e.g., Ti), detection limits are several times lower than those obtained using the most sensitive atomic absorption techniques. Certain nonmetals (e.g., P and B) can be easily determined directly in solution with good sensitivity, thus eliminating the need for time-consuming colorimetric methods. In addition, many of the matrix or interelement effects noted in other techniques are absent, and high dynamic concentration ranges are usually obtained.

These advantages may make atomic emission spectroscopy a singularly time- and cost-effective approach to the anlaysis of certain complex samples.

r.   Atomic Absorption Spectroscopy (AAS)

Atomic absorption spectroscopy is one of the most widely used analytical techniques when information on one or a few specific elements is required. Approximately 70 elements can be determined. With appropriate modifications, such as sample preconcentration, electro-thermal atomization, or atomization of gaseous hydrides, very low detection limits have been obtained for many elements.

The advantages of AAS which, in addition to broad applicability and low detection limits for many elements, include relative ease of sample preparation, standardization, operation, and moderate costs, have led to its widespread acceptance as a standard method for application to quantitative analysis of many element.

Some examples of these types of analyses are:

- Combustion Analysis - Classical combustion analysis for the determination of elements such as C, H, N, S, etc., requires large amounts of sample (10 - 50 mg) compared with the instrumental methods. However, it is probably the best means of obtaining the data needed for a total mass balance analysis of a sample. This could be supplemented by an ignition analysis (loss on ignition at 550°C, volatiles content) for inorganic content, but this latter method suffers from inaccuracies in unknown environmental samples.

- Thermal Analysis (TA, TGA, DTA) - Instrumental thermal analyses techniques have substantial sensitivity for the characterization of the thermal behavior of small amounts of sample. Although these procedures lack the specificity of other methods in the determination of elements or compounds, they can be of use in qualitatively and quantitatively comparing environmental samples with regard to parameters such as ash content.

- Spark Source Mass Spectroscopy (SSMA) - Spark Source excitation is usually used for elemental analysis of environmental samples because of the relatively uniform ion formation for all elements. Sample preparation, including removal of organic matter and mixing with graphite to prepare electrodes, is required prior to analysis. Standards prepared in a matrix similar to the sample are required for quantitative analysis. High molecular weight organics can interfere in this analysis.

- X-ray Fluorescence (XRF) - All elements with an atomic number greater than 11 can be analyzed by XRF, although satisfactory data are lacking for elements with atomic numbers greater than 20. Quantitative comparisons of energy discriminating and wavelength dispersive techniques have been published. XRF is dependent on particle size effects and sample thickness. For airborne particulate analysis, XRF works best when samples are deposited on surfaces. Only ngs of elements/cm^2 of deposit are required. Little or no sample pretreatment is required, and the technique is nondestructive.

- Neutron Activation - Neutron activation analysis has excellent sensitivity for certain elements and is applicable to a wide variety of matrices. It is especially convenient for analyzing filters in which sample deposition is not confined to the surface. Extensive auxiliary facilities and long turnaround time are required for certain elements. No sample pretreatment is required.

- Flame Emission Spectroscopy - Lower sensitivities than those obtainable by flame atomic absorption and flurescence techniques may be obtained for certain elements by flame emission methods. Simultaneous multielement analysis is possible, since a separate source is not required for each element (or groups of a few elements) as in atomic absorption. However, flame emission methods are critically dependent on flame conditions, and optimum excitation conditions vary for different elements. Many modern AA instruments have flame emission capability, and it is likely that such hybrid instruments may be required for situations calling for the widest possible scope at the highest possible sensitivity.

- Differential Pulse Anodic Stripping Voltammetry – Sensitivities comparable to nonflame AA techniques are obtained for certain metals with DPASV. The method is limited to the determination of approximately twenty metals which are soluble in mercury; the majority of applications have dealt with about ten of these. Approximately 4 - 6 elements may be determined in a single sample. The technique may be nondestructive.

- Microscopy – Microscopic techniques are especially useful for the examination of particulate matter. Information on compounds present, elemental composition, and crystalline form may be obtained rapidly on very small samples. Lower detection limits are 10^{-12} g with the light microscope, 10^{-15} g with the scanning electron microscope, and 10^{-18} g with the transmission electron microscope. Energy dispersive X-ray flurescence and X-ray diffraction analysis may be used in conjunction with the latter two microscopic techniques to determine the chemical composition of tiny single particles. The light microscope is one of the most powerful simple techniques although frequently overlooked because of lack of personnel training.

- Electron Diffraction – This technique, used in conjunction with a transmission electron microscope, is extremely powerful for cyrstal identification when only minute amounts of solid are available. However, extensive sample preparation is necessary, and the method is not likely to be viable for most environmental samples. One severe problem is that many analyses are necessary to represent a heterogeneous sample.

- X-ray Diffraction – This method is very powerful for the positive identification of crystalline species. Computer programs have recently been written which allow the simultaneous identification of up to 10 compounds in a mixture if the patterns of those compounds are present in the data file. Unfortunately, many environmental samples suffer the problem of having compounds present in amorphous forms and at various degrees of hydration. While the technique is sensitive and precise, it will probably only be applicable to a limited range of samples.

- Anion Measurements - For a screening program of the
  type in which an accuracy of a factor of 2 or so is
  acceptable for the analysis of samples of widely
  varying composition, the use of specific ion
  electrodes is strongly recommended wherever
  practicable. Since the electrodes give a
  (generally) linear response over several orders of
  magnitude, the need for further sample manipulation
  of off-scale samples (i.e., dilution [taking a
  different aliquot] changing cells), such as is
  experienced in narrow-band width measurement
  techniques (i.e., colorimetry, titrimetry) is
  eliminated. Colorimetric techniques may be required
  in some cases where highly precise determinations of
  trace anions are required. Acid-base and redox
  titrations can be applied in specific instances, but
  these procedures are generally not well suited for
  trace analyses.

s. Total Organic Carbon (TOC) and Total Organic Halogens
    (TOH)

The total organic content in groundwater samples may be determined
by a number of different commercially available total organic carbon
analyzers (total organic halogens may be of use in those cases where
halogenated species are of interest). The organic carbon present in a
sample is typically converted to methane (CH_4) or carbon dioxide (CO_2)
using reduction, oxidation, or pyrolysis techniques. The products are
subsequently measured using nondispersive infrared, flame ionization,
conductivity, or similar techniques. Detection limits vary from low ppb
to low ppm levels.

4. Prediction of Contaminant Transport

1. PHYSICAL, CHEMICAL, AND BIOLOGICAL PARAMETERS AND CONSTANTS
    APPLICABLE TO ORGANIC CONTAMINANTS AND PHYSICAL SYSTEMS OF CONCERN

    a. Overview

    Models designed to predict the fate and transport of organic
chemicals in soil/groundwater systems will usually require a variety of
parameters and constants to properly address two questions:

> (1) <u>Mobility</u>: How easily is the chemical mobilized (i.e.,
>   transported) through the important subcompartments, e.g.,
>   soil, soil air, and soil water?
>
> (2) <u>Persistence</u>: How easily is the chemical degraded by such
>   processes as biodegradation, hydrolysis, or oxidation?

    Associated with the answers to these questions are a number of
chemical-specific and environment-specific properties. Tables 24 and 25
provide a summary list of the most important chemical-specific and
environment-specific properties.

    Many of the chemical-specific parameters (or "constants") are
functions of one or more of the environment-specific parameters (and the
nature of the waste) and thus are not true constants. For example, the
rate of hydrolysis may be strongly affected by temperature, pH, the
presence of catalysts (e.g., certain heavy metal cations), and chemical
concentration. Similarly, there are some environment-specific
parameters that will be functions of the chemical, primarily when the
chemical (or waste) is present in significant concentrations. For
example, the soil porosity, groundwater pH and soil microbiological
population will be affected by the presence of many chemicals.

    The list of important properties provided in Tables 23 and 24
would, along with the requirement that their variability be known,
appear to place an excessive burden on a rigorous modeling effort.
There are, however, a few ways in which the burden can be reduced.
First, an initial prescreening of the site and chemical(s) of concern
will frequently allow a determination that one or more transport or
degradation pathways will be of little concern. For example, if the
contamination incident involves low concentrations (in groundwater) of a
highly soluble, refractory compound, then the volatilization and
biodegradation pathways might be neglected in the model. This, in turn,
removes the requirement for obtaining the chemical- and
environment-specific properties associated with these pathways (e.g.,
Henry's law constant, diffusion coefficient in air, rate of
biodegradation, and wind/air parameters).

TABLE 23. IMPORTANT CHEMICAL-SPECIFIC PROPERTIES

Bulk (Condensed Phase) Properties Affecting Mobility[a]

    Physical state (liquid or solid) of waste
    Chemical composition of waste
    Density (liquid)
    Viscosity (liquid)
    Interfacial tension (with water and minerals) (liquid)
    Water solubility
    Vapor Pressure

Properties to Assess Mobility of Low Concentrations[b]

    Soil adsorption coefficient
    Henry's law constant (or vapor pressure and water solubility)
    Diffusion coefficient (in air)
    Acid dissociation constant

Properties to Assess Persistence[c]

    Rate of biodegradation (aerobic and anaerobic)
    Rate of hydrolysis (and/or elimination)
    Rate of oxidation or reduction

Notes to Table 23

a. These properties will be important when it is known or
    suspected that a separate organic phase exists in the
    soil/groundwater system.

b. These properties are important in assessments of the mobility
    of chemicals present in low concentrations (i.e., not as a
    separate phase) in the soil/groundwater system.

c. For these properties it is generally important to know: (1)
    the effects of key parameters on the rate constants (e.g.,
    temperature, concentration, pH); and (2) the identify of the
    reaction products.

TABLE 24. IMPORTANT ENVIRONMENT-SPECIFIC PROPERTIES

Soil Properties[a]

    Porosity (air filled and total)
    Moisture content (in unsaturated zone)
    Particulate surface area (area per unit weight)
    Organic carbon content (weight percent basis)
    pH
    Cation exchange capacity
    Temperature
    Microbiological population density (and type(s))
    Nutrient availability
    Gas generation rate (esp. for landfills)

Leachate/Groundwater Properties

    pH
    Total dissolved solids, total dissolved carbon, and
      concentration of other major constituents and/or
      potential catalysts
    Groundwater flow rates

Meteorological Factors

    Infiltration rate
    Evapotranspiration rate
    Windspeed and direction (wind rose)
    Air temperature and pressure

Notes for Table 24

a.   If possible, values of these properties should be available
    for various depths, locations, seasons, etc., in the area of
    concern.

A second way to reduce the data generation burden (for chemical-specific properties) is to use estimates (see, for example, Lyman et al., 1982) or surrogates. In either case, sensitivity analyses could be run on the selected model to determine if the output was significantly affected by the uncertainty in the inputs.

A detailed discussion of all of these properties, their environmental significance and variability, common methods of representation and units, and measurement methods is beyond the scope of this section.

2.   CHARACTERISTICS OF INITIAL MIXING

The initial mixing of an (organic) pollutant with groundwater will have characteristics that will depend in complex ways on several variables including: (1) the time and space scales associated with the source release; (2) chemical-specific parameters; and (3) environment-specific parameters. We know of very few studies of this subject (especially for large, rapid releases into the soil/groundwater system) and the field engineer can only expect limited help from the available data generated by models, or from case histories of similar incidents.

Key questions associated with the characterization of initial mixing are:

a.   Was enough material released to result in the probable existence of a distinct organic phase in the soil/groundwater system?

b.   If so, where is this "second phase" material likely to be?

c.   How long will the "second phase" material persist?

d.   To what extent will groundwater be contaminated by this material?

Partial answers to these questions are given in the discussion below.

Time and space scales associated with the pollutant's release will be partial determinants of how much groundwater (area) is affected and to what degree. Time scales may range from minutes to hours on the short end (e.g., a one-time spill or leak from a container) to years at the long end (e.g., a slow leak through a liner or pipe). Space scales may range from a few square meters (for a point source leak) to several hectares (for large lagoons, pesticide application areas, or runoff). Clearly, a large release over a small time and/or space scale will have a greater potential for the formation of an distinct organic (second) phase in the soil/groundwater system immediately beneath the release site.

How quickly pollutants are transported down through the unsaturated soil zone to groundwater will depend on a variety of chemical- and environment-specific factors (see Section V). The rate of transport can probably be predicted reasonably well only when a second phase does <u>not</u> exist so that percolation with water is the dominant transport mechanism. In such cases, when the percolating water reaches the groundwater, mixing involves simple dilution.

The spill or release of bulk quantities of any liquid organics will, in most cases, have a high probability of resulting in the formation of a second (organic) phase near the groundwater table. Numerous case histories (e.g., from the rupture of storage tanks and leakage from unlined chemical dumpsites) have demonstrated this phenomenon.

There may be, in some cases, sufficient chemical- and environment-specific data to allow a prediction of the relative probability (or speed) of such second phase movement. Some of the more important factors are listed below.

Factors Enhancing the Mobility of Second Phase
Organic Material Through the Unsaturated Zone.

Chemical-Specific	Environment-Specific
High density	High soil porosity or fractures
Low viscosity	Low organic matter content of soil
Low interfacial tension	Low particle surface area
Persistent (i.e., resistant to degration by hydrolysis, etc.)	Shallow depth to groundwater
Low soil adsorption coefficient	

If and when a second phase of organic material does reach the groundwater table, the initial mixing characteristics are likely to be a function of the following parameters:

- Density difference between the organic material and groundwater;

- Water solubility of the second phase material;

- Slope of groundwater table and flow rate of groundwater; and

- Reactivity of organic material with water (esp. rate of hydrolysis)

Material that is denser than water (e.g., chlorinated solvents) may continue to sink down through the saturated soil zone leaving a near-vertical trail of contaminated soil and groundwater. A plume of contaminated groundwater deeper than usual is the likely result.

Material that is less dense than water (e.g., most hydrocarbon fuels) will tend to spread out over the groundwater table and, in many cases can flow in directions different from the local groundwater flow.

Irrespective of the density, the rate of dissolution of the chemical in groundwater will depend primarily on the chemical's water solubility, the contact area, and the groundwater flow in the area. In areas where groundwater movement is especially slow, molecular diffusion (of the chemical through water) may limit the rate of the dissolution process. A separate second phase of organics could persist for months to years in situations where the quantity released was large, the chemicals' solubility was low, the groundwater movement slow, and/or the area of contact was small.

In those cases where the material initially released to the soil/groundwater environment contained a mixture of different chemicals, one should expect a chromatographic effect. That is, chemicals with higher water solubility and lower soil adsorption coefficients will be transported more rapidly through both the unsaturated and saturated soil zones. Some of the more volatile components of the mixture may be transported up (through the unsaturated zone) to the soil surface. Such effects have been demonstrated in laboratory tests using kerosene and gas oil (van der Waarden et al., 1977 and 1971).

3. MATHEMATICS OF DISPERSION, DIFFUSION, AND DILUTION

a. Introduction

When two miscible fluids are brought into contact, there is an initial sharp interface between them which vanishes into a transition zone between the two fluids. As a solute is transported in a groundwater flow system, it gradually spreads and occupies an ever increasing portion of the flow domain. This spreading phenomenon that causes dilution of the contaminant is called hydrodynamic dispersion. Hydrodynamic dispersion occurs because of mechanical mixing during fluid advection and because of molecular diffusion due to the thermal-kinetic energy of the solute particles. Figure 84 illustrates the dilution process caused by mechanical dispersion. At high groundwater velocities, such as would occur in relatively permeable material (i.e., sand or fractured bedrock), mechanical mixing is principally responsible for dispersion. At low groundwater velocities, such as would exist in a clay or shale, molecular diffusion is principally responsible for fluid mixing. The two processes which make up hydrodynamic dispersion,

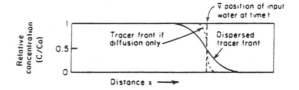

Figure 84.   Schematic Diagram Showing the Contribution of
              Molecular Diffusion and Mechanical Dispersion
              to the Spread of an Originally Sharp
              Concentration Front.

Source: Freeze and Cherry, 1979.

mechanical mixing and diffusion, will be discussed separately, then the combined effect of hydrodynamic dispersion and groundwater flow in the transport of solutes will be discussed.

(1) Mechanical Mixing

Mechanical mixing, or mechanical dispersion as it is sometimes called, is a mixing process, and it has a similar effect as turbulence in a surface-water regime. On a microscopic scale, it is caused by three mechanisms:

- dispersion due to differential velocity across a pore channel due to irregular pore surfaces,

- dispersion due to different pore velocities resulting from differences in pore size, and

- dispersion due to the tortuosity, branching and interfingering of pore channels.

(2) Diffusion

Diffusion is the process whereby ionic or molecular constituents move under the influence of their kinetic activity in the direction of their concentration gradient. It can occur in conjunction with mechanical dispersion or it can occur in the absence of any bulk hydraulic movement of the solution. Diffusion ceases only when concentration gradients are nonexistent. This process is sometimes referred to as molecular diffusion or ionic diffusion.

Fick's first law states that the mass of a diffusing substance passing through a given cross-section per unit time is proportional to the concentration gradient and is expressed as:

$$F = -D \frac{dC}{dX} \tag{11}$$

where:

F is the mass flux per unit area per unit time,

D is the diffusion coefficient,

C is the solute concentration,

$\frac{dC}{dX}$ is the concentration gradient.

In porous media, the diffusion coefficient, D, is replaced by an empirically derived "apparent diffusion coefficient," D^*, which is represented by the relation

$$D^* = AD \qquad (12)$$

where A is an empirical coefficient, less than unity, that takes into account the effect of the solid phase of the porous medium on the diffusion.

Fick's second law relates the concentration of a diffusion substance to space and time and is given by

$$\frac{\partial C}{\partial t} = D^* \frac{\partial^2 C}{\partial X^2} \qquad (13)$$

Solution to equation (12) is given by

$$C_i(X,t) = C_o \, erfc \left(\frac{X}{2} \sqrt{D^* t} \right) \qquad (14)$$

where:

$C_i(x,t)$ is the concentration of species i at location x and time t,

C_o is the source concentration of species i, and

erfc is the complementary error function

b. Contaminant Transport Processes

The principles involved in developing the fundamental differential equation for contaminant transport are described in Bear (1972), Freeze and Cherry (1979), and Fried (1975). This transport equation is developed by applying the principles of conservation of mass to the convection of a contaminant in a dispersive porous medium in which chemical sorption occurs.

For an elemental volume, this conservation of mass statement can be written as

net rate of   flux of   flux of   loss or gain
change of mass   =   solute out   −   solute   ±   of solute mass
of solute within   of the   into the   due to
the element   element   element   reactions

Mathematically, this can be written in one-dimensional form as

$$\frac{\partial C}{\partial t} = \frac{\partial}{\partial X}\left(D \frac{\partial C}{\partial X} \right) - V \frac{\partial C}{\partial X} + \frac{1}{n} \frac{\partial q^*}{\partial t} \qquad (15)$$

where:

C is the solute concentration
D is the dispersion coefficient
V is the average linear pore velocity
n_* is porosity, and
q is the mass of solute transferred to or from
   the solid phase

For nonreactive species, the last term on the right-hand side of equation (15) is zero, and the equation is called the advection-dispersion equation. Analytical solutions for one-, two-, and three-dimensional representations of the advection-dispersion equation are given by Bear (1972), Fried (1975), and many others. The use of numerical methods to solve this equation is reviewed by Anderson (1979).

Understanding contaminant hydrology requires obtaining quantitative or semiquantitative estimates for each of the terms in equation (15). Simulation of real contaminant migration patterns using mathematical models requires representation of Equation (15) in two or three dimensions in a formal sense. For persons confronted with the field problem of defining the areal extent of contaminant plumes using test drilling and piezometers or monitoring wells, it is the combined result of all of the processes described in Equation (15) that is important.

The remainder of this section will focus on the physical significance of the terms in Equation (14), excluding the chemical reaction term. To do this, we will follow the description in Freeze and Cherry (1979) and focus on homogeneous and heterogeneous media.

(1) Nonreactive Species in Homogeneous Media

The one-dimensional form of the advection-dispersion equation can be written as

$$\frac{\partial C}{\partial t} = \frac{\partial}{\partial X} \left(D \frac{\partial C}{\partial X} \right) - V \frac{\partial C}{\partial X} \qquad (16)$$

where all terms are as defined previously.

As discussed earlier, the coefficient of hydrodynamic dispersion can be expressed in terms of two components, mechanical dispersion, and molecular diffusion. Laboratory experiments (Cherry et al., 1975; Bear, 1972) have shown that the dispersion coefficient can be represented by

$$D = a_L V^m + D^* \qquad (17)$$

where:

D is the dispersion coefficient
a_L is the longitudinal dispersivity of the medium
V is the average linear pore velocity
m is an experimentally derived parameter, usually close to unity,
D^* is the apparent diffusion coefficient

The effect of the dispersion coefficient is to cause some of the contaminant to move faster than the average linear velocity of the groundwater and some of the contaminant to move slower than the average linear velocity. This results in a smearing out and dilution of an originally sharp boundary between contaminated and noncontaminated water. Figure 84 illustrates the combined effect of molecular diffusion and mechanical mixing to the spread of a concentration front.

The dispersion process causes spreading of the contaminant species in directions transverse to the flow path, as well as parallel to it. Figure 85 illustrates the lateral spreading due to transverse dispersion.

Solutions to one-dimensional flow problems are useful in the interpretation of laboratory column experiments, but are somewhat limited in application to field problems. Baetsle (1969) describes a method for obtaining preliminary estimates of the migration patterns that may arise from small contaminant spills or from leaching of buried wastes.

Assuming the contaminant originated as an instantaneous slug at a point source, the concentration at time t is given by

$$C(X,y,z,t) = \frac{M}{8(\pi t)^{1.5}\sqrt{D_x D_y D_z}} \exp\left(-\frac{x^2}{4D_x t} - \frac{y^2}{4D_z t} - \frac{z^2}{4D_z t}\right) \quad (18)$$

where:

M is the mass of contaminant introduced at the point source
D_x, D_y, D_z are directional dispersion coefficients, and
X, Y, Z are distances from the center of gravity of the contaminant mass

The peak concentration that occurs at the center of gravity of the contaminant plume is given by

$$C_{max} = (C_o V_o/(8(\pi t)^{1.5}\sqrt{D_x D_y D_z}) \quad (19)$$

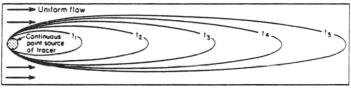

(a)

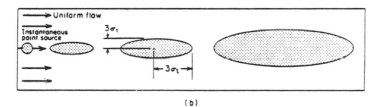

(b)

Figure 85.   Two-Dimensional Spreading of a Contaminant
              in a One-Dimensional Flow Field in a Uniform
              Sand. (a) Continuous Contaminant Source,
              (b) Instantaneous Contaminant Source.

Source: Freeze and Cherry, 1979.

where:

C_{max} is the maximum concentration at center of plume

C_o is the initial concentration, and

V_o^o is the initial volume

The other terms are as defined previously. The zone in which 99.7 percent of the contaminant mass occurs is described by the ellipsoid with dimensions $3\sigma_x = \sqrt{2D_x t}$, $3\sigma_y = \sqrt{2D_y t}$, and $3\sigma_z = \sqrt{2D_z t}$, where σ is the standard derivation of concentration (see Figure 85).

In complex hydrogeologic environments, these equations are of limited value because they do not consider the heterogeneity of the system. In relatively simple settings, however, they can provide preliminary estimates of migration patterns.

(2) Nonreactive Species in a Heterogeneous Media

The heterogeneity of natural materials complicates the problem of predicting and detecting contaminant behavior in groundwater flow systems. Contaminant transport is affected by both large scale and small scale heterogeneities. The larger scale heterogeneities, such as are caused by variations in geologic units, affect the general groundwater flow patterns. The smaller scale heterogeneities affect the dispersion of the contaminant.

Groundwater flow follows the path of least resistance and in a multilayered flow system most of the fluid flow will occur in the more permeable units with less flow occurring in the less permeable units.

Figure 86 illustrates the effect on the large-scale flow system and, hence on contaminant transport of stratification. These relatively simple variations can cause complex variations in the contaminant distribution pattern. Actual conditions are generally more complicated than these simple examples, and the end result is that one should expect the contaminant plume to have a very complex geometry.

On a smaller scale, the variations between different zones of the same layer can create a significant amount of dispersion of the contaminant front. Variations in the relative percentages of sand, silt, clay, and cementation or fracturing create local variations in permeability. These local scale variations are sufficient to cause fingering and spreading of the contaminant front. Figure 87 illustrates the effects of small scale heterogeneities on the pattern of contaminant migration in granular porous media. In Figure 87, the continuous higher permeability layers have allowed the contaminant front to advance more rapidly than in the lower permeability layers. In Figure 87, the

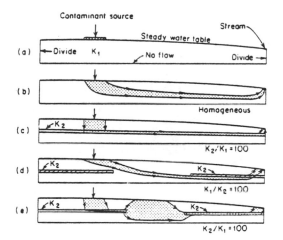

Figure 86. Effect of Layers and Lenses on Flow Paths
in Shallow Steady-State Groundwater Flows
Systems. (a) Boundary Conditions,
(b) Homogeneous Case, (c) Single Higher-
Conductivity Layer, (d) Two Lower-Conductivity
Lenses, (e) Two Higher Conductivity Lenses.

Source: Freeze and Cherry, 1979.

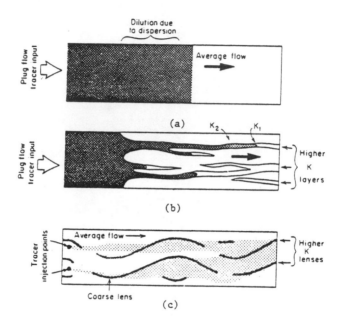

Figure 87.   Comparison of Advance of Contaminant Zones
Influenced by Hydrodynamic Dispersion.
(a) Homogeneous Granular Medium, (b) Fingering
caused by Layered Beds and Lenses, (c) Spreading
Caused by Irregular Lenses

Source: Freeze and Cherry, 1979.

discontinuous high permeability layers have resulted in a greater mixing of the contaminant front. This tends to create a more homogeneous concentration pattern in the down gradient direction.

Most laboratory studies of dispersion have dealt with relatively homogeneous material, and the experimental results have generally indicated small dispersivity values. Experiments which have tried to approximate actual conditions have generally resulted in calculating values of dispersivity which are much larger than were calculated for homogeneous material. Numerical analyses of the effects of dispersion on contaminant transport (Pickens and Lennox, 1976) have indicated that large values of dispersivity cause more mixing of the contaminant than do smaller values of dispersion.

4.   RELIABILITY OF PREDICTIVE METHODS

Many types of predictive methods are available which can be used to analyze the problem of groundwater contamination by organic chemicals. These methods range from simple analytical expressions for one-dimensional groundwater flow to multidimensional, multiphase, multicomponent reservoir simulation models. The simpler models are generally useful to a wider group of people and under a wider variety of circumstances than the more complex numerical simulation models. The more complex models can provide more detailed and precise answers to complex problems, but these advantages are frequently offset by the increased expenses related to data requirements and computer equipment, as well as the specialized training required for model application. The purpose of model application is to integrate several factors and to produce information on the basis of which intelligent decisions can be made. For the purpose of this project, these decisions are related to determining the source and/or extent of groundwater contamination, designing and executing the field data collection program to quantify the extent of contamination, and selecting the appropriate remedial action.

Classical approaches to problem solving have been to formulate the problem and then make as many simplifying assumptions as possible to produce a new problem which is manageable. For groundwater contamination problems, this may mean that a complex geometry is represented by a simpler geometry, that spatially or temporally varying properties are assumed to be constant, or that reacting chemical species are assumed to be nonreacting. The utility or reliability of the model results is dependent upon how well the simplified model represents actual conditions. When these conditions are closely approximated, then model results are directly applicable. Model results can also be useful when the limitations of model representations of actual conditions are known. The knowledge that the model results are not an accurate representation does not prevent use of the results to develop a better understanding of the system. Very few cases exist in nature where no answer is better than an approximate one, provided that the model limitations are known and understood.

Figure 88 illustrates how initial application of simplified models can aid the design of the field investigation program which, in turn, provides information which can be used to improve the model representation of the actual condition and lead to a better understanding of the physical system. The development of a better understanding of the physical environment and the simplified representation of this system leads to a better interpretation of model results and, makes the results more reliable.

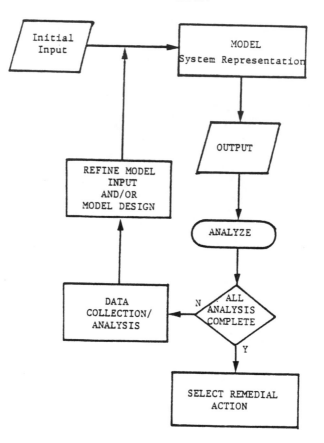

Figure 88.   Schematic Representation of Interaction Between Model Analyses and Data Collection Program Leading to Selection of Remedial Action.

5. Groundwater Treatment Control and Containment Methods

1. GRONDWATER TREATMENT METHODS

    a. Introduction

    Organic solvents, hydrocarbon fuels, pesticides, and other organics in groundwater present a number of issues for treatment selection. These compounds can usually be considered as members of one of the following chemical classes:

    - alcohols
    - ethers (halogenated)
    - aliphatic hydrocarbons
    - aromatics
    - halocarbons
    - pesticides
    - phenols

Because alcohols are readily biodegradable, these compounds are not a major groundwater contamination problem. The other chemical classes are generally persistent in the environment and pose acute and/or chronic chemical toxicity hazards if ingested. Table 25 (from CEQ, 1981) summarizes the highest concentration level measured, and the results of carcinogenic testing of the 33 organic compounds most commonly found in drinking water wells.

    Municipal drinking water wells typically yield 100 to 1,000 gallons per minute. A treatment facility to decontaminate groundwater would have to be capable of treating on the order of 100,000 to 1,000,000 gallons per day. The treatment capacity required could be increased by an order of magnitude if a major well field is contaminated.

    b. Overview of Treatment Technologies

    Groundwater contamination can be controlled by containment and treatment. Treatment is the focus of this section. Treatment technologies fall into one of the following classes:

    - biological
    - chemical
    - physical

Biological processes uses microbes to metabolize the contaminants. Biological treatment methods include:

    - activated sludge
    - aerated surface impoundments
    - trickling filters

253

TABLE 25.   33 ORGANIC COMPOUNDS MOST COMMONLY FOUND IN DRINKING
WELLS (CEQ 1981)

Chemical	Chemical Class	Highest Concentration (ppb)	Carcinogen Status*
Trichloroethylene	Halocarbons	27,300	CA
Toluene	Aromatics	6,400	NTA
1,1,1-Trichloroethane	Halocarbons	5,440	NA
Acetone	Ketone	3,000	–
Methylene chloride	Halocarbons	3,000	NTA
Dioxane	Ether	2,100	CA
Ethyl benzene	Aromatic	2,000	–
Tetrachloroethylene	Halocarbon	1,500	CA
Cyclohexane	Aliphatic hydrocarbon	540	NTA
Chloroform	Halocarbons	490	CA
Di-n-butyl-phthalate	Phthalates	–	NTA
Carbon tetrachloride	Halocarbons	400	CA
Benzene	Aromatic	330	H
1,2-Dichloroethylene	Halocarbons	323	NTA
Ethylene dibromide	Halocarbons	300	CA
Xylene	Aromatics	300	NTA
Isopropyl benzene	Aromatics	290	NTA
1,1-Dichloroethylene	Halocarbons	280	NTA
1,2-Dichloroethane	Halocarbons	250	CA
Bis(2-ethylhexyl)phthalate	Phthalates	170	NTA
Dibromochloropropane	Pesticide	137	CA
Trifluorotrichloromethane	Halocarbons	135	NTA
Dibromochloromethane	Halocarbons	55	NTA
Vinyl chloride	Halocarbons	50	H, CA
Chloromethane	Halocarbons	44	NTA
Butyl benzyl-phthalate	Phthalates	38	NTA
gamma-BHC(Lindane)	Pesticide	22	CA
1,1,2-Trichloroethane	Halocarbon	20	CA
Bromoform	Halocarbon	20	NTA
1,1-Dichloroethane	Halocarbon	7	SA
alpha-BHC	Pesticide	6	CA
Parathion	Pesticide	4.6	SA
delta-BHC	Pesticide	3.8	–

*H   = Confirmed human carcinogen
 CA   = Confirmed animal carcinogen
 SA   = Suggested animal carcinogen
 NA   = Negative evidence of carcinogenicity from animal bioassay
 NTA   = Not tested in animal bioassay
 Blank = No information found

- anerobic treatment
- land treatment

Activated sludge is the most commonly used biological process for wastewater treatment. Microbes oxidize or hydrolize organic compounds in an aerated tank. The activated sludge is separated in a clarifier and a portion is recycled to the aeration tank. Biological processes are capable of treating a wide range of organics but cannot effectively destroy refractory organics, such as PCBs, polynuclear aromatics, and halocarbons.

Chemical treatment destroys or immobilizes a contaminant by bringing it into contact with a chemical reactant. Chemical treatment methods include:

- alkaline chlorination (especially for cyanide)
- precipitation   (primarily for selected inorganics, e.g., heavy metals)
- ion exchange
- chemical reduction (especially for chromium)
- ozonation
- wet air oxidation
- hydrolysis

The first four processes are applicable to inorganic contaminants rather than organic. Ozonation and wet air oxidation are the chemical methods most capable of destroying organic contaminants. Hydrolysis, especially when used with an accelerating catalyst, may be an important process for the destruction of some pesticides.

Physical treatment methods generally separate contaminants from waste streams based on such physical properties of the contaminant as density, molecular weight, solubility, etc. Physical treatment methods include:

- carbon adsorption
- stripping
- flotation
- sedimentation
- reverse osmosis

Carbon adsorption is applicable to a wide range of organic compounds which include organic solvents, hydrocarbons, and pesticides. Stripping and reverse osmosis may be applicable to some organic solvents and pesticides, respectively.

The applicability of treatment methods to the chemical classes cited herein is summarized in Table 26.

Application of treatment technologies to groundwater contamination problems has been limited. In a survey of 169 uncontrolled hazardous

TABLE 26. TREATMENT PROCESS APPLICABILITY MATRIX
(USEPA 1980c)

Treatment Technology	Alcohols	Aliphatics	Aromatics	Ethers	Halocarbons	Pesticides	Phenols	Phthalates
Biological treatment								
Activated sludge	E	V	V	G	P	N,P	G	G
Rotating biological disc	E	V	V	G	P	N,P	G	G
Trickling filter	E	V	V	G	P	N,P	G	G
Surface impoundment	E	V	V	G	P	N,P	G	G
Land treatment	E	V	V	G	P	N,P	G	G
Chemical treatment								
Alkaline chlorination	N	N	N	N	N	N	N	N
Ozonation	G,E	P	F,G	—	F,G	E	E	—
Chemical reduction	N	N	N	N	N	N	N	N
Precipitation	—	—	F	—	—	—	—	G
Ion exchange	—	—	—	—	—	—	—	—
Wet air oxidation	X	X	X	X	X	X	X	X
Physical treatment								
Carbon adsorption	V	V	G,E	V	G,E	E	E	E
Sedimentation	—	—	—	—	—	—	—	—
Flotation	—	—	—	—	—	—	—	—
Filtration	—	—	—	—	—	—	—	—
Reverse osmosis	V	V	V	—	—	E	V	—
Stripping	—	—	—	—	—	—	—	—

Key:
E = Excellent performance likely
G = Good performance likely
F = Fair performance likely
P = Poor performance likely
N = Not applicable
V = Variable performance reported for different compounds in the class
X = Treatment is applicable but not specified in the source reference
— = A dash indicated no data available

waste sites, groundwater leachate was reported to have been contaminated at 110 sites (65 percent of the sites) (Neely, N. et al., 1981). The survey results revealed some form of treatment had been undertaken, at only 24 sites (~ 1/5 of those reporting groundwater contamination).*

The treatment technologies reported as being used (or proposed for use) were as follows:

Treatment Technology	No. of Sites Using Technology
Activated carbon or charcoal	8
Leachate recirculation through soil/landfill or process	4
Aeration†	4
Chemical treatment for removal of inorganics (e.g., precipitation, oxidation)	2
Biostimulation	1
Unspecified treatment method	5
	24

† Aeration is discussed under Activated Sludge Treatment in the next subsection.

    c. Process Treatment Trains

    Contaminated groundwater generally contains a mixture of contaminants, all of which can seldom be successfully treated by one technology. Therefore, it is usually necessary to employ more than one technology. Based on laboratory results, effective treatment methods and a sequential process treatment can be chosen. Data requirements necessary to select a process treatment train are summarized in Table 27. An example process treatment train to treat an aqueous mixture of metal salts and chlorinated solvents is shown in Figure 89. A detailed analysis of process treatment train selection can be found in USEPA 1980c.

    d. Treatment Technologies

    This subsection contains a detailed discussion of technologies which are applicable to organic contaminants in groundwater. These technologies are (USEPA 1981c):

---

* "Treatment" not considered to include diversion, isolation, simple pumping, or purging.

TABLE 27. DATA REQUIREMENTS TO SELECT TREATMENT PROCESS

TREATMENT TECHNOLOGY	Volume	TKN	Flow	pH	Acidity Expressed as $CaCO_3$ Equivalent	BOD	COD	TOC	TSS	TDS	Metals	Cyanide	Temperature	Complexing Agents	Viscosity	Climate	Soil Permeability	Soil CEC	Oxidation Reduction Potential	Leachate Variability	Phosphorus Content	Particle Size of Suspended Solids	Presence of Interfering Species
Biological Treatment																							
Activated Sludge	X	X	X	X	X	X	X	X	X		X	X	X			X				X	X		X
Rotating Biological Dis	X	X	X	X	X	X	X	X	X		X	X	X			X				X	X		X
Trickling Filter	X	X	X	X	X	X	X	X	X		X	X	X			X				X	X		X
Surface Impoundment	X	X	X	X	X	X	X	X	X		X	X	X			X				X	X		X
Land Treatment	X	X	X	X	X	X	X	X	X		X	X	X			X	X	X		X	X		X
Chemical Treatment																							
Alkaline Chlorination	X		X	X	X							X											X
Ozonation	X		X	X																			X
Chemical Reduction	X		X	X							X	X											X
Precipitation	X		X	X							X	X		X									
Ion Exchange	X		X	X					X	X	X												
Wet Air Oxidation	X		X	X				X	X	X													
Physical Treatment																							
Carbon Adsorption	X		X	X				X	X	X					X					X			X
Sedimentation	X		X	X					X													X	
Flocation	X		X	X					X													X	
Filtration	X		X	X					X													X	
Reverse Osmosis	X		X	X					X	X					X					X			X
Stripping	X		X	X																			X

Source: ADL, 1982

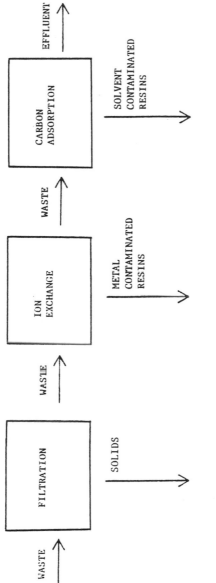

Figure 89. Process Treatment Train For An Aqueous Mixture of Metals and Chlorinated Solvents

- activated sludge
- carbon adsorption
- ozonation/UV
- stripping
- reverse osmosis/wet air oxidation

Selection of these technologies was based on the availability, experience, and range of applicability of the technology. Activated sludge is not usually an effective method to destroy halogenated solvents but was included because activated sludge sewage treatment systems are available at many military bases and would certainly be considered as a treatment option. Reverse osmosis is considered in combination with wet air oxidation as a potential method to concentrate and subsequently destroy pesticides. The remaining technologies are generally suitable for the removal or destruction of organic solvents and/or pesticides.

Technology discussions will all follow the same format and address:

- Technology Description
- Operating Characteristics
- Design Considerations

(1) Activated Sludge

(a) Description

Bacteria are utilized by the activated sludge process to oxidize and hydrolize organic waste in aqueous waste streams. The bacteria become acclimated to the wastewater environment through continuous recycle as shown in Figure 90.

The process includes an aeration basin, a clarifer, and equipment to recycle a portion of the activated sludge from the clarifer to the aeration basin. An aeration system releases either air or pure oxygen into the aeration tanks. Aeration methods are summarized in Table 29. Equalization, neutralization, and/or primary sedimentation may precede activated sludge processing. Activated sludge treatment yields a treated effluent and a residual sludge. Sludge disposal options include landfill, incineration, and composting.

(b) Operation Characteristics

Performance criteria for activated sludge processing systems are usually based on BOD removal efficiency. Mean BOD removal efficiency for 92 industrial wastewater streams, which were studied by the USEPA, was 86 percent (USEPA, 1980d). Mean influent and effluent levels were 1310 and 184 mg/l respectively. In contaminated groundwater treatment applications, removal efficiencies for specific compounds may be a much more important measure of the technology effectiveness. Activated sludge is not very effective for removal of halogenated solvents or pesticides from aqueous streams. For example, for five industrial

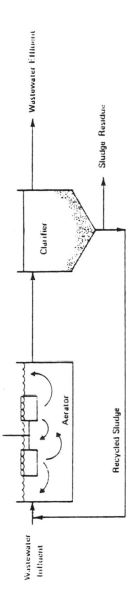

Figure 90. Typical Activated Sludge

Source: ADL, 1976

TABLE 28. SUMMARY OF AERATION METHODS

Method	Description	Application
Extended Aeration	Longer wastewater retention times in the aeration basin.	Low organic loading and reduced sludge quantities desired.
Pure Oxygen Aeration	Wastewater aeration with pure oxygen in a closed aeration tank.	High organic and/or metal loading.
Contact Stabilization	Aeration of recycled sludge on its return to the aeration tank.	Sludge removes BOD rapidly by biosorption. Contact stabilization decomposes the sorbed organics.

Source: ADL, 1976

wastewater streams containing a 26 mg/l mean concentration of methlyene chloride studied by the USEPA, removal efficiency was only 34 percent (USEPA 1980d). Removal was probably due to volatilization of the methylene chloride rather than biodegradation.

Effective operation of activated sludge systems requires that the influent pH level be near neutral and the process loading be constant. A pH adjustment system and an equalization tank are pretreatment processes usually employed to ensure an acceptable influent for activated sludge processing. If toxic species, such as heavy metal ions exist, then pretreatment may be required to remove them. Threshold toxicity concentrations for some metals are summarized in Table 29.

(c) Design Considerations

The major design factors are (Adams & Eckenfelder, 1974):

- retention time
- oxygen requirements
- food to microrganism ratio
- nutrient requirements
- sludge production

Based on these factors, the aeration tank(s) and clarifer can be sized, horsepower requirements for the aeration system determined, quantities of phosphorous and nitrogen needed to satisfy nutrient requirements determined, and sludge disposal needs defined. Design equations can be found in Adams and Eckenfelder (1974).

To generate quantitative design information, the following site specific information is necessary (Adams and Eckenfelder, 1974):

- biodegradable sludge fraction
- total Kjedahl nitrogen in influent (TKN)
- total phosphorous concentration in influent
- BOD or specific contaminant removal rate
- oxygen requirements
- sludge generation rate
- presence of interfering pollutants in influent

The last four information requirements listed are best determined by treatability studies. Removal rates are typically temperature dependent and should be studied over the expected operating temperatures at the site.

(2) Carbon Adsorption

(a) Technology Description

Carbon adsorption removes contaminants from an aqueous waste stream by binding the contaminants to the surface of a solid, activated carbon adsorbent. The carbon adsorbent is generally in a granualar form, but a

TABLE 29.   POTW THRESHOLD INTERFERENCE LEVELS* OF SOME
AEQUOUS CONTAMINANTS

Contaminant	Threshold Concentration (mg/l)	
	Chronic	+Slug Dose
Cyanide	1^1	40^1
Copper	1^2	75^2
Nickel	$1-2.5^2$	$50-200^2$
Zinc	$5-10^2$	160^2
Chromium	10^2	$>500^2$
Lead	$*1^1$	-
Cadmium	-	0.5^3
Silver	5^{23}	-

---

*Threshold level: 2-10% increase in BOD or COD of waste
water effluent

+Slug dose: 4-hour exposure which causes a significant
impact on the POTW for a 24-hour period

[1] Federal Guidelines: State and Local Pretreatment Programs,
Vol. 1, EPA-430/9-76-017a, p. E1-E26, 1977.

[2] Barth, E.F., et al, "Summary Report on the Effects of
Heavy Metals on the Biological Treatment Process,"
J. WPCF, Vol. 37, p. 86, 1965.

[3] Cenci, G., et al, "Evaluation of the toxic effect of Cd^{+2}
and $Cd(Cn)_4^{-2}$ on the growth of mixed microbial population
of activated sludges," The Science of the Total Environment 7
p. 131-143, 1977.

Source: Arthur D. Little, Inc., 1982.

powdered form is also used. Several operating configurations for granular activated carbon systems are shown in Figure 91. Carbon adsorption is capable of removing a wide range of organic compounds and some inorganic species such as: antimony, arsenic, chromium, mercury, and silver (USEPA, 1982b). A treated effluent and a contaminated carbon are the products of carbon adsorption treatment. The contaminated carbon can either be disposed of by landfilling or incineration, or reactivated by thermal methods.

(b) Operating Characteristics

Carbon adsorption is typically used to remove organic compounds which are not treatable by biological methods. Organics can be reduced to very low concentration levels with a well-designed and well-operated system. Carbon adsorption is most applicable to nonpolar, high molecular weight, slightly soluble organics (USEPA 1980d). Results of an EPA study indicated that 51 of the 60 toxic organic compounds tested could be removed by carbon adsorption. For further discussion on operating characteristics of carbon adsorption systems see: Symons, 1978; EPA, 1978a; Symons, 1979; Symons, 1980; Wood, 1978; and Demarco, 1978.

Several carbon adsorption contacting methods are used (see Figure 91). In granular activated carbon systems, the aqueous stream contacts the carbon as it flows through a fixed or moving bed. As the carbon adsorption capacity becomes spent, it is replaced with new or regenerated carbon. In powdered carbon systems, finely ground carbon is mixed with the aqueous stream and after sufficient contact time is removed and generally disposed. A typical mode of operation is to add the powdered carbon to the clarifier of an activated sludge system. Contacting methods and corresponding application conditions are summarized in Table 30.

Effective operation requires a fairly uniform influent. Because a system is sensitive to changes in the influent character, systems are generally oversized to prevent contaminant breakthrough in the case of increased flow rates and/or higher influent contaminant levels. If influent quantity or character vary widely, an equalization tank should precede the carbon adsorption system. Influent levels of suspended solids, and oil and grease greater than 50 ppm and 10 ppm, respectively, interfere with granular carbon adsorption systems (ADL, 1976). Biological activity sometimes occurs in the carbon system and can contribute positively via biodegradation or negatively via clogging.

(c) Design Considerations

The major design factors are:

- carbon system contacting configuration
- type of activated carbon
- carbon usage rate
- carbon regeneration carbon

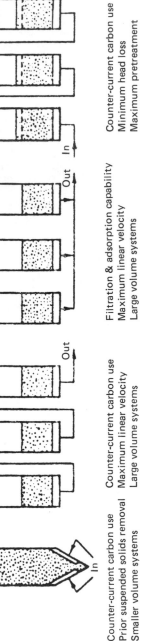

Figure 91.   Granular Activated Carbon System Configuration

Source: ADL, 1976

TABLE 30.   CONTACTING SYSTEMS

Method	Application Conditions	Comments
Single or parallel	Pollutant breakthrough curve is steep Carbon recharge interval is long Volume flow is high Influent is viscous	Typical flows are 1 to 4 gpm/ft^2 Parallel system is usually selected if pressure drop problems are expected for the system Moderate adsorbent expense
Adsorbers in series	Pollutant breakthrough curve is gradual Uninterrupted operation is necessary Relatively low effluent concentration is required Carbon recharge interval is short	Typical flows are 3–7 gpm/ft^2 High adsorbent expense
Expanded upflow adsorber(s)	For high flows and high suspended solids concentrations	Typical flows are 5–9 gpm/ft^2 Suspended solids are passed through the column and not separated
Moving bed	For systems requiring efficient use of carbon (i.e., carbon adsorption capacity is exhausted before removal from column)	Influent must contain less than 10 mg/ℓ TSS, and not biologically active. Either parameter will cause a pressure drop in the system and necessitate removal of carbon prior to exhaustion of its adsorption capacity
Powdered carbon with subsequent clarifier and/or filter	Carbon usage higher than for series of fixed-bed adsorbers Influent concentration of pollutants should be relatively constant to avoid frequent sampling and adjustment of carbon dosage	No restrictions or suspended solids or oil and grease in influent Capital equipment costs relatively low Simple to operate
Powdered activated carbon with activated sludge	For activated sludge systems receiving toxic or shock organic loadings	Protects the biological system from toxic organics and shock loadings Generally improves effluent quality

Source: ADL, 1976

The contacting configuration is based on the application conditions summarized in Table 30.

Carbon selection is an important consideration for optimizing the treatment process for the specific aqueous stream. Laboratory testing is required to determine the appropriate carbon for each specific waste stream. Properties of several commercially available carbons are summarized in Table 31. The general laboratory test procedure is to mix different quantities of activated carbon with batches of the contaminated aqueous stream and analyze the equilibrium conditions to generate adsorption isotherms. Based on the adsorption isotherms, a carbon type is selected. For a more detailed discussion of carbon adsorption isotherms, see Schweitzer (1979) or Adam and Eckenfelder (1974).

Because carbon usage cannot be determined by the carbon adsorption isotherm results, laboratory testing a flow-through system is required to size a system. A common test method used is known as the bed-depth/service time analysis (BDST) (Adams and Eckenfelder, 1974). Three to four columns are connected in series under hydraulic loads which simulate field conditions. Effluent from each column is analyzed for the chemicals of concern. The effluent-to-influent ratio for the chemicals measured is plotted against the total bed-depth. Based on these results and the carbon contacting system to be used, the carbon usage can be calculated. For more details on BDST analysis and carbon usage calculations, see Adams and Eckenfelder (1974).

Carbon regeneration rates for industrial waste carbon treatment systems in a multiple hearth furnace are approximately 3.5 lb/hr-ft² (Adams and Eckenfelder, 1974). Generally, a carbon usage rate of 1,000 pounds per day is considered the cutoff point for economic regeneration versus disposal.

The following site-specific data are required to design a carbon adsorption system:

- influent contaminant levels
- influent flow rate
- presence of interfering contaminants in influent
- carbon adsorption isotherms
- bed-depth/service time analysis

(3) Ozonation/Ultraviolet

(a) Technology Description

Ozone (O_3) is produced in a generator (ozonator) and introduced into a contacting chamber where it oxidizes a wide range of contaminants. Ozone is a strong oxidizing agent capable of oxidizing many refractory organic compounds in the following chemical classes:

TABLE 31. PROPERTIES OF SEVERAL COMMERCIALLY AVAILABLE CARBONS*

PHYSICAL PROPERTIES	ICI AMERICA HYDRODARCO 3000	CALGON FILTRASORB 300 (8x30)	WESTVACO NUCHAR WV-L (8x30)	WITCO 517 (12x30)
Surface area, m^2/gm (BET)	600–650	950–1050	1000	1050
Apparent density, gm/cc	0.43	0.48	0.48	0.48
Density, backwashed and drained, lb/cu ft	22	26	26	30
Real density, gm/cc	2.0	2.1	2.1	2.1
Particle density, gm/cc	1.4–1.5	1.3–1.4	1.4	0.92
Effective size, mm	0.8–0.9	0.8–0.9	0.85–1.05	0.89
Uniformity coefficient	1.7	1.9 or less	1.8 or less	1.44
Pore volume, cc/gm	0.95	0.85	0.85	0.60
Mean particle diameter, mm	1.6	1.5–1.7	1.5–1.7	1.2

SPECIFICATIONS

Sieve size (U.S. std. series)				
Larger than No. 8 (max. %)	8	8	8	—
Larger than No. 12 (max. %)	—	—	—	5
Smaller than No. 30 (max. %)	5	5	5	5
Smaller than No. 40 (max. %)	—	—	—	—
Iodine No.	650	900	950	1000
Abrasion No., minimum	**	70	70	85
Ash (%)	**	8	7.5	0.5
Moisture as packed (max. %)	**	2	2	1

* Other sizes of carbon are available on request from the manufacturers.
** No available data from the manufacturer.
— Not applicable to this size carbon.

Source: ADL, 1976

- chlorinated hydrocarbons
- chlorinated aromatics
- pesticides

The oxidation of refractory organic compounds is improved by combining ultraviolet light with ozonation as shown in Figure 92. Because ozone is acutely toxic, systems are equipped with automated devices which measure ozone levels in the gaseous effluent and reduce the ozonator voltage and frequency if gaseous levels exceed a preset limit (generally 0.05 ppm). These systems are also equipped with air monitors which sound an alarm and shut off the ozonator if an ozone leak occurs.

(b) Operating Characteristics

Ozonation is typically used to treat waste streams containing less than 1 percent oxidizable material. Because ozone is not a selective oxidizer, the presence of oxidizable materials other than target pollutants will increase the cost of the treatment. A list of organics which are economically treatable by O_3/UV is presented in Table 32.

Ozonators produce low concentrations of ozone in air (less than 2 percent). The contact chambers are large (typical depths of 5 meters) because reaction rates are limited by mass transfer. Ultraviolet lamps, when utilized, are operated in the contactor chamber. Typical UV operating power levels range from 1 to 10 watts per liter (Prengle, 1975).

Because ozone is corrosive, ozonation equipment requires special construction materials which include:

- stainless steel
- unplasticized PVC
- aluminum
- teflon
- chromium plated brass or bronze

(c) Design Considerations

Key design factors are:

- ozone dose rate
- retention time
- ultraviolet light dosage

Ozone dose rate is expressed as either ppm ozone or pounds of ozone per pound of aqueous constituents oxidized. Dose rates usually range from 10 to 40 ppm for the former and 1.5 to 3.0 pounds per pound of contaminants oxidized for the latter (ADL, 1976). Dosage rates are usually set based on laboratory studies.

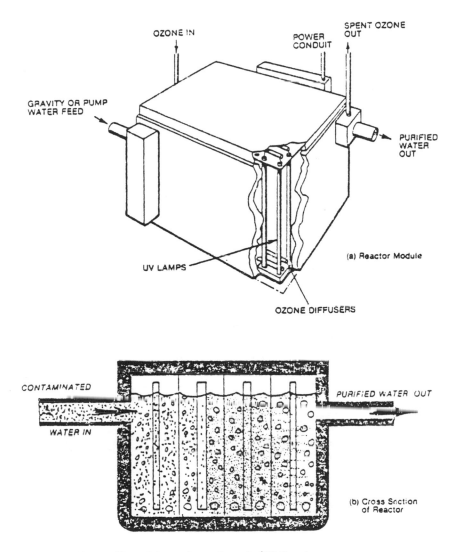

Figure 92.   Schematics of an O_3/UV Reactor

Source: Ghassemi, 1981.

TABLE 32. LIST OF ORGANICS DETERMINED TO BE ECONOMICALLY
TREATABLE BY THE O_3/UV PROCESS

Acetaldehyde	Glycols
Acetic acid	Hydroquinone
Alcohols	Kepone
Aldrin	Methylene chloride
Amines	Nitrobenzene
Anisole	Nitrophenol
Benzoic acid	Organic phosphates
Chelating compounds	Organosulfur compounds
Chlorinated phenols	Organo-tin compounds
Chlorobenzene	PCB's
Detergents	Phenol
Dieldrin	Phthalic acid
Dioctylpthalate	RDX
Endrin	Sodium acetate
Ethylene dichloride	Styrene
Formaldehyde	Sugars
Formic acid	TNT
Glycerols	Vinyl chloride
Glycine	Xylenol

Source: Ghassemi, 1981

Retention times generally range from 10 minutes to one hour in a multistage application. The design parameter is generally set based on laboratory studies. If ultraviolet light is used, the dose rate is also determined by laboratory studies.

Data requirements for system design include:

- treatment stream flow rate
- concentration of oxidizable constituents
- ozone dosage
- retention time
- ultraviolet light dosage
- operating temperature

(4) Stripping

(a) Technology Description

Stripping removes volatile components from an aqueous waste stream by passing air or steam through the waste stream. The "stripped" constituents must be removed from the gas stream by subsequent treatment, such as carbon adsorption or condensation. Stripping has been mainly used to remove ammonia from water, but is applicable to volatile organics (Kelleher, 1981; Shuckrow et al., 1981). Figures 93 and 94 show typical air and steam stripping process designs.

(b) Operating Characteristics

Air stripping is capable of achieving ammonia removal efficiencies of greater than 90 percent for aqueous streams containing less than 100 ppm (ADL, 1976). An EPA study (USEPA 1980d) reported that steam stripping removal efficiencies for volatile organics range from 10 to 97 percent. For example, the average removal efficiencies for chloroform and 1,1,2,2-tetrachloroethane were 89 percent and 40 percent. These results represent 5 data points for each pollutant. The average influent concentration was 65,000 ug/l and 78,000 ug/l for chloroform and 1,1,2,2-tetrachloroethane. Application of air stripping to remove 1,1,1-trichloroethane (TCEA) from a contaminated groundwater drinking supply resulted in the reduction of TCEA concentrations from 1200 ppb to 10 ppb (Kelleher, 1981). Other studies also indicate that air stripping is suitable for removing volatile organic compounds (Shuckrow, 1981; James, 1981). Laboratory studies of removal efficiencies for a particular organic constituent are necessary because removal efficiencies are dependent on the characteristics of each aqueous stream (Gosset, 1981).

In an air stripping process, wastewater flows counter current to an air stream. For ammonia removal, the influent pH is maintained in the 11-12 range. Typical operating parameter values for ammonia stripping are 40 L/(min-m^2) hydraulic load and a 400 ft^3/gal air flow rate (ADL, 1976).

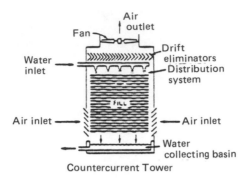

Countercurrent Tower

Source: USEPA 1982b

Figure 93. Air Stripping Tower

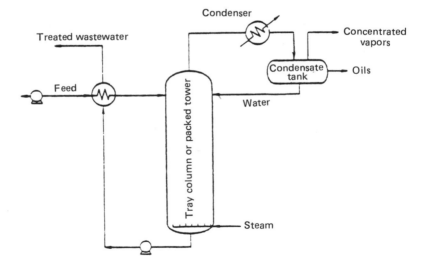

Figure 94. Typical Steam Stripping System

Source: ADL, 1976

In a steam stripping operation, the wastewater is typically introduced at the top of a distillation column and steam at the bottom. The steam and stripped wastewater components flow through a condenser. Condensed water is recycled and the pollutants are collected. Typical operating conditions are 0.6–2.0 pounds of steam per gallon of water treated and a wastewater flow of 200 gallons per minute.

(c) Design Considerations

For air stripping, the hydraulic load, air flow, depth of stripping tower, operating temperature and aqueous stream pH (if NH_3 is the stripping target) are important design factors (ADL, 1976).[3] For steam stripping, the treatment stream flow, steam flow, and column size are also important design factors (USEPA 1980d).

Information required to design a stripping system includes (Adams and Eckenfelder, 1978):

- aqueous treatment stream flow
- water temperature
- cold weather wet bulb temperature
- pH (for ammonia)
- influent concentration

(5) Reverse Osmosis/Wet Air Oxidation

(a) Technology Description

Reverse osmosis separates contaminants from wastewater by forcing water through a membrane which is impermeable to most soluble inorganic species and some organic compounds. Wet air oxidation is a thermal oxidation process which oxidizes waste in the liquid phase at high pressure and high temperature. A wet air oxidation process schematic is shown in Figure 95. These two technologies are considered together because, used in consort, they could be effective for treating groundwater contaminated with high molecular weight organic compounds (300–500 g/mol). Reverse osmosis would be utilized to concentrate the organic constitutients followed by wet air oxidation of the concentrate.

(b) Operating Characteristics

Reverse osmosis systems are typically operated in the 200–400 psi range. The operating characteristics of the three basic reverse osmosis module configurations are summarized in Table 33.

Reverse osmosis removal efficiencies are species and waste stream specific. An EPA study (USEPA 1980d) reported on average TOC removal efficiency of 84 percent for 13 data points.

Wet air oxidation is typically applied to waste streams containing 1 to 15 percent organic material. These concentrations are too high for conventional biological treatment or carbon adsorption, and generally

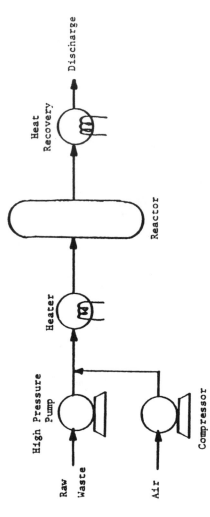

After: Ghassemi, 1981

Figure 95.   Schematic of Wet Air Oxidation

TABLE 33. COMPARISON OF REVERSE OSMOSIS MODULE CONFIGURATION

	Spiral wrap	Tubular	Hollow fine fiber
Membrane surface area per volume, ft^2/ft^3	100 - 300	40 - 100	5,000 - 10,000
Product water flux, gpd/ft^2	8 - 25†	8 - 25	0.1 - 2
Typical module factors:			
Brine velocity, ft/sec	†	1.5	0.04
Brine channel diameter, in	0.03§	0.5	0.004
Method of membrane replacement	As a membrane module assembly - on site	As tubes - on site	As entire pressure module - on site, module returned to factory
Membrane replacement labor	Low	High	Medium - requires equipment
High pressure limitation	Membrane compaction	Membrane compaction	Fiber collapse
Pressure drop, product water side	Medium	Low	High
Pressure drop, feed to brine exit	Medium	High	Low
Concentration polarization problem	Medium	High	High
Membrane cleaning - mechanical	No	Yes	No
Membrane cleaning - chemical	Yes - pH and solvent limited	Yes - pH and solvent limited	Yes - less restricted
Particulate in feed	Some filtration required	No problem	Filtration required

†Product flux varies with the net driving pressure and temperature; a flux of 10-25 gpd/ft^2 is typical at a pressure of about 400 psi.

‡It is difficult to define velocity in a spiral element since the space between membrane is filled with a polypropylene screen which acts as a spacer and turbulent promoter.

§Height of brine channel (not diameter).

**Permissible pH and temperature ranges dependent primarily on membrane type and not on module configuration; for example, polyamide hollow fine fiber is pH limited from 4 to 11, cellulose acetate from 3 to 7.5, thin film composite (TFC) spirals have been operated and cleaned at pH levels ranging from 1 to 12.

Source: Ghassemi, 1981

too low for cost-effective incineration. A summary of destruction efficiencies for ten priority pollutants is presented in Table 34. Units are typically operated at pressures of 24 atmospheres and temperatures of 300 degrees centigrade.

The effective utilization of reverse osmosis and wet air oxidation to remove and destroy organic compounds is limited by the concentration of inorganic salts in the groundwater. If the inorganic salt concentration is too high in relation to the organic concentration, then reverse osmosis treatment will not be capable of producing a sufficiently concentrated solution for economical destruction by wet air oxidation.

(c) Design Considerations

Flow, solvent flux, and solute flux are key design considerations for designing a reverse osmosis treatment system. For a discussion on calculating solvent flux and solute flux, see Ghassemi (1981). Reactor pressure, operating temperature, and retention time are key design considerations for wet air oxidation. A more detailed discussion of these design parameters can be found in Ghassemi, 1981.

Information required to design a reverse osmosis system includes:

- solvent flux
- solute flux
- hydraulic load
- total dissolved solvents in influent
- influent pH
- operating temperature

Information required to design a wet air oxidation system includes:

- treatment flow rate
- concentration of oxidizable materials
- laboratory studies to determine optimum operating pressure and temperature

2.   GROUNDWATER CONTROL AND CONTAINMENT METHODS

a. Introduction

In certain circumstances, groundwater contamination can be effectively controlled by containing, rather than treating, the source of contamination or the contaminated plume. The purpose of leachate treatment methods is to remove the hazardous constituents from groundwater after it has been removed from the ground. Control and containment methods, on the other hand, are designed to prevent hazardous constituents from entering groundwater or to restrict the movements of contaminated groundwater. In some cases, both treatment and containment may be appropriate.

TABLE 34.   WAO EFFICIENCY FOR TEN PRIORITY POLLUTANTS
(1-HOUR DETENTION TIMES)

Compound	Starting con-centration (g/1)	% Starting material destroyed	
		320°C	275°C
Acenaphthene	7.0	99.96	99.99
Acrolein	8.41	>99.96*	99.05
Acrylonitrile	8.06	99.91	99.00†
2-Chlorophenol	12.41	99.86	94.96†
2,4-Dimethylphenol	8.22	99.99	99.99
2,4-Dinitrotoluene	10.0	99.88	99.74
1,2-Diphenylhydrazine	5.0	99.98	00.08
4-Nitrophenol	10.0	99.96	99.60
Pentachlorophenol	5.0	99.88	81.96†
Phenol	10.0	99.97	99.77

* The concentration remaining was less than the detection limit of 3 mg/1.

† The percent destruction from acrylonitrile, 2-chloropenol, and pentachlorophenol at 275°C was increased to 99.50, 99.88, and 97.3 by addition of cupric sulfate (catalyst).

Source: Ghassemi, 1981

Groundwater control and containment measures are not limited to methods which act on the groundwater alone. The entire water balance at the site, as well as the location of the source of contamination with respect to the water table, must be considered. Cutting off groundwater flow upgradient of a site, for example, would be ineffective where the source of contamination is above the water table, where infiltration of surface water through the source is a problem, or where the source contains liquid wastes. These require other source control measures, such as surface water control methods or excavation. Upgradient groundwater control, however, would be appropriate if groundwater flow through the source is the primary means of release of hazardous constituents into the groundwater. The types of control and containment measures which will be most effective should be determined on a site-by-site basis. Data requirements for evaluating groundwater control and containment technologies are summarized in Table 35.

b. Control and Containment Technologies

This subsection contains a detailed discussion of technologies applicable to the control and containment of contaminated groundwater. Categories of technologies discussed are:

- Impermeable Barriers
- Groundwater Pumping
- Leachate Collection
- Removal
- Surface Water Control

Discussions include a description of the technologies in the category and engineering (design) information necessary to determine a technology's applicability at a given site.

(1) . Impermeable Barriers

(a) Description

Impermeable barriers are underground physical barriers designed to restrict groundwater flow. The most common impermeable (actually low permeability) barriers in use today are slurry walls, grout curtains, and sheet piles.

Slurry walls are constructed by digging a trench using a soil or cement bentonite and water mixture (slurry) to maintain the trench during excacation. For soil-bentonite (SB) slurry walls, the trench is then backfilled with carefully engineered, low-permeability soil. For cement-bentonite (CB) slurry walls, the slurry sets to form a low permeability barrier. Effectiveness of the slurry wall depends both on the characteristics of the SB backfill or set CB slurry, as well as on the formation of a slurry filter cake on the sides of the trench during construction.

TABLE 35.   GROUNDWATER CONTROL AND CONTAINMENT TECHNOLOGY DATA REQUIREMENTS

x = useful data

✓ = primary data required

Data Requirement	Slurry Walls	Grout Curtains	Sheet Piles	Pumping	Subsurface Drains & Ditches	Liners	Surface Water Technologies
Soil							
Topography	x	x	x	x	x	x	✓
Site of Material Accessibility	✓	✓	x	x	x	x	x
Vegetation						x	
Depth to Impermeable Stratum	✓	✓	✓	✓	✓	x	
Soil Type	✓	x	x	✓	x		x
Grain Size Distribution	✓	✓	✓	✓	x	x	x
Compaction	x	x	✓	x	x	x	x
Moisture Content	✓	✓			x	x	x
Permeability	✓	✓	x	x	✓	x	x
Porosity	x	✓	x				
Depth	x	x	x	x	x	x	x
Chemistry	x	✓	x			✓	
Groundwater							
Climatic Conditions						x	
Depth to Water Table	✓	✓	✓	✓	✓	✓	
Potentiometric Surfaces				✓		x	
Direction/Rate of Flow	✓	✓	x	x	x		
Recharge				✓			
Aquifer Characteristics	x	x	x	✓	x		
Chemistry	✓	✓	✓	x	✓	x	
Infiltration	x	x	x	x	x	x	✓
Runoff							✓
Drainage Area					✓		✓
Waste Characteristics	✓	✓	✓	✓	✓	✓	x

Source: Arthur D. Little, Inc., 1982.

Grout curtains are installed principally to seal voids in porous or fractured rock in situations where other groundwater controls are impractical. Solutions or water-solid suspensions are injected under pressure into the ground, filling voids and reducing groundwater flow. Soil and geologic conditions determine the type of grout applied: cement, bentonite, and chemicals are common types; grouts specific to soil and other local conditions are epoxy resins, silicone rubbers, lime, fly ash and bituminous compounds. A grout curtain designed to eliminate groundwater flow is installed by first placing two or three rows of pipes in a grid pattern (as shown in Figure 96. Grout is then injected through the pipes at successive depths to fill pore spaces in the surrounding areas.

Sheet piling cutoff walls are constructed by driving web sections of steel sheet piling permanently into the ground. Each sheet pile is interlocking at its end by either a socket or a bowl and ball end. Sheet piles vary in length and shape, as shown in Figure 97. Sections are assembled before they are driven into the ground. Initially, they are not watertight due to rough fitting of the interlocking edges; however, in predominantly fine to medium grained soils, the joint connections soon fill with soil particles to reduce groundwater flow.

(b) Engineering Considerations

Slurry Walls

The most important engineering factors which influence the effectiveness of slurry walls are the characteristics of the slurry and the characteristics of the backfill (for SB walls). Viscosity is the most important property of the slurry. A viscosity of 40 seconds Marsh (i.e., 40 seconds for a volume of slurry to pass through a standard Marsh Funnel) is recommended for trench stability and proper filter cake formation (D'Appolonia, 1980). In addition, unit weight of the slurry is important. For SB slurry, a unit weight of 240 Kg/m^3 (15 lb/ft^3 is recommended to allow proper displacement by the backfill. For CE slurry, a unit weight of approximately 1920 kg/m^3 (120 lb/ft^3) is recommended (Sommerer and Kitchens, 1980). Slurry should be designed according to in situ soil and groundwater characteristics. Varying slurry additives are available to counteract adverse in situ conditions (e.g., to increase viscosity).

Permeability is the most important backfill characteristic. For SB walls, a minimum bentonite content of 1 percent, with at least 20 percent fines content, is recommended to give minimum permeability. In addition, using plastic fines, as opposed to nonplastic or low plasticity fines, can decrease permeability an additional two orders of magnitude. With plastic fines, permeabilities as low as 10^{-8} cm/sec are possible (D'Appolonia, 1980). As with slurry materials, compatibility of backfill materials with contaminated groundwater should be tested. Adjusting fines content or using already contaminated soils may help the wall resist attack of contaminated groundwater. The effect of various pollutants on SB backfill is shown in Table 36.

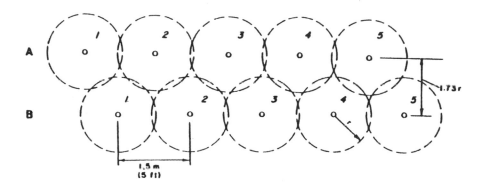

Figure 96. Typical Two-Row Grid Pattern for Grout Curtain

Source: EPA, 1982b.

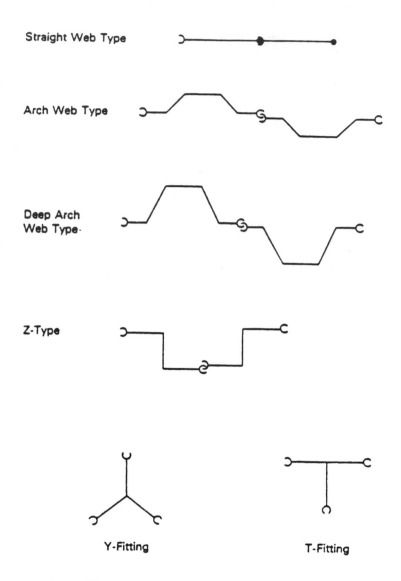

Figure 97.   Steel Piling Shapes and Interlocks

Source: EPA, 1982 b.

TABLE 36. PERMEABILITY INCREASE DUE TO LEACHING WITH VARIOUS POLLUTANTS

Pollutant (1)	Filter Cake (2)	SB Backfill (silty or clayey sand) 30% to 40% Fines (3)
Ca++ or Mg++ at 1,000 ppm	N	N
Ca++ or Mg++ at 10,000 ppm	M	M
NH$_4$NO$_3$ at 10,000 ppm	M	M
HCL (1%)	N	N
H$_2$SO$_4$ (1%)	M	N
HCL (5%)	M/H[a]	M/H[a]
NaOH (1%)	M	M
CaOH (1%)	M	M
NaOH (5%)	M	M/H[a]
Sea water	N/M	N/M
Brine (SG = 1.2)	M	M
Acid mine drainage (FeSO$_4$ = pH ~ 3)	N	N
Lignin (in Ca++ solution)	N	N
Alcohol	H (failure)	M/H

[a]Significant dissolution likely.

Note: N = no significant effect, permeability increase by about a factor of 2 or less at steady state; M = moderate effect, permeability increase by factor of 2 to 5 at steady state; H = permeability increase by factor of 5 to 10.

Source: D'Appolonia, 1980

Grout Curtains

The effectiveness of a grout curtain and selection of grout materials depends primarily on soil permeability and grain size, chemical characteristics of the soil and groundwater, and grout strength requirements. Soils are considered unsuitable for grouting if more than 20 percent of the soil passes through a #200 sieve (Sommerer and Kitchens, 1980). Important properties of available grout materials are given in Table 37. Grout curtains are useful only under certain site specific conditions. In addition, they tend to be more expensive than slurry walls, and it is difficult to verify whether a contiguous impermeable curtain has actually been formed.

Sheet Piles

Steel sheet piles are typically useful to depths up to 15 meters in soils which are loosely packed and consist of primarily sand and gravel. Piling lifetime depends on groundwater characteristics. The pH of the groundwater is of particular concern. For steel pilings, pH of 5.8 to 7.8 is best, allowing a lifetime of up to 40 years, while a pH as low as 2.3 can reduce effective lifetime to 7 years or less (USEPA, 1982a).

(c) Groundwater Pumping

Description

Groundwater pumping utilizes one or more pumps to draw groundwater to the surface through a series of wells, forming a cone of depression in the groundwater table. Shallow well points or deep well systems may be used, dependent on the depth of the aquifer. Well point systems are used in shallow, unconfined aquifers. They consist of a series of riser pipes connected to a common header pipe and a centrifugal pump. A typical well point-dewatering system is shown in Figure 98. Deep well systems can be used in confined or unconfined aquifers up to depths of several hundred meters.

Groundwater pumping can be used to lower the water table, contain a plume, or remove contaminated groundwater. It is often used in conjunction with other groundwater controls, such as impermeable barriers, for more effective groundwater control.

Engineering Considerations

The effect of a well or well system on a water table can be very difficult to predict. The following general equations can be used for estimating drawdown (s) under certain conditions (Freeze and Cherry, 1979).

TABLE 37. GROUT PROPERTIES

Grout Material	Catalyst Material	Unconfined Compressive Strength (psi) of Grouted Soil	Viscosity (cp)	Setting Time (min)	Toxicant*	Pollutant**
Silicate Base						
Low concentration	Bicarbonate	10-50	1.5	0.1-300	No	No
Low concentration	Halliburton Co., material	10-50	1.5	5-300	No	No
Low to high concentration	Siroc-Diamond Shamrock Chemical Co.	10-500	4-40	5-300	No	No
Low to high concentration	Chloride-Jooster Process	10-1000	30-50	0	No	No
Low to high concentration	Ethyl acetate-Soletanche & Halliburton	10-500	4-40	5-300	No	No
Low to high concentration	Rhone-Progil 600	—	—	—	—	—
Low to high concentration	Geloc-3, H. Baker Co.	10-500	4-25	2-200	No	No
Low to high concentration	Geloc-3X	10-250	4-25	0.5-120	No	No
Lignin Base						
Blox-All	Halliburton Co., material	5-90	8-15	3-90	Yes	Yes
Tom	Cementation Co., material	50-500	2-4	5-120	Yes	Yes
Terra-firma	Intrusion Co., material	10-50	2-5	10-300	Yes	Yes
Lignosol	Lignosol Co., material	10-50	50	10-1000	Yes	Yes
Formaldehyde Base						
Urea-formaldehyde	Halliburton Co., material	$\geq$1000	10	4-60	Yes	Yes
Urea-formaldehyde	American Cyanamid Co. material	$\geq$500	13	1-60	Yes	Yes
Resorcinol formaldehyde	Cementation Co., material	>500	3.5	—	Yes	Yes
Tannin-paraformaldehyde	Borden Co., MQ-8	—	—	—	—	—
Geoseal MQ-4 & MQ-5	Borden Co., material	—	—	—	—	—
Unsaturated Fatty Acid Base						
Polythixon FRD	Cementation Co., material	>500	10-80	25-360	No	No

*A material which must be handled using safety precautions and/or protective clothing.
**Pollutant to fresh water supplies contacted.
Source: Halliburton Services, 1976.

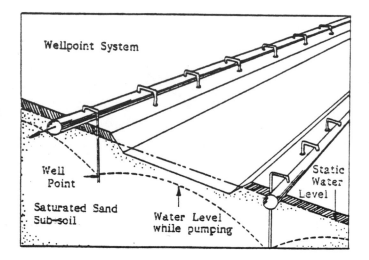

Figure 98.   Well Point Dewatering System

Source: Sommerer and Kitchens, 1980

- In a confined, isotropic aquifer:

$$s = \frac{Q_w}{4 \pi T} W(u) \tag{20}$$

- In an unconfined aquifer at early time (t < a few minutes):

$$s = \frac{Q_w}{4 \pi T} W(u_A, n) \tag{21}$$

- In an unconfined aquifer at later time (t > a few minutes):

$$s = \frac{Q_w}{4 \pi T} W(u_B, n) \tag{22}$$

where:

$$u = u_A = \frac{r^2 S}{4 T t}$$

$$u_B = \frac{r^2 S_y}{4 T t}$$

$$n = \frac{r^2}{b^2} \quad \text{in an anisotropic aquifer}$$

$$n = \frac{r^2 k_1}{b^2 k_2} \quad \text{in an isotropic aquifer}$$

and:

Q_w	=	pumping rate of the well
T	=	transmissivity of the aquifer
$W(u)$	=	well function for confined aquifers
$W(u_A, n)$	=	type A well function
$W(u_B, n)$	=	type B well function
r	=	radial distance from the well where drawdown is measured
t	=	time from initial pumping at which drawdown is measured
S	=	storativity of the aquifer
S_y	=	specific yield of the aquifer
b	=	depth of the aquifer before pumping
k_1	=	vertical hydraulic conductivity
k_2	=	horizontal hydraulic conductivity

Values for $W(u)$, $W(u_A,n)$ and $W(u_B,n)$ can be found in standard hydrology texts or engineering manuals. For $\mu < .01$, $W(u)$ can be approximated as:

$$W(u) = \ln \frac{2.246 \, T \, t}{r^2 \, S} \tag{23}$$

The previous equations are based on the following assumptions:

- the aquifer is homogeneous
- the aquifer is not leaky
- the well penetrates and is screened over the entire depth of the aquifer
- pumping rate is uniform over time
- only one aquifer is affected by the well
- there are no barriers or rivers within the radius of influence of the well
- flow to the well remains saturated for confined aquifers

If any of these assumptions is not valid for a particular well system, the simple drawdown equations are not valid. A hydrologist should be consulted to determine drawdown on a site-specific basis.

For a multiple well system, total drawdown at a given place and time is simply the added drawdown of each individual well such that (Freeze and Cherry, 1979):

$$S_{total} = \sum_{i=1}^{n} S_{well(i)} \tag{24}$$

Well point systems are practical at depths up to 10 meters, and most effective at 4.5 meters. Spacing is typically 1 to 2 meters and well points should be close enough together to maintain sufficient drawdown between the wells. Spacing and effectiveness will depend on site specific conditions (Sommerer and Kitchens, 1980).

Deep wells must be of sufficient diameter (at least 10 cm) to house a submersible pump and handle expected flow (Sommerer and Kitchens, 1980). Construction of deep wells is similar to the construction of monitoring wells, which is discussed in detail in Section ~ IV.

(d) Leachate Collection

Description

Leachate collection technologies include:

- subsurface drains;
- drainage ditches;
- liners; and
- wells

Subsurface drains are constructed by placing tile or perforated pipe in a trench, surrounding it with gravel envelope, and backfilling with topsoil or clay. They have been used extensively to dewater construction sites. At remedial action sites, subsurface drains can be installed to collect leachate, as well as lower the water table for site dewatering. They have been used at Love Canal and other major remedial action sites for leachate collection.

Drainage ditches are open trenches designed to collect surface water runoff, collect flow from subsurface drains, or intercept lateral seepage of water or leachate through the site. They are considered preferable to subsurface drains when the slope of the flow is steep. Drainage ditches are generally simple to construct, but may require extensive maintenance to maintain operating efficiency (USEPA, 1982b).

Liners, although mentioned in the National Contingency Plan as a leachate control technique, are not generally applicable to remedial action sites. Liners are useful at newly constructed disposal facilities to protect groundwater from leachate migration. Construction of liners at existing, unlined sites, however, is extremely difficult. Bottom sealing techniques, such as pressure-injection grouting, are practical over a very limited range of hydrogeologic conditions and are generally undemonstrated as a remedial action technique. In addition, costs for bottom sealing are expected to be extremely high, and no method is currently available to determine whether the seal is complete (USEPA 1982b). In view of this, liners will not be further considered as a numerical action technique.

(e)   Engineering Considerations

Subsurface Drains

Drain pipe material should be compatible with groundwater and leachate characteristics. In general, fired clay is more suitable for corrosive or high strength chemical wastes than plastic or metal pipe. In addition, an envelope of permeable material (typically gravel) should surround the drain pipe. Recommended minimum thickness of the drain envelope is 8 to 10 centimeters (3 to 4 inches). A typical envelope thickness is 14 centimeters (6 inches) and can be much larger. For example, at Love Canal the gravel envelope was about 66 centimeters thick (26 inches). The envelope of permeable material may be wrapped with a fabric to prevent clogging with soil (USEPA 1982b).

Drain depth is determined based on site specific conditions. In general, the deeper the drain, the wider the spacing that is possible (and, therefore, the fewer drains that are required). However, the cost of deeper drains with larger design flow should be compared with shallower drains with smaller design flow to determine the optimal number and depth of drains.

The distance between adjacent drains is primarily a function of drain depth, design flow (hydraulic capacity) of the drain, and soil permeability. The equation normally used to determine drain spacing is (Linsley and Frazini, 1979):

$$L = \frac{4k \ (b^2 - a^2)}{Q} \qquad (25)$$

where (see Figure 99):

L = distance between adjacent drains (m)
k = soil permeability (m/sec)
Q = design flow per meter of drain (m^3/sec/m of drain)
a = height of drain above impermeable barrier (m)
b = maximum height of water table above impermeable barrier (m)

This equation assumes steady state, one-dimensional flow through homogeneous soil. If these assumptions are not valid, spacing may be determined experimentally based on soil properties. Determining spacing based on two-or three-dimensional flow becomes a differential boundary value problem based on Laplace's equation. This can be solved using computer generated or published solutions (USEPA 1982b).

Design flow per meter of drain can be determined by performing a water balance to estimate the amount of water a drain will need to be able to transport. Manning's formula can then be used to determine pipe size.

Inflow to a pipe can also be roughly estimated as (Frogge and Sanders, 1977):

$$Q = \frac{DA \ (k)}{10} \qquad (26)$$

where:

Q = inflow to pipe (m^3/sec)
DA = area drained by pipe (m^2)
k = soil permeability (m/sec)

This should be used as a rule of thumb only.

Draining Ditches

Water level and flow velocity in drainage ditches depend on site specific characteristics and the function of the ditch. Factors which determine flow velocity include soil type, channel shape, grade and roughness, and sediment loading. The size of the ditch required can be determined using a water balance and Manning's formula. Ditch side slopes depend on soil stability and hazard of scour. Stabilization of side slopes by compaction, vegetation, or fabric liner may be necessary. Trapezoidal and parabolic cross-sections are generally considered most

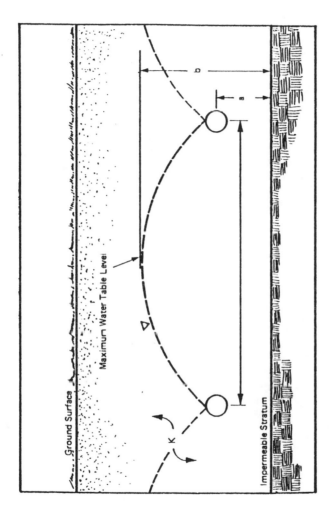

Figure 99. Spacing Equation Diagram

Source: Frogge and Sanders, 1977

stable. Maintenance of vegetation, side slopes, and depth (removal of
obstruction or sediments) may be required (USEPA 1982a; USEPA 1982b).

(f) Removal

Description

Removal of the source of contamination may involve excavation of
loose, drummed or tanked waste and contaminated soil, pumping of
impounded liquids, and similar measures.   Removal may be appropriate
when the source is small, such as a leaky 500-gallon underground storage
tank, when containment or treatment measures are not appropriate or cost
effective in comparison, or when the hazards of not removing the source
are acute.

Excavation of waste is the major removal technology.   Although
excavation at construction sites is a well-demonstrated technology, the
application to hazardous waste sites presents some unique problems. The
load-bearing capacity and fill density, which may be affected by the
buried waste, should be considered before deciding to operate heavy
equipment at the site. Landfilled drums must be handled with caution.
If drums are punctured or already leaking, additional soil at the site
can be contaminated. Sparks created by drum contact with grappling
hooks can ignite flammable waste. Typically, drums are moved to a
staging area for transfer, if necessary, to a secure drum or a tank
truck.

Removal operations at a site may be hazardous and special
precautions should be taken. Operators of equipment may be exposed to
hazardous vapors and to direct contact with liquids, solids, and
contaminated surfaces. Protective clothing, including respirators, may
be required in some cases. Equipment may become contaminated and
require decontamination before it can be taken offsite and used
somewhere else.

Surface impoundment sludge bottoms and contaminated soils can be
removed by dredging techniques such as centrifugal pumping and hydraulic
pipeline dredges. Both methods are readily available and comparable in
cost. The waste can be pumped directly to tank trucks as a low solid
content sludge (less than 20 percent solids). If transport distances to
a dewatering facility are large, it may be cost effective to dewater on
site. Impoundments can be drained prior to sludge removal by pumping
the liquid phase to a tank or other receptor. The uncovered sludge
could present an odor problem. Dried sludges can be removed with the
backhoe or dragline equipment discussed previously.

Engineering Considerations

Excavation of landfilled waste typically employs backhoes or
dragline crane units. Backhoes are effective for removing compacted or
loosely packed materials up to a depth of 21 meters. They offer
accurate bucket placement and can be equipped for drum removal.

Dragline units are effective for the removal of unconsolidated materials to a depth of 18 meters. Optimal digging depth for both is considered to be 4.5 meters.

Removal of drummed or containerized waste poses special problems. Several approaches are available. Drums in good condition can simply be loaded on a truck and transported to an off site treatment, storage, or disposal (TSD) facility. The contents of corroded drums can be transferred to secure drums for offsite disposal and the former drums compacted for offsite disposal. A third alternative is to blend the contents of drummed waste in holding tanks and subsequently pump the blended materials into a tank truck for removal. Blending operations must be carefully monitored; extensive preblending and sampling of drums is necessary to screen for incompatible wastes.

(g) Surface Water Control

Many technologies are available to control surface water flow at a site. These include:

- dikes;
- terraces;
- channels;
- chutes and downpipes;
- grading;
- surface seals; and
- vegetation

These technologies perform five basic functions, which are summarized in Table 38. Management of surface water is important since the entire water balance can affect groundwater contamination problems. Remedial actions for groundwater contamination, therefore, may require one or more of these technologies to minimize the production of leachate and prevent off site contamination. Each of these technologies is described briefly below.

Dikes

Dikes are compacted earthen ridges designed to divert or retain surface water flow. They can be used to control floodwater or to control runoff. Design of flood control dikes (or levees) depends on the amount of flood protection required based on expected height of water and failure hazard. Runoff control dikes can be used either to intercept flow (with a 0-percent grade) or, with a positive grade, to divert flow to stabilized outlets. Dike height, spacing, and construction are the primary design parameters.

Terraces

Terraces are embankments or combinations of embankments and channels constructed across a slope. As shown in Figure 100, a variety of terrace cross-sections are possible depending on slope and site

TABLE 38. SURFACE WATER TECHNOLOGIES

Technology	Primary Function					
	Minimize Runon	Minimize Infiltration	Reduce Erosion	Protection from Flooding	Collect and Transfer Water	Discharge Water
Flood control dikes				X		
Runoff control dikes	X		X			
Terraces	X		X			
Channels			X		X	
Chutes			X		X	
Downpipes			X		X	
Grading	X		X			
Surface seals		X				
Vegetation	X	X	X			
Seepage basins						X
Seepage ditches						X

Source: Arthur D. Little, Inc., 1982.

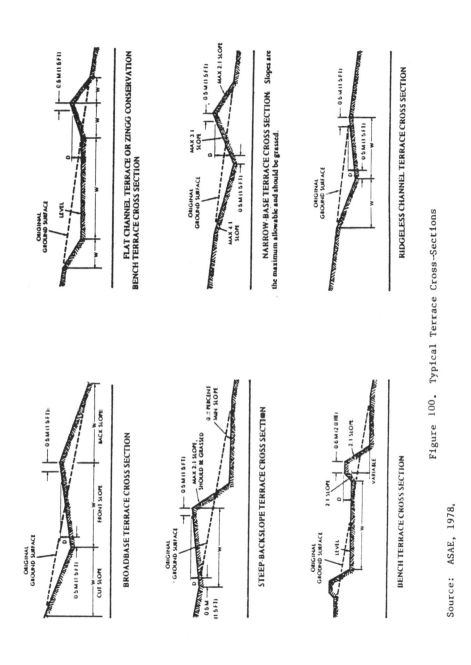

Figure 100. Typical Terrace Cross-Sections

Source: ASAE, 1978.

specific requirements. Terraces can be used to intercept and divert surface flow away from a site and to control erosion by reducing slope length.

Channels

Channels are wide and shallow excavated ditches with trapezoidal, triangular, or parabolic cross sections. Diversion channels are used primarily to intercept runoff or reduce slope length. They may or may not be stabilized. Channels stabilized with vegetation or stone riprap (waterways) are used to collect and transfer diverted water off site or to on site storage or treatment.

Chutes and Downpipes

Chutes (or flumes) are open channels normally lined with bituminous concrete, portland cement, concrete, grouted riprap, or similar non-erodible material.

Downpipes (or downdrains) are drainage pipes constructed of rigid piping (such as corrugated metal) or flexible tubing of heavy duty fabric. They are installed with prefabricated entrance sections. Downpipes can also be open structures constructed by joining half sections of bituminous fiber or concrete pipe.

Chutes and downpipes are useful in transferring concentrated flows of surface runoff from one level of a site to a lower level without erosive damage. They generally extend downslope from earthen embankments and convey water to stabilized waterways or outlets located at the base of the slope. Downpipes are particularly useful in emergency situations since they can be quickly constructed during severe storms to handle excess flow when downslope waterways overflow and threaten the containment of hazardous waste (USEPA 1982b).

Grading

Grading is the general term for technologies used to modify the natural topography and runoff characteristics of a waste site. Grading primarily involves the use of heavy equipment (such as dozers, loaders, scrapers, and compacters) to spread and compact loose soil, roughen and loosen compacted soil, and modify the surface gradient. There are six basic grading techniques described in Table 39.

Grading has two primary applications:

- Slope Grade Construction: Excavation, spreading, compaction, and hauling are used to optimize the slope at a waste site such that surface runoff increases and infiltration and ponding decrease without significantly increasing erosion. This is of primary importance in the construction of surface seals and other waste covers.

- Preparation for Revegetation: Roughening techniques
  (scarification, tracking, and contour furrowing) are
  used to reduce runoff, thereby increasing
  infiltration, and make the soil receptive to seed or
  seedlings. This is an important aspect of offsite
  revegetation once an effective surface seal has been
  applied. These techniques can also be used offsite
  in conjunction with surface water diversion
  technologies to control runon.

Surface Seals

Surface seals (caps or covers) are impermeable barriers placed over waste disposal sites to:

- reduce surface water infiltration
- reduce water erosion
- reduce wind erosion and fugitive dust emissions
- contain and control gases and odors
- provide a surface for vegetation and other
  postclosure uses

Various impermeable materials may be used, including soils and clays, admixtures (e.g., asphalt concrete, soil cement), and polymeric membranes (e.g., rubber and plastic linings).

Typical surface seals are composed of several layers, including:

- barrier layer to restrict the passage of water or
  gas. The barrier has low permeability and is
  usually composed of clayey soil or a synthetic
  membrane;

- buffer soil layer above and/or below the barrier
  layer to protect the barrier layer from cracking,
  drying, tearing, or from being punctured. It is
  usually a sandy soil;

- filter layer, made of intermediate grain sizes, to
  prevent fine particles of the barrier from
  penetrating and sifting through the coarser buffer
  layer;

- gas channeling layer of sand and gravel placed
  immediately above the waste to allow generated gases
  to escape or be collected. Pipe and trench vents
  can be used in conjunction with this layer for gas
  and odor control;

- top soil layer for growth of vegetation.

TABLE 39.   GRADING TECHNIQUES

Technique	Description	Use	Equipment
Excavation	soil removal	slope grade construction	dozer, loader, scraper
Spreading	soil application smoothing	slope grade construction	dozer, loader, compactor,
Compaction	compacts soil increases density	slope grade construction	dozer, loader, compactor
Scarification	roughening technique loosens soil	preparation for revegetation increases infiltration	loader, tractor, harrow
Tracking	roughening technique grooves soil along contour	preparation for revegetation increases infiltration	cleated crawler tractor
Contour Furrowing	roughening technique creates small depressions in soil along contour	preparation for revegetation increases infiltration	dozer

Source: Arthur D. Little, Inc., 1982.

Vegetation

Vegetation can perform four basic functions:

- it can stabilize soil and earthen structures against wind and water erosion by intercepting rainfall, slowing runoff, and holding soil together with a tight root system;

- it can reduce the quantities of water available for runoff through interception, infiltration, uptake, and transpiration;

- it can treat contaminated soil and leachate through the uptake and removal of waste constituents, nutrients, and water from the soil;

- it can improve the aesthetic appearance of the site.

Plants used for revegetation include various types of grasses, legumes, shrubs, and trees. A revegetation program involves careful plant selection, land preparation (such as increasing soil depth, grading, fertilizing and tilling), seeding, and maintenance.

6. Information Sources

The information which needs to be evaluated during analysis of a groundwater contamination problem includes a description of the physical framework through which water is moving, a description of the hydrologic system and documentation of present and past base operations, including construction details.

Data and interpretive reports regarding the physical framework and hydrologic system would generally be available from federal, state, or local governmental agencies that are responsible for natural resource or environmental studies. Information may also be available from engineering or consulting firms that may have performed local specialized studies. Site use and construction information should be available through the office of the Base Engineer.

1. GEOLOGIC FRAMEWORK AND HYDROLOGIC SYSTEM

Information regarding the geologic framework and hydrologic system is generally available from federal and state geologic or natural resource surveys. The type of information which can be obtained from each agency will range from fundamental data, such as well logs and water level measurements, to specialized technical reports. In some states, the local state agency will be the principal information source, and in other states, the federal agency will provide the most information. Tables 40 and 41 list the addresses of the state geologists and the district offices of the Water Resources Division of the U.S. Geological Survey. In addition to providing in-house data and reports regarding local hydrogeologic conditions, the U.S. Geological Survey district offices provide access to national water-related databases, such as NAWDEX and WATSTORE, as well as providing a convenient access to other geological survey reports and information.

NAWDEX (National Water Data Exchange) is a computerized data system that can identify sources of water data, locations of sites at which water data are being collected, and assist users of water data in locating and procuring data that meet their specific criteria and are in the geographic area of interest. WATSTORE (National Water Data Storage and Retrieval System) provides access to streamflow, water quality and groundwater data. Access to this data base can be made through computer facilities of the U.S. Geological Survey or by establishing a separate user account. U.S. Geological Survey Circular 777, entitled "Guide to Obtaining Information from the U.S. Geological Survey," provides more detailed instructions about obtaining information from the U.S. Geological Survey.

Local engineering and/or consulting firms may have performed specialized studies in the area of interest and may have geologic or hydrologic data available. The Base Engineer may have information

302

TABLE 40. ADDRESSES OF STATE GEOLOGISTS

(from the 1980 Directory of the Association of American State Geologists)

ALABAMA (205)349-2852
Thomas J. Joiner
Geol. Survey of Alabama
P. O. Drawer O
University, AL 35486

ALASKA (907)279-1433
Ross G. Schaff
Div. of Geology and
 Geophysical Surveys
3001 Porcupine Drive
Anchorage, AK 99501

ARIZONA (602)626-2733
Larry D. Fellows
Bureau of Geology and
 Mineral Technology
Geol. Survey Branch
845 N. Park Avenue
Tucson, AZ 85719

ARKANSAS (501)371-1488
Norman F. Williams
Arkansas Geol. Commission
Vardelle Parham Geol. Center
3815 W. Roosevelt Road
Little Rock, AR 72204

CALIFORNIA (916)445-1923
F. Davis
Div. of Mines & Geology
Calif Dept of Conservation
1416 9th Street, Room 1341
Sacramento, CA 95814

COLORADO (303)839-2611
John W. Rold
Colorado Geological Survey
1313 Sherman St., Room 715
Denver, CO 80203

CONNECTICUT (203)566-3540
Hugo F. Thomas
Conn. Geol. & Natural
 History Survey
State Office Bldg., Room 553
165 Capitol Avenue
Hartford, CT 06115

FLORIDA (904)488-4191
Charles W. Hendry, Jr.
Bureau of Geology
903 W. Tennessee St.
Tallahassee, FL 32304

GEORGIA (404)656-3214
William McLemore
Geol. and Water
 Resources Division
Dept. of Natural Resources
19 Dr. Martin Luther King,
 Jr. Drive, S.W.
Atlanta, GA 30334

HAWAII (808)548-7533
Robert T. Chuck
Div. of Water & Land Dev.
P. O. Box 373
Honolulu, HI 96809

IDAHO (208)885-6785
Maynard M. Miller
Idaho Bur. of Mines
 and Geology
Moscow, ID 83843

ILLINOIS (217)333-5111
Jack A. Simon
Illinois State Geological
 Survey
121 Natural Resource Bldg
Urbana, IL 61801

INDIANA (812)337-2862
John B. Patton
Dept. of Natural Resources
Indiana Geological Survey
611 North Walnut Grove
Bloomington, IN 47401

IOWA (319)338-1173
Stanley C. Grant
Iowa Geological Survey
123 N. Capitol
Iowa City, IA 52242

KENTUCKY (606)622-2270
Donald C. Haney
Kentucky Geol. Survey
University of Kentucky
311 Breckinridge Hall
Lexington, KY 40506

LOUISIANA (504)342-6754
Charles G. Groat
Louisiana Geol. Survey
Box G, Univ. Station
Baton Rouge, LA 70893

MAINE (207)289-2801
Walter Anderson
Maine Geological Survey
State Off Bldg, Rm 211
Augusta, ME 04330

MARYLAND (301)235-0771
Kenneth N. Weaver
Maryland Geol. Survey
Merryman Hall
Johns Hopkins University
Baltimore, MD 21218

MASSACHUSETTS
Joseph A. Sinnott
Dept. of Environ.
 Quality Engineering
Div of Waterways, Rm 532
100 Nashua Street
Boston, MA 02114

MICHIGAN (517)373-1256
Arthur E. Slaughter
Mich. Dept. of Nat. Res.
Geological Survey Div.
P. O. Box 30028
Lansing, MI 48909

MINNESOTA (612)373-3372
Matt Walton
Minnesota Geol. Survey
1633 Eustis Street
St. Paul, MN 55108

TABLE 40. (CONTINUED)

DELAWARE (302)738-2833
Robert R. Jordan
Delaware Geological Survey
University of Delaware
Newark, DE 19711
MISSOURI (314)364-1752
Wallace B. Howe
Div. of Geol. & Land Survey
P. O. Box 250
Rolla, MO 65401

MONTANA (406)792-8321
Sid Groff
Mont. Bureau of Mines
& Geology
Montana College of Mineral
Science & Technology
Butte, MT 59701

NEBRASKA (402)472-3471
Vincent H. Dreeszen
Conservation & Survey Div.
University of Nebraska
Lincoln, NE 68508

NEVADA (702)784-6691
John Schilling
Nevada Bureau of Mines
& Geology
University of Nevada
Reno, NV 89557

NEW HAMPSHIRE (603)862-1216
Glenn W. Stewart
Office of State Geologist
James Hall
Univ. of New Hampshire
Durham, NH 03824

NEW JERSEY (609)292-2576
Kemble Widmer
New Jersey Bureau of
Geol. & Topography
P. O. Box 1390
Trenton, NJ 08625

KANSAS (913)864-3965
William W. Hambleton
State Geol. Survey of Kansas
Raymond C. Moore Hall
1930 Avenue A, Campus West
Lawrence, KS 66044

NORTH DAKOTA (701)777-2231
Lee C. Gerhard
N. Dakota Geological Survey
University Station
Grand Forks, ND 58202

OHIO (614)466-5344
Horace R. Collins
Ohio Div. of Geol. Survey
Fountain Square, Bldg. B
Columbus, OH 43224

OKLAHOMA (405)325-3031
Charles J. Mankin
Oklahoma Geol. Survey
830 Van Vleet Oval, Rm 163
Norman, OK 73019

OREGON (503)229-5580
Donald A. Hull
State Dept. of Geology &
Mineral Industries
1069 State Office Bldg.
1400 S.W. Fifth Avenue
Portland, OR 97201

PENNSYLVANIA (717)787-2169
Arthur A. Socolow
Bureau of Topo. & Geol.
Survey
Dept. of Environ. Resources
P. O. Box 2357
Harrisburg, PA 17120

PUERTO RICO
Director
Servicio Geologico de P.R.
Dept. de Recursos Naturales
Apartado 5887, Puerta de
Tierra
San Juan, PR 00906

MISSISSIPPI
(601)354-6228
William H. Moore
Miss. Geol., Econ., &
Topo. Survey
P. O. Box 4915
Jackson, MS 39216

TENNESSEE (615)741-2726
Robert E. Hershey
Dept. of Conservation
Division of Geology
G-5 State Office Bldg.
Nashville, TN 37219

TEXAS (512)471-1534
W. L. Fisher
Bureau of Econ. Geology
Univ. Station, Box X
Austin, TX 78712

UTAH (801)581-6831
Donald T. McMillan
Utah Geol. & Min. Survey
606 Black Hawk Way
Salt Lake City, UT 84108

VERMONT (802)828-3357
Charles A. Ratte
Agency of Environmental
Conservation
5 Court Street
Montpelier, VT 05602

VIRGINIA (804)293-5121
Robert C. Milici
Virginia Div. of Univ. of
Mineral Resources
P. O. Box 3667
Charlottesville, VA 22903

WASHINGTON (206)753-6183
Vaughn E. Livingston, Jr.
Dept of Natural Resources
Geology & Earth Resources
Division
Olympia, WA 98504

TABLE 40. (CONCLUDED)

NEW MEXICO (505)835-5420
Frank E. Kottlowski
Mexico Bureau of Mines
 & Mineral Resources
New Mexico Tech
Socorro, NM 87801

NEW YORK (518)474-5816
Robert H. Fakundiny
N.Y. State Geol. Survey
State Education Bldg.
Albany, NY 12234

N. CAROLINA (919)733-3833
Stephen G. Conrad
N. Carolina Dept. of Nat.
 Res. & Community Develop.
P. O. Box 27687
Raleigh, NC 27611

RHODE ISLAND
Robert L. McMaster
Assoc. State Geologist for
 Marine Affairs
Grad. Schl. of Oceanography
Kingston, RI 02881

S. CAROLINA (803)758-6431
Norman K. Olson
S. Carolina Geol. Survey
State Development Board
Harbison Forest Road
Columbia, SC 29210

SOUTH DAKOTA (605)624-4471
Duncan J. McGregor
S.D. State Geol. Survey
Science Center
Univ. of South Dakota
Vermillion, SD 57069

W. VIRGINIA (304)292-6331
Robert B. Erwin
W. Va. Geol. & Econ. N.
P. O. Box 879
Morgantown, WV 26505

WISCONSIN (608)262-1705
Meredith E. Ostrom
Wisc. Geol. & Natural
History Survey
1815 University Ave.
Madison, WI 53706

WYOMING (307)742-2054
Daniel N. Miller, Jr.
Wyoming Geol. Survey
Box 3008, Univ. Station
Laramie, WY 82071

TABLE 41.   ADDRESSES OF THE DISTRICT OFFICES OF THE U.S. GEOLOGICAL
SURVEY WATER RESOURCES DIVISION

University of Alabama
Oil & Gas Bldg. - Room 202
P. O. Box V
Tuscaloosa, ALABAMA 35486
(205) 752-8104

218 E. Street
Anchorage, ALASKA 99501
(907) 271-4138

Federal Building
301 W. Congress Street
Tucson, ARIZONA 85701
(501) 378-6391

855 Oak Grove Avenue
Menlo Park, CALIFORNIA 94025
(415) 323-8111

Building 53
Denver Federal Center
Lakewood, COLORADO 80225
(303) 234-5092

135 High Street - Room 235
Hartford, CONNECTICUT 06103
(203) 244-2528

Subdistrict - District Office/MD
Federal Building - Room 1201
Dover DELAWARE 19901
(302) 734-2506

325 John Knox Road - Suite F-240
Tallahassee, FLORIDA 32303
(904) 386-1118

Suite B
6481 Peach Tree, Indust. Blvd.
Doraville, GEORGIA 30360
(404) 221-4858

Field Headquarters
4398D Loke St., P. O. Box 1856
Lihue, Kauai, GUAM 96766

Subdistrict
U.S. Navy Public Works Center
FPO S.F. 96630 - P. O. Box 188
Agana, GUAM 96910

P. O. Box 50166
300 Ala Moana Blvd. - Room 6110
Honolulu, HAWAII 96850
546-8331

P. O. Box 2230
Idaho Falls, IDAHO 83401
(208) 526-2438

P. O. Box 1026
605 N. Nek Street
Champaign, ILLINOIS 61820
(217) 398-5353

1819 North Meridan Street
Indianapolis, INDIANA 46202
(317) 269-7101

Federal Building - Room 269
P. O. Box 1230
Iowa City, IOWA 52244
(319) 337-4191

University of Kansas
Campus West
1950 Avenue A
Lawrence, KANSAS 66045
(913) 864-4321

Table 41. (CONTINUED)

Federal Building - Room 572
600 Federal Place
Louisville, KENTUCKY 40202
(502) 582-5241

6554 Florida Boulevard
Baton Rouge, LOUISIANA 70896
(504) 389-0281

District Office in Mass.
26 Ganneston Drive
Augusta, MAINE 04330
(207) 623-4797

208 Carroll Building
8600 Lasalle Road
Towson, MARYLAND (301) 828-1535

150 Causeway Street, Suite 1001
Boston, MASSACHUSETTS 02114
(617) 223-2822

6520 Mercantile Way - Suite 5
Lansing, MICHIGAN 48910
(517) 372-1910

702 Post Office Building
St. Paul, MINNESOTA 55101
(612) 725-7841

Federal Building, Suite 710
100 West Capitol Street
Jackson, MISSISSIPPI 39201
(601) 969-4600

Mail Stop 200
1400 Independence Road
Rolla, MISSOURI 65401
(314) 341-0824

Federal Building, Drawer 10076
Helena, MONTANA 59601
(406) 559-5263

Fed. Bldg/Courthouse - Rm 406
100 Centennial Mall North
Lincoln, NEBRASKA 68508
(402) 471-5082

Federal Building - Room 227
705 North Plaza Street
Carson City, NEVADA 89701
(702) 882-1388

Subdistrict - Dist. Off./Mass
Federal Building - 210
55 Pleasant Street
Concord, NEW HAMPSHIRE 03301

Federal Building - Room 436
402 E. State St. - P. O. Box 1238
Trenton, NEW JERSEY 08607
(609) 989-2162

Western Bank Building
505 Marquette, NW
Albuquerque, NEW MEXICO 87125

236 U.S. Post Office/Courthouse
P. O. Box 1350
Albany, NEW YORK 12201
(518) 472-3107

Century Station - Room 436
Post Office Building
P. O. Box 2857
Raleigh, NORTH CAROLINA 27602
(919) 755-4510

821 E. Interstate Avenue
Bismarck, NORTH DAKOTA 58501
(701) 255-4011

975 West Third Avenue
Columbus, OHIO 43212
(614) 469-5553

215 NW 3rd - Room 621
Oklahoma City, OKLAHOMA 73102
(405) 231-4256

(Mail) P. O. Box 3202
Ship-830 NE Holladay St., 97232
Portland, OREGON 97208
(503) 231-5242

Table 41. (CONCLUDED)

Federal Building - 4th Floor
P. O. Box 1107
Harrisburg, PENNSYLVANIA 17108
(717) 782-4514

Building 652, Ft. Buchanan
G.P.O. Box 4424
San Juan, PUERTO RICO 00936
(809) 783-4660

District Office in Mass.
Federal Bldg. & U.S. P. O.
Room 224
Providence, RHODE ISLAND 02903
(401) 528-4655

Strom Thurmond Federal Building
1835 Assembly Street - Suite 658
Columbia, SOUTH CAROLINA 29201
(803) 765-5966

Federal Building - Room 308
200 4th Street, S.W.
Huron, SOUTH DAKOTA 57350
(605) 352-8651

U.S. Courthouse
U.S. Federal Building-A-413
Nashville, TENNESSEE 37203
(615) 251-5424

Federal Building - 649
300 East 8th Street
Austin, TEXAS 78701
(512) 397-5766

Administration Building - 1016
1745 West 1700 South
Salt Lake, UTAH 84104
(801) 524-5663

District Office in Mass.
U.S. Post Office/Courthouse
Rooms 330B and 330C
Montpelier, VERMONT 05602
(802) 229-4500

200 West Grace Street - Room 304
Richmond, VIRGINIA 23220
(804) 771-2427

1201 Pacific Avenue - Suite 600
Tacoma, WASHINGTON 98402
(206) 593-6510

Federal Bldg./U.S. Courthouse
500 Quarrier St., East - Room 3017
Charlestown, WEST VIRGINIA 25301
(304) 343-6181

1815 University Building
Madison, WISCONSIN 53706
(608) 262-2488

P. O. Box 1125
J. C. O'Mahoney Federal Center
2120 Capitol Avenue - Room 5017
Cheyenne, WYOMING 82001
(307) 778-2220

regarding any specialized engineering or consulting services that may have been performed on the base.

Soils information, including soil maps, types, physical characteristics, and depths is available from the Soil Conservation Service of the U.S. Department of Agriculture. Table 42 lists the addresses of various offices of the Soil Conservation Service.

2. SITE INFORMATION

Site information includes site construction information, as well as site use information. Construction information, such as the location of buried utilities, storage tanks, and pipelines should be available from the Base Engineer or local utility companies. This information is necessary to identify potential sources of contamination, to identify areas where drilling is precluded, and to identify those areas where construction activities may have altered the water-bearing characteristics of the natural geologic materials sufficiently to affect the groundwater flow direction.

Information regarding past operations on the base needs to be evaluated to identify possible past sources of contamination, such as abandoned landfills, evaporation pits, or storage areas. Comparison of current and older base engineering records and plans may be useful in interpreting changes in base operation. Aerial photographs taken at different times may also provide information regarding past changes in base operations, particularly with respect to changes in disposal operations. Aerial photographs and other remotely sensed imagery are generally available from the EROS Data Center and the National Cartographic Information Center. General assistance and identification of local sources of information can be obtained from:

National Cartographic Information Center (NCIC)
Headquarters
U.S. Geological Survey
507 National Center   (703) 860-6045
Room 1-C-107
12201 Sunrise Valley Drive
Reston, VA 22092

EROS Data Center
User Services Unit   (605) 594-6511
Sioux Falls, SD 57198

TABLE 42. ADDRESSES OF STATE OFFICES OF THE SOIL
CONSERVATION SERVICE

ALABAMA, Auburn 36830
Wright Building
138 South Gay Street
P. O. Box 311
(205) 821-8070

ALASKA, Anchorage 99504
Suite 129, Professional Bldg.
2221 E. Northern Lights Blvd.
(907) 276-4246

ARIZONA, Phoenix 85025
230 N. 1st Avenue
3008 Federal Building
(602) 261-6711

ARKANSAS, Little Rock 72203
Federal Building, Room 5029
700 West Capitol Street
P. O. Box 2323
(501) 378-5445

CALIFORNIA, Davis 95616
2828 Chiles Road
(916) 758-2200, ext. 210

COLORADO, Denver 80217
2490 W. 26th Avenue
P. O. Box 17107
(303) 837-4275

CONNECTICUT, Storrs 06268
Mansfield Professional Park
Route 44A
(203) 429-9361/9362

DELAWARE, Dover 19901
Treadway Towers, Suite 2-4
9 East Loockerman Street
(302) 678-0750

FLORIDA, Gainsville 32602
Federal Building
P. O. Box 1208
(904) 377-8732

GEORGIA, Athens 30603
Federal Building
355 E. Hancock Avenue
P. O. Box 832
(404) 546-2274

HAWAII, Honolulu 96850
300 Ala Moana Blvd.
Room 4316
P. O. Box 5004
(808) 546-3165

IDAHO, Boise 83702
304 North 8th Street, Rm 345
(208) 384-1601, ext. 1601

ILLINOIS, Champaign 61820
Federal Building
200 W. Church Street
P. O. Box 678
(217) 356-3785

INDIANA, Indianapolis 46224
Atkinson Square W. Suite 2200
5610 Crawfordsville Road
(317) 269-3785

IOWA, Des Moines 50309
693 Federal Building
210 Walnut Street
(515) 862-4260

KANSAS, Salina 67401
760 South Broadway
P. O. Box 600
(913) 825-9535

KENTUCKY, Lexington 40504
333 Waller Avenue
(606) 233-2749, ext. 2749

LOUISIANA, Alexandria 71301
3737 Government Street
P. O. Box 1630
(318) 448-3421

MAINE, Orona 04473
USDA Building
University of Maine
(207) 866-2132/2133

MARYLAND, College Park 20740
Room 522, Hartwick Building
4321 Hartwick Road
(301) 344-4180

Table 42. (CONTINUED)

MASSACHUSETTS, Amherst 01002
29 Cottage Street
(413) 549-0650

MICHIGAN, East Lansing 48823
1405 South Harrison Road
Room 101
(517) 372-1910, ext. 242

MINNESOTA, St. Paul 55101
200 Federal Bldg. & U.S. Courthouse
316 North Robert Street
(612) 725-7675

MISSISSIPPI, Jackson 39205
Milner Building, Room 590
210 South Lamar Street
P. O. Box 610
(601) 969-4330

MISSOURI, Columbia 65201
555 Vandiver Drive
(314) 442-2271, ext. 3155

MONTANA, Bozeman 59715
Federal Building
P. O. Box 970
(406) 587-5271, ext. 4322

NEBRASKA, Lincoln 68508
Federal Building
U.S. Courthouse, Room 345
(402) 471-5301

NEVADA, Reno 89505
U.S. Post Office Bldg., Room 308
P. O. Box 4850
(702) 784-5304

NEW HAMPSHIRE, Durham 03824
Federal Building
(603) 868-7581

NEW JERSEY, Somerset 08873
1370 Hamilton Street
P. O. Box 219
(201) 246-1205, ext. 20

NEW MEXICO, Albuquerque 87103
517 Gold Avenue, SW
P. O. Box 2007
(505) 766-2173

NEW YORK, Syracuse 13260
U.S. Courthouse & Federal Bldg.
100 S. Clinton Street, Room 771
(315) 423-5493

NORTH CAROLINA, Raleigh 27611
310 New Bern Avenue, Federal Bldg.
Room 544, P. O. Box 27307
(919) 755-4165

NORTH DAKOTA, Bismarck 58501
Federal Bldg. - Rosser Ave & 3rd St
P. O. Box 1458
(701) 255-4011, ext. 421

OHIO, Columbus 43215
200 No. High Street, Room 522
(614) 469-6785

OKLAHOMA, Stillwater 74074
Agriculture Building
Farm Road & Brumley Street
(405) 624-4360

OREGON, Portland 97209
Federal Office Building
1220 SW 3rd Avenue
(503) 221-2751

PENNSYLVANIA, Harrisburg 17108
Federal Bldg. & Courthouse
Box 985, Federal Square Station
(717) 782-4403

PUERTO RICO, Hato Rey 00918
Federal Office Bldg., Room 633
Mail: GPO Box 4868
Puerto Rico, San Juan 00936
(809) 753-4206

RHODE ISLAND, West Warwick 02893
46 Quaker Lane
(401) 828-1300

TABLE 42. (CONCLUDED)

SOUTH CAROLINA, Columbia 29210
240 Stoneridge Drive
(803) 765-5681

SOUTH DAKOTA, Huron 57350
Federal Bldg., 200 4th St., SW
P. O. Box 1357
(605) 352-8651

TENNESSEE, Nashville 37203
675 U.S. Courthouse
(615) 749-5471

TEXAS, Temple 76501
W. R. Poage Federal Building
101 S. Main Street, P. O. Box 648
(817) 773-1711, ext. 331

UTAH, Salt Lake City 84138
4012 Federal Bldg - 125 S. State S
(801) 524-5051

VERMONT, Burlington 05401
1 Burlington Square, Suite 205
(802) 862-6501, ext. 6261

VIRGINIA, Richmond 23240
Federal Bldg., Room 9201
400 N. 8th Street - P. O. Box 10026
(804) 782-2457.

WASHINGTON, Spokane 99201
360 U.S. Courthouse
W. 920 Riverside Avenue
(509) 456-3711

WEST VIRGINIA, Morgantown 26505
75 High Street, P. O. Box 865
(304) 599-7151

WISCONSIN, Madison 53711
4601 Hammersley Road
(608) 252-5351

WYOMING, Casper 82601
Federal Office Bldg., P.O. Box 2440
(307) 265-5550, ext. 3217

TECHNICAL SERVICE CENTERS

MIDWEST
NEBRASKA, Lincoln 68508
    Federal Bldg.-U.S. Courthouse, Rm 393
    Phone: 541-5346 (FTS)
           402-471-5361 (CML)

WEST
OREGON, Portland 97209
    511 N.W. Broadway
    Phone: 423-2824 (FTS)
           503-221-2824 (CML)

NORTHWEST
PENNSYLVANIA, Broomall 19008
    1974 Sproul Road
    Phone: 596-5783 (FTS)
            215-596-5710 (CML)

SOUTH
TEXAS, Ft. Worth 76115
    Ft. Worth Federal Center
    P. O. Box 6567
    Phone: 817-334-5456 (FTS & CML)

CARTOGRAPHIC UNITS
(Not located at TSC)
MARYLAND, Lanham 20782
    10,000 Aerospace Road
    Phone: 301-436-8756 (FTS & CML)

References

Adams, C.E., Jr. and W.W. Eckenfelder, Jr. Process Design Techniques for Industrial Waste Treatment. Enviro Press. Nashville, Tennessee. 1974.

American Petroleum Institute. The Migration of Petroleum Products in Soil and Ground Water - Principles and Countermeasures. Publication No. 4149. 1972.

American Society of Agricultural Engineers (ASAE), ASAE Standard: ASAE 5268.2. Design, Layout, Construction and Maintenance of Terrace Systems, MI. 1978.

Anderson, M.P. Using Models to Simulate the Movement of Contaminants Through Groundwater Flow Systems: CRC Critical Reviews in Environmental Control. Volume 9, Number 2, pp. 97-156. 1979.

Arthur D. Little, Inc. (ADL). Physical, Chemical, and Biological Treatment Techniques for Industrial Wastes. Report to U.S. Environmental Protection Agency, Office of Solid Waste Management Programs. PB-275-054/56A (Volume I) and PB-275-278/1GA (Volume II). November 1976.

Arthur D. Little, Inc. (ADL). (Lagace, R.L. and R.H. Spencer). Assessment of the Operational Utility of a Developmental Ground Penetration Radar for Detecting Buried Ordnance. Prepared for U.S. Army Mobility Equipment (R&D Command). 1980.

Baetsle, L.H. Migration of Radionuclides in Porous Media. Progress in Nuclear Energy. Series XII. Health Physics (A.M.F. Duhamel, ed.). Pergamon Press. pp. 707-730. 1969.

Barth, E.F. et al. Summary Report on the Effects of Heavy Metals on the Biological Treatment Process. J. WPCF. Volume 37, p. 86, 1965.

Bartlet, R.E. State Groundwater Protection Programs - A National Summary. Groundwater. Volume 17, Number 1, pp. 89-93. 1979.

Bear, J., Dynamics of Fluids in Porous Media. American Elsevier Publishing Co. New York. 1972.

Bear, J., Hydraulics of Groundwater. McGraw-Hill Book Company, New York, 1979.

Campbell, M.D. and J.H. Lehr. Water Well Technology. McGraw-Hill Book Co. New York. 1973.

Cenci, G. et al. Evaluation of the Toxic Effect of Cd^{+2} and $Cd(CN)_4^{-2}$ on the Growth of Mixed Microbial Population of Activated Sludges. The Science of the Total Environment. Volume 7, pp. 131-143. 1977.

313

Cherry, J.A. Groundwater Contamination. Part B. Chemistry and Field Sampling. Groundwater Hydrology. Boston Society of Civil Engineers Lecture Series, 1981.

Council on Environmental Quality (CEQ). The Eleventh Annual Report of the Council on Environmental Quality. December 1980.

Council on Environmental Quality (CEQ). Contamination of Groundwater by Toxic Organic Chemicals. U.S. Government Printing Office, Washington, D.C. January 1981.

D'Appolonia, D.J., Soil Bentonite Slurry Trench Cutoffs. J. Geotech, Eng. Div. ASCE, Vol. 106 4:399-418. April 1980.

Davis, S.M. and R.J.M. DeWiest. Hydrogeology. John Wiley and Sons, Inc. 1966.

DeMarco, J. and P.R. Wood. Design Data for Organics Removal by Carbon Beds. Presented at the Environmental Engineering Division National Conference on Engineering Design, ASCE, Kansas City, Missouri, July 10-12, 1978.

Docre, J.C. et al. Preliminary Pollutant Limit Values for Human Health Effects. Environ. Sci. Technol. Volume 14, Number 7, pp. 778-784. 1980.

Domenico, P.A. Concepts and Models in Groundwater Hydrology. McGraw-Hill Book Company. 1972.

Donigan, A.S., Jr. Agricultural Runoff Management (ARM) Model Version II: Refinement and Testing. U.S. Environmental Protection Agency. EPA-600/3-77-098. 1977.

Environmental Science and Technology (ES&T). Safeguards for Groundwater. Special Report. Volume 14(1). January 1980.

Farmer, W.J. et al. Land disposal of Hexachlorobenzene Wastes, Controlling Vapor Movement in Soil. U.S. Environmental Protection Agency. EPA-600/2-80-119. 1980.

Fiksel, J. and M. Segal. An Approach to Prioritization of Environmental Pollutants: The Action Alert System. U.S. Environmental Protection Agency. 1980.

Flint, R.F. and B.J. Skinner. Physical Geology. John Wiley and Sons. 1974.

Freeze, R.A. and J.A. Cherry. Groundwater. Prentice-Hall, Inc. 1979.

Fried, J.J. and M.A. Combarnous. Dispersion in Porous Media. Advances in Hydroscience. Volume 7, pp. 169-282. 1971.

Fried, J.J. Groundwater Pollution: Elsevier Scientific Publishing Company, New York. 1975.

Frogge, R.R. and G.D. Sanders. USBR Subsurface Drainage Design Procedure. Water Management for Irrigation and Drainage, pp. 30-46. July 20-22, 1977.

Geer, R.D. Predicting the Anaerobic Degradation of Organic Chemical Pollutants in Waste Water Treatment Plants from their Electrochemical Reduction Behavior. Montana State University, Waste Resources Research Center. MUJWRRC-95, W79-01-OWRT-A-097-MONT(1). 1978.

Ghassemi, M. et al. Feasibility of Commercialized Water Treatment Techniques for Concentrated Waste Spills. TRW Environmental Engineering Division. PB 82-108440 or EPA 600/2-81-213. September 1981.

Ginilka, A. and J.E. Harwood. Groundwater Monitoring Program Duke Power Company. US/USSR Joint Group on Design and Operation of Air Pollution Reduction and Waste Disposal Systems for Thermal Power Plants Symposium. Harpers Ferry, West Virginia. June 1979.

Gosset, J.M. and A.H. Lincoff. Solute-Gas Equilibria in Multi-Organic Aqueous Systems. Prepared for the U.S. Air Force Office of Scientific Research (AFSC), Directorate of Chemical and Atmospheric Sciences. November 1981.

Grant, F.S. and G.F. West. Interpretation Theory in Applied Geophysics. McGraw Hill. 1965.

Halliburton Services. Grouting in Soils. Vols. 1 and 2. Federal Highway Administration. NTIS No. PB-259-043, 4. 1976.

Hamaker, J.W. The Interpretation of Soil Leaching Experiments in Environmental Dynamics of Pesticides. (R. Haque and V.H. Freed eds.). Plenum Press. New York. 1975.

Harris, J.C. Rate of Hydrolysis in Handbook of Chemical Property Estimation Methods. McGraw-Hill. New York. 1982.

Hillerich, M.S. Air Driven Piston Pump for Sampling Small Diameter Wells. WRD Bulletin. October - December 1976.

Inside PEA, Vol. 3, Number 27, p. 8. July 9, 1982.

James, S.C. History and Bench Scale Studies for the Treatment of Contaminated Groundwater at the Ott/Story Chemical Site, Muskegon, Michigan. National Conference on Management of Uncontrolled Hazardous Waste Sites, sponsored by the USEPA and the Hazardous Materials Control Research Institute, October 28-30, 1981.

Johnson Division. Groundwater and Wells. Universal Oil Producers Company. 1972.

Kelleher, D.L. et al. Investigation of Volatile Organics Removal. J. of the New England Water Works Association. June 1981.

Keys, W. Scott and L.M. MacCary. Techniques of Water Resources Investigations of the United States Geologic Survey: Application of Borehole Geophysics to Water-Resources Investigations. Book 2, Chapter E. 1971.

Kolmer, J.R. Investigation of the Lipari Landfill Using Geophysical Techniques. EPA-600/9-81-002/0. March 1981.

Leistra, M. Computation Models for the Transport of Pesticides in Soil. Residue Rev. Volume 49, pp. 87-131. 1973.

LeRoy, L.W. et al. Subsurface Geology - Petroleum Mining Construction. Colorado School of Mines. 4th edition. 1977.

Letey, J. and J.K. Oddson. Mass Transfer, in Organic Chemicals in the Soil Environment. Volume 1. (C.A.I. Goring and J.W. Hamaker eds.). Marcel Dekker. New York. 1972.

Letey, J. and W.J. Farmer. Movement of Pesticides in Soil, in Pesticides in Soil and Water, Soil Science Society of America, Inc., Madison, Wisconsin. 1974.

Linsley, R.K. and J.B. Franzini. Water Resources Engineering. McGraw-Hill, Inc. New York. 1972.

Linsley, R.K., Jr. et al. Hydrology for Engineers. McGraw-Hill, Inc. New York. 1975.

Lyman, W.J., W.F. Reehl and D.H. Rosenblott (eds). Handbook of Chemical Property Estimation Methods. McGraw-Hill Book Co. New York. 1982.

Lyman, W. Adsorption Coefficient for Soils and Sediments. In the Handbook of Chemical Property Estimation Methods. McGraw-Hill. New York. 1982.

Mabey, W. and T. Mill. Critical Review of Hydrolysis of Organic Compounds in Water Under Environmental Conditions. J. Phys. Chem. Volume 7, pp. 383-415. 1978.

Mackay, D. and W.Y. Shiu. A Critical Review of Henry's Law Constants for Chemicals of Environmental Interest. J. Phys. Chem. Volume 10, Number 4, pp. 1175-1199. 1982.

Matis, J.R. Petroleum Contamination of Groundwater in Maryland. Groundwater. Volume 9, Number 6. November - December 1971.

McCarty, P.L. et al. Trace Organics in Groundwater. Environ. Sci. Technol. Volume 15, Number 1, pp. 40-49. 1981.

National Academy of Sciences (NAS). Water Quality Criteria 1972. 1972.

National Academy of Sciences (NAS). Drinking Water and Health. Federal Register. Volume 42, Number 132, pp. 35763-35779. July 11, 1977.

Neely, N. et al. Survey of On-Going and Completed Remedial Action Projects. U.S. Environmental Protection Agency. EPA-600/2-81-246. 1981.

New York Department of Environmental Conservation (NYDEC). Toxic Substances in New York's Environment. May 1979.

Page, G.W. Comparison of Groundwater and Surface Water for Patterns and Levels of Contamination by Toxic Substances. ES&T. Volume 15, Number 12. 1981.

Parizek, R.R. and B.E. Lane. Soil-Water Sampling Using Pan and Deep Pressure-Vacuum Lysimeters. J. of Hydrology, Volume 11, p. 1-21, 1970.

Pettyjohn, W.A. et al. Sampling Groundwater for Organic Contaminants. Groundwater. Volume 19, Number 2, pp. 180-189. March - April 1981.

Pickens, J.F. and W.C. Lennox. Numerical Simulation of Waste Movement in Steady Groundwater Flow Systems. Water Resources Res. Volume 12, No. 2, pp. 171-180. 1976.

Prengle, H.W., Jr. et al. Ozone/UV Process Effective Wastewater Treatment. Hydrocarbon Processing. October 1975.

Press, F. and R. Siever. Earth. Second Edition. W.A. Freeman and Company. 1978.

Rao, P.S.C. and J.M. Davidson. Estimation of Pesticide Retention and Transformation Parameters Required in Non-Point Source Pollution Models. In: Environmental Impact of Non-Point Source Pollution, Ann Arbor Science Publishers, Inc. Ann Arbor, Michigan. 1980.

Richter, R.O. Adsorption of Trichloroethylene by Soils from Dilute Aqueous Systems. Prepared for the U.S. Air Force Office of Scientific Research. 1981.

Schlumberger. Log Interpretation. Volume I - Principles (1972); Volume II - Applications (1974); Volume III - Charts (1979).

Schweitzer, P.A. (ed.). Handbook of Separation Techniques for Chemical Engineers. McGraw-Hill Book Company, New York, 1979.

Scow, K.M. Rate of Biodegradation, in Handbook of Chemical Property Estimation Methods. McGraw-Hill. New York. 1982.

Shen, T.T. Estimating Hazardous Air Emissions from Disposal Sites. Pollution Engineering. August 1981.

Shuckrow, A.J. et al. Bench Scale Assessment of Concentration Technologies for Hazardous Aqueous Waste Treatment. In: Proceedings of the Seventh Annual Research Symposium on Land Disposal: Hazardous Waste. Sponsored by the U.S. Environmental Protection Agency. March 16-18, 1981.

Sommerer, S. and J.F. Kitchens. Engineering and Development Support of General Decon Technology for the DARCOM Installation and Restoration Program, Task 1: Literature Review on Groundwater Containment and Diversion Barriers. Draft Report by Atlantic Research Corp. to U.S. Army Hazardous Materials Agency, Aberdeen Proving Ground, Contract NO. DAAK 11-80-C-0026. October 1980.

Sommerfeldt, T.G. and D.E. Campbell. A Pneumatic System to Pump Water from Piezometers. Groundwater. Volume 13, Number 3, p. 293. 1975.

Sowers, G.F. Soil Mechanics and Foundations: Geotechnical Engineering. MacMillan Publishing Co., Inc. New York. 1970.

Symons, J.M. Interim Treatment Guide for Controlling Organic Contaminants in Drinking Water Using Granular Activated Carbon. Prepared for U.S. Environmental Protection Agency. January 1978.

Symons, J.M. Removal of Organic Contaminants from Drinking Water Using Techniques Other than Granular Activated Carbon Alone. Prepared for U.S. Environmental Protection Agency. May 1979.

Symons, J.M. Utilization of Various Treatment Unit Processes and Treatment Modification for Trihalomethane Control. In: Proceedings - Control of Organic Chemical Contaminants in Drinking Water. Sponsored by U.S. Environmental Protection Agency, Office of Drinking Water. January 1980.

Telford, W.M. et al. Applied Geophysics. Cambridge University Press. 1976.

Thibodeaux, L.J. Chemodynamics. Environmental Movement of Chemicals in Air, Water, and Soil. John Wiley and Sons. New York. 1979.

Thibodeaux, L.J. Estimating the Air Emissions of Chemicals from Hazardous Waste Landfills. J. Hazardous Materials. Volume 4, pp. 235-244. 1981.

Thomas, R.G. Volatilization from Soil. In the Handbook of Chemical Property Estimation Methods. McGraw-Hill. New York. 1982a.

Thomas. R.G. Volatilization from Water. In the Handbook of Chemical Property Estimation Methods. McGraw-Hill. New York. 1982b.

Tinsley, I.J. Chemical Concepts in Pollutant Behavior. John Wiley & Sons. New York. 1979.

Todd, D.K. Groundwater Hydrology. John Wiley & Sons, Inc. New York. 1959.

Tomson, M.B. et al. A Nitrogen Powered Continuous Delivery All Glass Teflon Pumping System for Groundwater Sampling from Below 10 Meters. Groundwater. Volume 18, Number 5, pp. 444-446. September - October 1980.

Trescott, P.C. and G.F. Pinder. Air Pump for Small Diameter Piezometer. Groundwater. Volume 8, Number 3, pp. 10-15. 1970. U.S. Environmental Protection Agency. Quality Criteria for Water. 1976.

U.S. Environmental Protection Agency (U.S.EPA). Handbook for Evaluation of Remedial Action Technology Plans; Draft Report; USEPA Municipal Environmental Research Laboratory, Concinnati, Ohio. June 1982a.

U.S. Environmental Protection Agency (U.S.EPA). Handbook for Remedial Action at Waste Disposal Sites. Report to OERR, ORD, MERL, EPA Report No. EPA-625/6-86-006. June 1982b.

U.S. Environmental Protection Agency (U.S.EPA). Water Quality Criteria Documents; Availability. Federal Register. Volume 45, Number 231, pp. 79317-79379. November 28, 1980a.

U.S. Environmental Protection Agency (U.S.EPA). Procedures Manual for Ground Water Monitoring at Solid Waste Disposal Facilities. EPA 530/SW-611. 1980b.

U.S. Environmental Protection Agency (U.S.EPA). Technical Resource Document #5: Management of Hazardous Waste Leachate. September 1980c.

U.S. Environmental Protection Agency (U.S.EPA). Treatability Manual, Volumes I, II, III, IV, and V. EPA-600/8-80-042 a-e. July 1980d.

U.S. Environmental Protection Agency (U.S.EPA). National Secondary Drinking Water Regulations. Federal Register. Volume 44, Number 240, pp. 42195-42202. July 19, 1979.

U.S. Environmental Protection Agency (U.S.EPA). IERL-RTP Procedures Manual, Level 1, Environmental Assessment. EPA-600/7-78-201. 1978.

U.S. Environmental Protection Agency (U.S.EPA). The Report to Congress: Waste Disposal Practices and Their Effects on Goundwater. 1977a.

U.S. Environmental Protection Agency (U.S.EPA). Federal Guidelines: State and Local Pretreatment Programs. Volume 1. EPA-430/9-76-017a. 1977b.

U.S. Environmental Protection Agency (U.S.EPA). Quality Criteria for Water. USEPA, Washington, D.C. 1976a.

U.S. Environmental Protection Agency (U.S.EPA). National Interim Primary Drinking Water Regulations. EPA-570/9-76-003. 1976b.

van der Waarden, J. et al. Transport of Mineral Oil Components to Groundwater - I. Model Experiments on the Transfer of Hydrocarbons from a Residual Oil Zone to Trickling Water. Water Res. Volume 5, pp. 213-226. 1971.

van der Waarden, M. et al. Transport of Mineral Oil Components to Groundwater - II. Influence of Lime, Clay and Organic Soil Components on the Rate of Transport. Water Res. Volume 11, pp. 359-365. 1977.

Walton, W.C. Groundwater Resource Evaluation. McGraw-Hill Book Co., New York. 1970.

Williams, D.G. and D.G. Wilder. Gasoline Pollution of a Groundwater Reservoir - A Case History. Groundwater. Volume 9, Number 6, pp. 50-56. November - December 1971.

Wood, W.W. A Technique Using Porous Cups for Water Sampling at any Depth in the Unsaturated Zone. Water Resources Research. Volume 9, Number 2, pp. 486-488. 1973.

Wood, P.R. and J. DeMarco. . Effectiveness of Various Adsorbents in Removing Organic Compounds from Water. Presented at the 176th ACS Meeting. Activated Carbon Adsorption of Organics from the Aqueous Phase. Miami Beach, Florida. September 10-15, 1978.

Wyss, A.W. et al. Closure of Hazardous Waste Surface Impoundments. Prepared for U.S. Environmental Protection Agency. SW-873. September 1980.

Zohdy, A.A.R. et al. Techniques of Water Resources Investigations of the United States Geodetic Survey: Application of Surface Geophysics to Groundwater Investigations. Book 2, Chapter D1. 1974.

Bibliography

Selected manuals, books, and reports used in planning and implementing groundwater contamination assessment, prevention and spill control measures:

ACS Committee on Environmental Improvement. Guidelines for Data Acquisition and Data Quality Evaluation in Environmental Chemistry. Anal. Chem. Volume 52, Number 14. December 1980.

Alexander, M. Introduction to Soil Microbiology. 2nd ed. John Wiley & Sons. New York. 1977.

American Institute of Chemical Engineers. Chemical Engineering Applications of Solid Waste Treatment. Report No. S-122. New York, New York. 1972.

American Society of Agronomy. Chemistry in the Soil Environment. ASA Special Publication No. 40. Madison, New Jersey. 1981.

Arthur D. Little, Inc. Report of a Limited Survey of Industry Capacity to Respond to Environmental Emergencies Arising from the Release of Hazardous Chemicals. Report to the U.S. Environmental Protection Agency. 1978.

Arthur D. Little, Inc. User's Guide for the Evaluation of Remedial Action Technologies. Draft Report for the EPA MERL, Cincinnati, Ohio. Contract #68-01-5949. 1982.

Arthur D. Little, Inc. Guide to Water Cleanup Materials and Methods. Learning Systems. Cambridge, Massachusetts. 1974.

Atlantic Research Corporation. Engineering and Development Support of General Decon Technology for the DARCOM Installation and Restoration Program, Task 1. Literature Review on Groundwater Containment and Diversion Barriers. Draft Report to U.S. Army Hazardous Materials Agency, Aberdeen Proving Grounds. Contract No. DAAK11-80-C-0026. 1980.

Baldwin, H.L. and C.L. McGuiness. A Primer on Groundwater. U.S. Geological Survey. 1970.

Berkowitz, J. et al. Unit Operations for Treatment of Hazardous Industrial Wastes. Noyes Data Corporation. Park Ridge, New Jersey. 1978.

Bonazountas, M. and J. Wagner. SESOIL, A Seasonal Soil Compartment Model. Prepared for U.S. Environmental Protection Agency. 1981.

Boost, C.W. Modeling the Movement of Chemicals in Soil by Water. Soil Sci. Volume 115, Number 3, pp. 224-230. 1973.

Chemical Engineering. Liquids Handling Deskbook, 1978, and Materials Handling Deskbook, 1978. New York, New York.

Chemical Industries Association, Ltd. Transport Emergency Cards, Volumes 2-4. London. 1973-1976.

Cherry, J.A. et al. Contaminant Hydrogeology - Part 1, Physical Processes: Geoscience Canada. Volume 2, No. 2, pp. 76-84. 1975.

Conway, R.A. (ed). Environmental Risk Analysis for Chemicals. Van Nostrand Reinhold Co. New York. 1982.

Crichlow, H.B. Modern Reservoir Engineering - A Simulation Approach. Prentice-Hall, Inc. 1977.

D'Appolonia, D.J. Soil-Bentonite Slurry Trench Cutoffs. Journal of the Geotechnical Engineering Division, ASCE. Volume 106(4), pp. 399-418. 1980.

Davidson, J.M. et al. Use of Soil Parameters for Describing Pesticide Movement Through Soils. U.S. Environmental Protection Agency. EPA-660/2-75-009. 1974.

Dawson, G.W. et al. Control of Spillage of Hazardous Polluting Substances. Report No. 15090 in the Water Pollution Control Research Series. Department of the Interior. Washington, D.C. 1970.

Faust, C.R. et al. Computer Modeling and Groundwater Protection: Groundwater. Volume 19, Number 14, pp. 362-365. 1981.

Garfield, F.M., N. Palmer, and G. Schwartzman, eds. Optimizing Chemical Laboratory Performance Through the Application of Quality Assurance Principles. Proceedings of a Symposium, Association of Official Analytical Chemistry 94th Annual Meeting, Washington, D.C. October 22-23, 1980.

Giles, M.T. (ed.). Drinking Water Detoxification. Noyes Data Corporation. Park Ridge, New Jersey. 1978.

Gillett, J.W. et al. A Conceptual Model for the Movement of Pesticides Through the Environment: A Contribution of the EPA Alternative Chemicals Program. U.S. Environmental Protection Agency. EPA-660/3-74-024.

Goring, C.A.I. and J.W. Hamaker (eds). Organic Chemicals in the Soil Environment. Marcel Deker. New York. 1972.

Graphics Management Corporation. Control of Hazardous Material Spills. Proc. 1980 National Conference, Washington, D.C. 1980. [Others: 1972, 1974, 1976, 1978.]

Guenzi, W.D. (ed.). Pesticides in Soil and Water. Soil Science Society of America, Inc. Madison, Wisconsin. 1974.

Hackman, E., III. Toxic Organic Chemicals Destruction and Waste Treatment. Noyes, Data Corporation. Park Ridge, New Jersey. 1978.

Howard, P.H. et al. Determining the Fate of Chemicals. Environ. Sci. Technol. Volume 12, Number 4, pp. 398-407. 1978.

Huck, P.J. Assessment of Time Domain Reflectometry and Acoustic Emission Monitoring: Leak Detection Systems for Landfill Liners. In: Land Disposal of Hazardous Waste, Proc. of 7th Annual EPA Res. Symp. EPA-600/9-81-002/0. pp. 261-273. March 1981.

Hushon, J. et al. Information Required for Regulation of Toxic Substances. Volumes I (Survey of Data Requirements) and II (Survey of Test Methods). The MITRE Corporation. MTR-7887. 1978.

Jacob, C.E. Flow of Groundwater. In Engineering Hydraulics, H. Rouse, ed. John Wiley & Sons, New York. pp. 321-386.

Josephson, J. Groundwater Monitoring. Environmental Science and Technology. Volume 15, Number 9, pp. 993-996. September 1981.

Kolthoff, I.M. et al. Quantitative Chemical Analysis. MacMillan Company. New York. 1969.

Lee, G.F. and R.A. Jones. Interpretation of Chemical Water Quality Data. Aquatic Toxicology. (L.L. Marking and R.A. Kimerle, eds.). ASTM STP 667. pp. 302-231. 1979.

Leopold, L.B. and W.B. Langbein. A Primer on Water. U.S. Government Printing Office. 1960.

Lindorff, D.E. Groundwater Pollution - A Status Report. Groundwater. Volume 17, Number 1, p. 9-17. 1979.

Lyman, W.J. et al. Research and Development of Methods for Estimating Physicochemical Properties of Organic Compounds of Environmental Concern. Prepared for the U.S. Army Medical Bioengineering Research and Development Command. NTIS AD-074829. 1979.

Mackison, F.W., R.S. Stricoff and L.J. Partridge (eds.). Pocket Guide to Chemical Hazards (NIOSH/OSHA). DHEW (NIOSH). Publication No. 78-210. 1978.

Mandel, S. and Z.L. Shifton. Groundwater Resources Investigation and Development. Academic Press. 1981.

Metry, A.A. et al. The Handbook of Hazardous Waste Management. Technomic Publishing Company. Westport, Connecticut. 1979.

National Fire Protection Association. Fire Protection Guide on Hazardous Materials, 6th ed. Boston, Massachusetts. 1975.

National Research Council. Testing for Effects of Chemicals on Ecosystems. National Academy Press. Washington, D.C. 1981.

Oak Ridge National Laboratory. Material Safety Data Sheets - The Basis for the Control of Toxic Chemicals. Report No. ORNL/TM-6981/V1, V2, V3. Prepared for U.S. Department of Energy. 1979.

Ogata, A. Transverse Diffusion in Saturated Isotropic Granular Media. U.S. Geological Survey Professional Paper 411-B. 1961.

Ogata, A. The Spread of a Dye Stream in an Isotropic Granular Medium. U.S. Geological Survey Professional Paper 411-G. 1964.

Ogata, A. Theory of Dispersion in a Granular Medium. U.S. Geological Survey Professional Paper 411-I.

Pojasek, R.J. (ed.). Toxic and Hazardous Waste Disposal, Vols. 1-4. Ann Arbor Science. Ann Arbor, Michigan. 1979-1980.

Prickett, T.A. Modeling Techniques for Groundwater Evaluation. Advances in Hydrosciences. Volume 10, pp. 1-143. 1976.

Prickett, T.A. Groundwater Computer Models - State of the Art. Groundwater. Volume 17, Number 2, pp. 167-173. 1979.

Railway Systems and Management Association. Handling Guide for Potentially Hazardous Commodities. Chicago, Illinois. 1972.

Robinson, J.S. (ed.). Hazardous Chemical Spill Cleanup. Noyes Data Corporation. Park Ridge, New Jersey. 1979.

Selim, H.A. et al. Transport of Reactive Solutes Through Multi-layered Soils. Soil Sci. Soc. Amer. J. Volume 41, Number 1, pp. 3-10. 1977.

Skibitzke, H.E. and G.M. Robinson. Dispersion in Groundwater Flowing Through Heterogeneous Materials. U.S. Geological Survey Professional Paper 386-B. 1963.

Travis, C.C. Mathematical Description of Adsorption and Transport of Reactive Solutes in Soil: A Review of Selected Literature. Oak Ridge National Laboratory. ORNL 5403. 1978.

U.S. Army. State-of-the-Art Survey of Land Reclamation Technology. Report No. EC-CR-76076. 1976.

U.S. Coast Guard. Chemical Data Guide for Bulk Shipment by Water. Report No. CG-338. Washington, D.C. 1976.

U.S. Coast Guard. A Feasibility Study of Response Techniques of Hazardous Chemicals that Disperse Through the Water Column. Report No. CG-D-16-77. Washington, D.C. 1976.

U.S. Coast Guard. Agents, Methods and Devices for Amelioration of Discharges of Hazardous Chemicals on Water. Report No. CG-D-38-76. Washington, D.C. 1975.

U.S. Coast Guard. Survey Study of Techniques to Prevent or Reduce Discharges of Hazardous Chemicals. Report No. CG-D-1984-75. Washington, D.C. 1975.

U.S. Coast Guard. CHRIS Response Methods Handbook (Manual 4). Report No. CG-446-4. Washington, D.C. 1975.

U.S. Coast Guard. Influence of Environmental Factors on Selected Amelioration Techniques for Discharges of Hazardous Chemicals. Report No. CG-D-81-75. Washington, D.C. 1975.

U.S. Coast Guard. A Condensed Guide to Chemical Hazards (Manual 1). Report No. CG-446-1. Washington, D.C. 1974. [Updated as required.]

U.S. Coast Guard. CHRIS Hazardous Chemical Data (Manual 2). Report No. CG-446-2. Washington, D.C. 1974. [Updated as required.]

U.S. Coast Guard. Survey Study to Select a Limited Number of Hazardous Materials to Define Amelioration Requirements. Report No. CG-D-46-75. Washington, D.C. 1974.

U.S. Department of Transportation. Hazardous Materials, 1980 Emergency Response Guidebook. Report No. DOT-P 5800.2. Washington, D.C. 1980.

U.S. Environmental Protection Agency. Emission Monitoring: Leak Detection Systems for Landfill Liners. EPA-600/9-81-002/0. March 1981.

U.S. Environmental Protection Agency. Groundwater Protection-A Water Quality Management Report. 1980.

U.S. Environmental Protection Agency. The Quality Assurance Bibliography. EPA-600/4-80-009. 1980.

U.S. Environmental Protection Agency. EPA/IERL-RTP Procedures for Level 2 Sampling and Analysis of Organic Materials. EPA-600/7-79-03. 1979.

U.S. Environmental Protection Agency. An EPA Manual for Organic Analysis Using Gas Chromatography/Mass Spectrometry. EPA-600/8-79-006. 1979.

U.S. Environmental Protection Agency. Handbook for Analytical Quality Control in Water and Wastewater Laboratories. EPA-600/4-79-019. March 1979.

U.S. Environmental Protection Agency. Hazardous Materials Spill Monitoring Safety Handbook and Chemical Hazard Guide. PART A - Safety Handbook; PART B - Chemical Data. Report No. EPA-600/4-79-008a and b. Las Vegas, Nevada. 1979.

U.S. Environmental Protection Agency. Methods for Chemical Analysis of Water and Wastes. EPA-600/4-79-020. 1979.

U.S. Environmental Protection Agency. Manual of Treatment Techniques for Meeting the Interim Primary Drinking Water Regulations. EPA 600/8-77-005. April 1978.

U.S. Environmental Protection Agency. Pollution Prediction Techniques for Waste Disposal Siting, A State-of-the-Art Assessment. SW-162c. 1978.

U.S. Environmental Protection Agency. Procedure for the Evaluation of Environmental Monitoring Laboratories. EPA-600/4-78-017. 1978.

U.S. Environmental Protection Agency. Survey of States on Response to Environmental Emergencies. Report by Arthur D. Little, Inc. 1978.

U.S. Environmental Protection Agency. Symposium on Environmental Transport and Transformation of Pesticides. EPA-600/9-78-003. 1978.

U.S. Environmental Protection Agency. Emergency Collection System for Spilled Hazardous Materials. Report No. EPA/600/2-77/162. 1977.

U.S. Environmental Protection Agency. In Situ Treatment of Hazardous Material Spills in Flowing Streams. Report No. EPA/600/2-77/164. 1977.

U.S. Environmental Protection Agency. Manual for the Control of Hazardous Material Spills. Volume 1: Spill Assessment and Water Treatment Techniques. Report No. EPA-600/2-77-227. Oil and Hazardous Materials Spill Branch, IERL. Edison, New Jersey. 1977.

U.S. Environmental Protection Agency. Multipurpose Gelling Agent and Its Application to Spilled Hazardous Materials. Report No. EPA/600/2-77/151. 1977.

U.S. Environmental Protection Agency. Performance Testing of Spill Control Devices on Floatable Hazardous Materials. Report No. EPA/600/2-77/222. Edison, New Jersey. 1977.

U.S. Environmental Protection Agency. Quality Assurance Guidelines for Air Pollution Measurement Systems. Volume I, II, and III. EPA-600/9-76-005, EPA-600/4-77-027a, and EPA-600/4-77-027b. 1976 and 1977.

U.S. Environmental Protection Agency. Manual of Analytical Quality Control for Pesticides in Human and Environmental Media. EPA-600/1-76-017. 1976.

U.S. Environmental Protection Agency. Methods to Treat, Control and Monitor Spilled Hazardous Materials. Report No. EPA-670/2-75-042. Industrial Waste Treatment Research Laboratory. Edison, New Jersey. 1975.

U.S. Environmental Protection Agency. Spill Prevention Techniques for Hazardous Polluting Substances. Report No. OHM 7102 001 in the Oil and Hazardous Materials Program Series. 1971.

No.	Title	Report No.	Date
1.	Evaluating Cover Systems for Solid and Hazardous Waste	SW-867	1980
2.	Hydrologic Simulation on Solid Waste Disposal Sites	SW-868	1980
3.	Landfill and Surface Impoundment Performance Evaluation	SW-869	1980
4.	Lining of Waste Impoundment and Disposal Facilities	SW-870	1980
5.	Management of Hazardous Waste Leachate	SW-871	1981
6.	Guide to the Disposal of Chemically Stabilized and Solidified Wastes	SW-872	1981
7.	Closure of Hazardous Waste Surface Impoundment	SW-873	1981
8.	Design and Management of Hazardous Waste Land Treatment Facilities	SW-874	1981
9.	Soil Properties, Classification and Hydraulic Conductivity Testing	SW-875	1981
10.	Solid Waste Leaching Procedures Manual	Draft	1981
11.	Landfill Closure Manual	Draft	1981

U.S. Environmental Protection Agency Oil and Hazardous Materials Spills

Branch, Hazardous Spills Reports:

OHMSB Code	Date	Title	EPA Report Number / NTIS Number
HR-1	Nov. '70	Control of Spillage of Hazardous Polluting Substances	EPA – 15090 F OZ 10/70 NTIS – PB-197 596/0BA PC A18/MF A01
HR-2	Aug. '72	Rapid Detection System for Organophosphates and Carbamate Insecticides in Water	EPA – R2-72-010 NTIS – PB-214 764/3BA PC A04/MF A01
HR-3	Mar. '73	Control of Hazardous Chemical Spills by Physical Barriers	EPA – R2-73-185 NTIS – PB-221 493/0BA PC A05/MF A01
HR-4	May '73	Feasibility of Plastic Foam Plugs for Sealing Leaking Chemical Containers	EPA – R2-73-251 NTIS – PB-222 627/2BA PC GPO/MF A01
HR-5	Sep. '73	Treatment of Hazardous Materials Spills with Floating Mass Transfer Media	EPA – 670/2-73-078 NTIS – PB-228 050/AS PC GPO/MF A01
HR-6	Dec. '74	Evaluation of MTP for Testing Hazardous Material Spill Control Equipment	EPA – 670/2-74-073 NTIS – PB-240 762/5BA PC A14/MF A01
HR-7	Apr. '75	Feasibility of 5 gpm Dynactor/Magnetic Separator System to Treat Spilled Hazardous Materials	EPA – 670/2-75-004 NTIS – PB-241 080/1BA PC A03/MF A01
HR-8	Jun. '75	Methods to Treat, Control and Monitor Spilled Hazardous Materials	EPA – 670/2-75-042 NTIS – PB-243 386/0BA PC A07/MF A01

ID	Date	Title	Reference
HR-9	Jun. '75	Guidelines for the Disposal of Small Quantities of Unused Pesticides	EPA – 670/2-75-057 NTIS – PB-244 557/5BA PC A15/MF A01
HR-10	Jul. '76	Development of a Mobile Treatment System for Handling Spilled Hazardous Materials	EPA – 600/2-76-109 NTIS – PB-256 707/1BA PC A05/MF A01
HR-11	Sep. '76	Removal and Separation of Spilled Hazardous Materials from Impoundment Bottoms	EPA – 600/2-76-245 NTIS – PB-266 140/3BE PC A05/MF A01
HR-12	Dec. '76	Prototype System for Plugging Leaks in Ruptured Containers	EPA – 600/2-76-300 NTIS – PB-267 245/9BE PC A06/MF A01
HR-13	Aug. '77	Emergency Collection System for Spilled Hazardous Materials	EPA – 600/2-77-162 NTIS – PB-272 790/7BE PC A05/MF A01
HR-14	Aug. '77	Heavy Metal Pollution from Spillage at Ore Smelters and Mills	EPA – 600/2-77-171 NTIS – PB-272 639/6BE PC A06/MF A01
HR-15	Aug. '77	Multipurpose Gelling Agent and its Aplication to Spilled Hazardous Materials	EPA – 600/2-77-151 NTIS – PB-272 763/4BE PC A04/MF A01
HR-16	Oct. '77	In Situ Treatment of Hazardous Material Spills in Flowing Streams	EPA – 600/2-77-164 NTIS – PB-274 455/5 BE PC A04/MF A01
HR-17	Nov. '77	Evaluation of "CAM-1", a Warning Device for Organophosphate Hazardous Material Spills	EPA – 600/2-77-219 NTIS – PB-276 647/5 BE PC A04/MF A01
HR-18	Nov. '77	Performance Testing of Spill Control Devices on Floatable Hazardous Materials	EPA – 600/2-77-222 NTIS – PB-276 581/6BE PC A08/MF A01

HR-19	Nov. '77	Manual for the Control of Hazardous Materials Spills: Vol. I Spill Assessment and Water Treatment Techniques	EPA - 600/2-77-227 NTIS - 276 734/1BE PC A21/MF A01
HR-20	Dec. '77	Effects of Hazardous Material Spills on Biological Treatment Processes	EPA - 600/2-77-239 NTIS - PB-276 724/2BE PC A10/MF A01
HR-21	Mar. '78	Development of a Kit for Detecting Hazardous Material Spills in Waterways	EPA - 600/2-78-055 NTIS - PB-281 284/0BE PC A05/MF A01
HR-22	Apr. '78	Hazardous Material Spills: A Documentation and Analysis of Historical Data	EPA - 600/2-78-066 NTIS - PB-281 090/1BE PC A11/MF A01
HR-23	Jul. '78	System for Applying Powdered Gelling Agents to Spilled Hazardous Materials	EPA - 600/2-78-145 NTIS - PB286 308/2 BE PC A03/MF A01
HR-24	Mar. '79	Selected Methods for Detecting and Tracing Hazardous Material Spills	EPA - 600/2-79-064 NTIS - PB-298 054/8BE PC A04/MF A01
HR-25	Jul. '79	Comparison of Three Waste Leaching Tests	EPA - 600/2-79-071 NTIS - PB-299 259/2BE PC A11/MF A01
HR-26	May '79	Background Study on the Develpoment of a Standard Leaching Test	EPA - 600/2-79-109 NTIS - PB-298 280/9BE PC A12/MF A01
HR-27	Chem. Stf. Pub. '79	Publications on the Analysis of Spilled Hazardous and Toxic Chemicals and Petroleum Oils	EPA - OHMSB
HR-28	Aug. '79	Acoustic Monitoring to Determine the Integrity of Hazardous Waste Dams	EPA - 625/2-79-024 NTIS - PB-80-176787
HR-29	Jan. '80	CAM-4, Portable Warning Device for Organophosphate Hazardous Material Spills	EPA - 600/2-80-033 NTIS - PB-80-159494

IIR-30	May '80	Environmental Emergency Response Unit Capability	EPA - OHMBS
IIR-31	May '80	Application of Buoyant Mass Transfer to Hazardous Materials Spills	EPA - 600/2-80-078 NTIS - PB-80-198427
IIR-32	May '80	Alternate Enzymes for Use in Cholinesterase Antagonist Monitors ("CAM's")	EPA - 600/2-80-083 NTIS - PB-80-200835
IIR-33	Jul. '80	Hazardous Material Spills and Responses for Municipalities	EPA - 600/2-80-108 NTIS - PB-80-214141
IIR-34	1981	Detection and Mapping of Insoluble Sinking Pollutants	
IIR-35	1981	Use of Selected Sorbents and Aqueous Film Forming (AFF) Foams on Floatable Hazardous Material Spills	
IIR-36	May '81	Development of a System to Protect Groundwater Threatened by Hazardous Spills on Land	EPA - 600/2-81-085 NTIS - PB-81-209587 PC A07/MF A01
IIR-37	1981	Removal of Water-Soluble Hazardous Materials Spills from Waterways by Activated Carbon	
IIR-38	1981	Guidelines for the Use of Chemicals in Removing Hazardous Substance Discharge	
IIR-39	1981	Restoring Hazardous Spill-Damaged Areas: Identification and Assessment of Techniques	
IIR-40	1981	A Mobile Stream Diversion System for Hazardous Materia & Spills Isolation	
IIR-41	1981	Modifications of Spill Factors Affecting Air Pollution	
IIR-42	1981	Removal of Spilled Hazardous Materials from Bottoms of Flowing Watercourses: Phase I	
IIR-43	1981	Feasibility of Commercialized Water Treatment Techniques for Concentrated Waste Spills	

NATIONAL CONFERENCE ON CONTROL OF HAZARDOUS MATERIAL SPILLS

OHMSB Code	Title/Date/Place
HP-1	Proceedings of the 1972 National Conference on Control of Hazardous Materials Spills. March 21-23, 1972, Houston, Texas.
HP-2	Proceedings of the 1974 National Conference on Control of Hazardous Material Spills. August 25-28, 1974, San Francisco, California.
HP-3	Proceedings of the 1976 National Conference on Control of Hazardous Material Spills. August 25-28, 1976, New Orleans, Louisiana.
HP-4	Proceedings of the 1978 National Conference on Control of Hazardous Materials Spills. April 11-13, 1978, Miami Beach, Florida.
HP-5	Proceedings of the 1980 National Conference on Control of Hazardous Materials Spills. May 13-15, 1980, Louisville, Kentucky.
HP-6	Proceedings of the 1980 US EPA National Conference on Management of Uncontrolled Hazardous Waste Sites, October 15-17, 1980, Washington, D.C.

van Geruchten, M.T. and P.J. Wierenga. Mass Transfer Studies in Sorbing Porous Media I. Analytical Solutions. Soil Sci. Soc. Am. J. Volume 40, Number 4, pp. 473–480. 1976.

von Lehmden, D.J. and C. Nelson. Quality Assurance Handbook for Air Pollution Measurement Systems. Volume I. Principles. PB 254658. January 1976.

Witherspoon, J.P. et al. State-of-the-Art and Proposed Testing for Environmental Transport of Toxic Substances. U.S. Environmental Protection Agency. EPA-560/5-76-001. 1976.

Yaffe, H.J. et al. Application of Remote Sensing Techniques to Evaluate Subsurface Contamination and Buried Drums. Proceedings of 7th Annual EPA Res. Symposium. EPA-600/9-81-002/0. March 1981.

Xanthakos, P.O. Slurry Walls. McGraw-Hill Book Company. New York, New York. 1979.

Glossary

The geologic, hydrologic, and chemical terms pertinent to this report are defined as follows:

Aqueous Phase - water in the saturated or unsaturated zone, which may contain hydrocarbon compounds (see Hydrocarbon Fluid Phase; Hydrocarbon Solid Phase).

Aquifer - a formation, group of formations, or part of a formation that contains sufficient saturated permeable material to yield significant quantities of water to wells or springs.

Confined Ground Water - ground water under pressure significantly greater than atmospheric. Its upper limit is the bottom of a bed of distinctly lower vertical hydraulic conductivity than that of the material in which the confined water occurs (see "Confining Bed").

Confining Bed - a body of material with low vertical permeability stratigraphically adjacent to one or more aquifers. Replaces the terms "aquiclude," "aquitard," and aquifuge."

Desorption - the removal of contaminants from the solid matrix of the porous medium by fluids in the ground water system.

Diffusion - molecular movement of chemical constituents of ground water or hydrocarbon fluids in response to chemical-concentration gradients.

Dispersion, Mechanical - differences in the rate and direction of movement of individual tracer particles owing to variations in path lengths, and pore geometry or size.

Dispersion, Hydrodynamic - the combined effects of "Diffusion" and "Dispersion, mechanical."

Drawdown - the vertical distance between the static (nonpumping) water level and the level caused by pumping.

Ground Water - that part of subsurface water that is in the saturated zone.

Head, Static - the height above a standard datum of the surface of a column of water that can be supported by the static pressure at a given point.

Hydraulic Conductivity - capacity of a rock to transmit water under
pressure. It is the rate of flow of water at the prevailing
kinematic viscosity passing through a unit section of area,
measured at right angles to the direction of flow, under a unit
hydraulic gradient (see "Permeability, Intrinsic").

Hydrocarbon Fluid Phase - a liquid mixture of hydrocarbon compounds,
immiscible with water, that forms a fluid phase physically
distinct from the aqueous phase. It is distinctly denser and
more viscous and has a higher surface tension than the aqueous
phase. (See Aqueous Phase; Hydrocarbon Solid Phase.)

Hydrocarbon Solid Phase - hydrocarbon compounds sorbed onto the matrix
of the porous media. (See Aqueous Phase; Hydrocarbon Fluid
Phase; Sorption.)

Isopotential Line - line connecting points of equal static head.
(Head is a measure of the potential.)

Multiaquifer Well - any well that hydraulically connects more than one
aquifer. The connection may be due to original open-hole
construction or to deterioration of casing or grout seal.

Permeability, Intrinsic - a measure of the relative ease with which a
porous medium can transmit liquid under a potential gradient. It
is a property of the medium alone and is independent of the
nature of the liquid and of the force field causing movement. It
is a property of the medium that is dependent upon the shape and
size of the pores.

Piezometer - a small-diameter pipe placed in the ground in such a way
that the water level in the pipe represents the static head at
the very point in the flow field where the piezometer terminates.

Porosity - the property of a rock or soil to contain interstices or
voids. It may be expressed quantitatively as the ratio of the
volume of interstices to total volume of the rock. (See
"Porosity, Effective.")

Porosity, Effective - the amount of interconnected pore space
available for fluid transmission. It is expressed as a decimal
fraction or as a percentage of the total volume occupied by the
interconnecting interstices.

Potentiometric Surface - a surface that represents the static head.
As related to an aquifer, it is defined by the levels to which
water will rise in tightly cased wells. Where the head varies
appreciably with depth in the aquifer, a potentiometric surface
is meaningful only if it describes the static head along a
particular specified surface or stratum in that aquifer. More
than one potentiometric surface is then required to describe the
distribution of head. The water table is a particular
potentiometric surface. Replaces the term "Piezometric Surface."

Saturated Zone - zone in earth's crust in which all voids are ideally
    filled with water. The water table is the upper limit of this
    zone. Water in the saturated zone is under pressure equal to or
    greater than atmospheric.

Sorption - the removal of contaminant from fluids in the ground water
    system by the solid matrix of the porous medium.

Specific Capacity - the rate of discharge of water from a well divided
    by the drawdown of water level within the well. It varies slowly
    with duration of discharge, which should be stated when known.
    If the specific capacity is constant except for time variation,
    it is roughly proportional to the transmissivity of the aquifer.

Specific Yield - the ratio of the volume of water that a saturated
    rock or soil will yield by gravity to its own volume.

Storage Coefficient - the volume of water an aquifer releases from or
    takes into storage per unit surface area of the aquifer per unit
    change in head. In an unconfined aquifer, it is virtually equal
    to the specific yield.

Transmissivity - the rate at which water of the prevailing kinematic
    viscosity is transmitted through a unit width of an aquifer under
    a unit hydraulic gradient.

Unconfined Ground Water - water in an aquifer that has a water table.

Valley Fill - drift or alluvial sediments deposited in an erosional
    depression in the bedrock surface.

Water Table - that surface in an unconfined water body at which the
    pressure is atmospheric. It is defined by the levels at which
    water stands in wells that penetrate the water body just far
    enough to hold standing water. In wells that penetrate to
    greater depths, the water level will stand above or below the
    water table if an upward or downward component of ground water
    flow exists.

Well Field - as used in this report, any combination of wells
    withdrawing water from the same area and close enough to cause
    mutual drawdown effects.

Zone, Saturated - that part of the water-bearing material in which all
    voids, large and small, are ideally filled with water under
    pressure greater than atmospheric. The saturated zone may depart
    from the ideal in some respects. A rising water table may cause
    entrapment of air in the upper part of the zone of saturation,
    and some parts may include accumulations of other fluids.

Zone, Unsaturated - the zone between the land surface and the water table. It includes the capillary fringe. Characteristically, this zone contains liquid water under less than atmospheric pressure and water vapor and air or other gases generally at atmospheric pressure. In parts of the zone, interstices, particularly the small ones, may be temporarily or permanently filled with water. Perched water bodies may exist within the unsaturated zone and some parts may include accumulations of other fluids. Replaces the terms "zone of aeration" and "vadose zone."

Appendix

Costs Associated with
Waste Disposal Site Cleanup

The enclosed cost information describes the costs of various aspects of the contamination assessment and remedial action programs as well as some examples of costs associated with actual instances of groundwater contamination. Many of the costs are representative of rates prevailing in the northeastern United States in the mid 1970s to early 1980s and actual costs may be higher or lower depending on local conditions. The purpose of the cost information is to indicate the potential magnitude of the expense associated with site cleanup as well as the relative cost of the various phases.

COST ESTIMATES FOR VARIOUS MONITORING TECHNIQUES AND
CONSTRUCTION METHODS IN THE ZONE OF SATURATION (FROM
U.S. EPA, 1980)

Monitoring Techniques & Construction Method	. . . Price per Installation Well Diameter ($) . . .		
	51 mm (2 in.)	102 mm (4 in.)	152 mm (6 in.)
Screened over a single interval (plastic screen and casing)			
1. Entire aquifer	1,600–3,700	2,300–4,500	6,400–7,500
2. Top 3 m (10 ft) of aquifer	600–1,050	700–1,150	—
3. Top 1.5 m (5 ft) of aquifer with drive point	100–200	—	—
Piezometers (plastic screen & casing)			
1. Entire aquifer screened			
a. Cement grout	2,100–4,700	2,800–5,500	6,900–8,500
b. Bentonite seal	1,850–4,150	2,350–4,950	6,650–7,950
2. Top 3 m (10 ft) of aquifer screened			
a. Cement grout	1,150–2,050	1,200–2,150	—
b. Bentonite seal	900–1,500	950–1,600	—
Well clusters			
1. Jet-percussion			
a. Five-well cluster, each well with a 6 m (20 ft) long plastic screen	2,500–3,800	—	—
b. Five-well cluster, each well with only a 1.5 m (5 ft) long plastic screen	1,700–2,300	—	—
2. Augers			
a. Five-well cluster, each well with a 6 m (20 ft) long stainless steel wire-wrapped screen	4,600–5,300	—	—
b. Five-well cluster, each well with only a 1.5 m (5 ft) long gauze-wrapped drive points	1,800–2,600	—	—
3. Cable tool			
a. Five well cluster, each well with a 6 m (20 ft) long stainless steel, wire-wrapped screen	—	—	9,850–14,150
4. Hydraulic rotary			
a. Five-well cluster, each well with a 6 m (20 ft) long plastic screen, casing grouted in place	—	9,050–14,900	13,800–19,400
b. Five-well cluster, completed in a single large-diameter borehole 4.5 m (15 ft) long plastic screens, 1.5 m (5 ft) seal between screens	4,240–5,880	8,250–11,000	—
5. Single well/multiple sampling point			
a. 33.5 m (110 ft) deep well with 1 ft long screens separated by 1.2 m (4 ft) of casing starting at 3 m (10 ft) below ground surface	—	—	3,000–4,700
Sampling during drilling	—	3,000–4,700	3,300–5,200

*Cost estimates are for an aquifer composed of unconsolidated sand with a depth to water of 3 m (10 ft) and a total saturated thickness of 30 m (100 ft). Cost estimates are based on rates prevailing in the Northeast in Autumn, 1975. Actual costs will be lower and higher depending upon conditions in other areas. Therefore, while the costs presented will become outdated with time, the relative cost relationships among the monitoring techniques should remain fairly constant.

EXAMPLES OF UNIT COSTS FOR SITE CLEAN-UP

VEHICLES
- Front Loader: $500/day
- Backhoe: $225/day
- Small Bulldozer: $225/day
- Vacuum Trucks: $50/hr
- Light Trucks: $30/day
- Heavy Trucks: $100-200/day

SAMPLE COLLECTION FROM WELLS, SOILS, STREAMS
- 1-2 hrs/well
- 2-3 person crew
- Truck, pumps, containers
- $50-200/sample

SAMPLE ANALYSIS
- $30-60/sample/parameter for conventional pollutants
- GC/MS priority pollutants: $900-2000/sample
- GC/ECD pesticides: $100-350/sample
- ICAP metals: $100-350/sample

CLEAN-UP CREW (8-10)
- Supervisor (1), Chemist (1), Technician (1), Equipment Operators (2-3), Laborers (3-4)
- ~ $250-300/hr ($2000 - 2400/day)

BARRELS
- Disposal: $75-300 each
- Repackaging: $100 each

DAILY COSTS FOR SITE CLEAN-UP (EXAMPLE)
- Crew $2000
- Equipment $1000
- Sampling/Analysis $1000
- Barrel Costs $1000
                    $5000

YEARLY COST AT ABOVE RATE (EXAMPLE): $1,200,000 (240D)

SPECIAL EXPENSES
- Site Access/Preparation
- Utilities
- Fences, Security
- Administration
- Insurance
- Inspection

YEARLY COST + SPECIAL COSTS: $1,500,000 + (?)

TYPICAL REMEDIAL ACTION COSTS[*]

	ACTION	MED.SIZE LANDFILL	MED.SIZE IMPOUNDMENT	UNITS
		COST (MID 1980 $)		
1A.	Contour grading and surface water diversion	8,000	-	/AC
1B.	Pond closure and contour grading	-	20,000	/AC
2.	Surface sealing	34,000	25,000	/AC
3.	Revegetation	7,000	2,000	/AC
4.	Bentonite slurry-trench cutoff wall	9	9	/FT^2
5.	Grout curtain	150	57	/FT^2
6.	Sheet piling	9	11	/FT^2
7.	Bottom sealing	3,600,000	640,000	/AC
8.	Drains	120	560	/FT
9.	Well point system[+]	13	33	/FT^2
10.	Deep well system[+]	3.2	13	/FT^2
11.	Injection[+]	166	12	/FT^2
12.	Leachate handling by sub-grade irrigation	9,100	-	/AC
13.	Chemical fixation	52,000	-	/AC
14.	Chemical injection	0.1	-	/FT^3
15.	Excavation and disposal at secure landfill	3.5/FT^3	1.5/GAL	
16.	Ponding	370	-	/AC
17.	Trench construction	6	-	/FT
18.	Perimeter gravel trench	40	-	/FT
19.	Treatment of contaminated groundwater	15	27	/GAL
20.	Gas migration control[++] Passive	300	-	/FT
	Active	70	-	/FT
21.	Berm construction	-	4	/YD^3

[+]Units: $ per intercept face area

[++]Units: $/Length (ft) of site perimeter

[*]Source: U.S. EPA, Cost of Remedial Response Actions at Uncontrolled Hazardous Waste Sites. Report to MERL, Cincinnati, Ohio.

EXAMPLES OF CLEAN-UP COSTS AT ACTUAL SITES
(not all costs are final)

- OLIN CORP/REDSTONE ARSENAL MILITARY RESERVATION (AL)

  - Dichlorodiphenyltrichloroethane (DDT) contamination of surface waters used for water supply. Produced from 1947-1970. Sued in 1970.

  - Wastewater discharge was permitted.

  - High levels of DDT found in soil, water, fish, near site.

  - U.S. Army Corps of Engineers commissioned a $1.3 million study of problem and possible corrective action (1979).

  - Estimated 837 tons of DDT in sediments of stream.

  - Suggested rerouting stream; estimated cost: $88.9 million.

  - TVA was authorized to spend $1.5 million on special studies/local aid in 1981.

- CHEM-DYNE HAZARDOUS WASTE DUMP (OH)

  - Approximately 20,000 drums of chemicals at site.

  - Pollution of surface and groundwater, soil, air; fires.

  - EPA negotiated agreement with 109 companies for partial clean-up. Other generators identified.

  - Cost: $2.4 million ($250,000 by Velsicol).

  - Justice Department suing mine owner/operators plus 16 waste contributors.

- LOVE CANAL (NY)/HOOKER CHEMICAL CORPORATION

  - Abandoned canal used for disposal of chemicals (1942-1953) (21,000 tons)

  - Sold land to Niagara Falls Board of Education in 1953 for $1.00.

  - Problem identified in 1978 (surface breakthrough).

  - Settlement agreement for clean-up negotiated cleaned-up cost: $62 million.

  - Suits asking for compensatory and punitive payments total several billion dollars.

- W. R. GRACE & COMPANY, ACTON, MASS.

  - Organic chemicals contaminated Acton town wells.

  - Wells 1500 ft. from waste lagoons.

  - Initial hydrogeological study: $90,000 - $140,000.

  - Findings criticized by Grace; second hydrogeological study
    paid for by company ($?).

  - Consent decree signed with state in October 1980 for 20-30
    year clean-up of aquifer.

  - Company being sued for $22M by Acton Water Board.

- SUPERFUND ALLOCATIONS

  - Picillo Farm (RI): $4.9M

  - Kim-Buc Landfill (NJ): $2.44M (Removal)

  - Love Canal (NY): $5.3M

  - Commencement Bay (WA): $0.718M (Site investigations)

Part III

Emergency Response

The information in Part III is from *Rapid Assessment of Potential Ground-Water Contamination Under Emergency Response Conditions,* by A.S. Donigian, Jr.; T.Y.R. Lo; and E.W. Shanahan of Anderson-Nichols & Co., Inc., for the U.S. Environmental Protection Agency, November 1983.

Acknowledgments

This manual is the result of support and guidance from a number of groups and individuals. Financial support was provided by the EPA Office of Emergency and Remedial Response through the Exposure Assessment Group and the Environmental Research Laboratory (Athens, GA) of EPA's Office of Research and Development. Mr. Lee Mulkey (Athens ERL) was the Project Officer and Mr. John Schaum (Exposure Assessment Group) was the technical Project Monitor; the technical assistance and guidance provided by these individuals was instrumental to the successful completion of this manual.

Anderson-Nichols was assisted in this effort by Battelle, Pacific Northwest Laboratories, Richland, WA, and Dr. P.S.C. Rao of the University of Florida, Gainsville, FL. Battelle PNL provided technical review of reports and assistance in the compound and site characterization efforts; Dr. Rao assisted in the review of methods and development of procedures for estimating contaminant fate and transport in the unsaturated zone.

In addition, technical review comments were provided by Dr. Wayne Pettyjohn of Oklahoma State University and the EPA RSK Environmental Research Laboratory in Ada, OK. Peer review comments and suggestions on the draft manual were provided by Dr. Carl Enfield (EPA-RSKERL), Dr. Charles Faust of Geo-Trans, Inc., and Mr. Robert Carsel (EPA-Athens ERL) in addition to the project team members noted above. These reviews and suggestions were especially helpful in preparing this final manual.

Among the authors, Mr. Anthony Donigian was project manager responsible for the overall technical content of the manual, development of the methodology and parameter estimation guidelines, and preparation of the manual. Mr. T.Y. Richard Lo developed the assessment nomograph and application procedures, and prepared example applications. Mr. Edward W. Shanahan assisted in the methodology development, and the methods review and parameter estimation for the saturated-zone procedures.

Mr. John Imhoff was involved in the methodology review and development, and assisted in the preparation of the interim report. Guidance in chemical parameter estimation was provided by Mr. J. David Dean and Mr. Brian R. Bicknell; Mr. Bicknell also assisted in preparing the draft manual.

Technical support was provided by Mary Maffei, word processing was provided by Ms. Lyn Hiatt and Ms. Sandy Guimares, and the drafting and graphics were prepared by Ms. Virginia Rombach. The dedicated assistance of all these individuals allowed the successful completion of this project, and is sincerely appreciated.

1. Introduction

The purpose of this manual is to provide a methodology for estimating potential ground-water contamination, under emergency response conditions, at an abandoned hazardous waste or toxic chemical spill site. Specifically, this manual is designed for use by field personnel to quickly estimate how contaminant concentrations might change with time and distance from an emergency response site. The procedures include evaluation of critical contaminant and site charcteristics as input to an assessment methodology for analyzing the fate and movement of chemicals through both the unsaturated and saturated (i.e. ground water) soil zones. A graphical technique (i.e. nomograph) has been developed for contaminant movement through both the unsaturated and saturated (ground water) zones to provide a complete, integrated assessment methodology. Guidelines for evaluating critical waste and site characteristics are provided to allow estimation of needed nomograph parameters.

1.1 SCOPE AND LIMITATIONS OF THIS MANUAL

The phrase EMERGENCY RESPONSE is emphasized throughout this manual because it has been the over-riding criterion (and constraint) for selection, evaluation, and development of pollutant transport assessment methods and parameter evaluation techniques included herein. Emergency response situations require assessments of potential ground-water contamination to be completed in less than 24 hours. Consequently, extensive field sampling, laboratory analyses, data search and collection, and sophisticated computer analyses are generally impractical during this limited time frame. Although these extensive sampling and analysis activities may be initiated during the emergency response period, the results are not expected to be available for use in an emergency assessment.

The assessment procedures in this manual are designed to allow emergency response personnel to make a first-cut, order-of-magnitude estimate of the potential extent of contamination from a waste site or chemical spill within the 24-hour emergency response time frame. These procedures are not intended to provide a definitive, indepth analysis of the complex fate and transport processes of contaminants in the subsurface environment.

The primary goal of this manual is to provide the basis for determining the need for emergency actions, such as emergency sampling, containment/removal, drinking water restrictions, etc. in order to preclude or minimize human exposure from ground-water contamination at an emergency response site. Two specific emergency response situations are envisioned where the assessment procedures in this manual would be applied.

1. Discovery of an abandoned hazardous waste site where an assessment of the potential extent of the waste plume is needed within the emergency response time frame.

2. Spill (or leakage) of a toxic waste or chemical where the potential for ground-water contamination and/or the extent of contamination must be assessed within the emergency response time frame.

Time and resource limitations expected during an emergency response have required a number of simplifying assumptions in our assessment procedures; additional simplifications may be needed by the user due to limited data and information available at a particular emergency response site. The major assumptions incorporated into the assessment procedures in this manual are as follows:

1. Homogeneous and isotropic properties are assumed for both the unsaturated and saturated zones (or media).

2. Steady and uniform flow is assumed in both the unsaturated and saturated zones.

3. Flow and contaminant movement are considered only in the vertical direction in the unsaturated zone and the horizontal direction in the saturated zone.

4. All contaminants are assumed to be water-soluble and exist in concentrations that do not significantly affect water movement.

A variety of other assumptions and limitations in the procedures are further discussed in Section 3.5. The user should carefully review all the assumptions and limitations, and must make specific judgements as to their validity for the specific site, contaminant(s), and emergency situation being analyzed. Perhaps the most critical aspect of an emergency response situation will be the ability of the user to adequately characterize, within the 24-hour time frame, the subsurface media (e.g. heterogeneities, depth to ground water, soil/aquifer properties, aquifer thickness) through which the contaminants may move. Consequently, access and/or availability of data, expertise, and familiarity with local, site-specific soils and hydrogeologic conditions is critical to the successful application of the assessment procedures in this manual.

1.2 REQUIRED USER BACKGROUND, TRAINING, and PREPARATION

Effective use of this manual requires an understanding of a mix of disciplines, such as hydrology, hydrogeology, soil science, chemistry, on the part of the intended user, and sufficient familiarity or training with the techniques, procedures, and auxiliary sources of information described herein. Moreover, this manual is not intended to be a primer on pollutant fate and movement through soils and ground water; a variety of excellent introductory textbooks and reports in these areas are available to the potential user to provide the needed background (e.g., Freeze and Cherry, 1979; EPA, 1981; Thibodeaux, 1979).

Ideally, academic training in any one of the above disciplines supplemented with experience, job training, and/or exposure (e.g. short course attendance) in the other disciplines provides a profile of the recommended background for a user. Alternatively, an engineering or science undergraduate degree with appropriate training is acceptable as long as a basic understanding in the following areas is included:

a. the hydrologic cycle and its components

b. hydrogeologic concepts, processes, and terminology related to ground-water movement

c. soil science concepts related to soil processes and water movement

d. chemical processes, parameters, and terminology

e. mathematical capabilities and skills in the use of scientific hand calculators.

In many emergency response situations, the user will have access to experts in the above disciplines to provide guidance in parameter evaluation. Thus, the user must have sufficient comprehension of the appropriate terminology in order to communicate effectively with the experts and "ask the right questions!"

User training and preparation is needed to develop familiarity with the assessment procedures described in this manual and the wide range of auxiliary sources of information that supplement and complement the parameter evaluation guidelines in Section 4. In essence, the user should be able to ask and answer the question - "What information do I need and where can I get it?"

Training and/or familiarity with the specific procedures described herein is absolutely essential to effectively use this manual. Without prior study users cannot expect to use this manual for assessing potential ground-water contamination within a 24-hour period. Although every effort has been made to simplify the procedures and parameter evaluation guidelines, prior study is needed to become familiar with the assumptions/limitations, the step-by-step calculations, the application of the nomographs, the parameter evaluation guidelines, and the auxiliary sources of information. Also, knowledge of the most sensitive, critical parameters will allow the user to allocate data search efforts most effectively.

Familiarity with supplementary sources of information cannot be over-emphasized. Section 2.4 describes a variety of handbooks and data bases from which contaminant characteristics (and input parameters) can be evaluated or estimated. Precious time can be saved if the user is knowledgeable about which sources are most likely to contain the information he is seeking.

Since site characterization may require the greatest effort during an emergency assessment, preparation of a regional or local data base on meteorology, hydrologic characteristics, soils/aquifer properties, ground-water characteristics, prior hydrology/hydrogeologic studies, and local experts (i.e. contacts and phone numbers) could considerably shorten the time needed to obtain data and improve the resulting parameter estimates. A similar, regional data base for the characteristics of wastes and chemicals produced in, or transported through, the region would be extremely valuable. Recommendations for the contents and format of such a regional data base have been developed for EPA (Battelle PNL, 1982a).

1.3 FORMAT OF THE MANUAL

Section 2 describes the types of hazardous waste and spill situations for which the assessment procedures are designed, and provides a methodology flow chart to guide an application. An overview of critical compound and site characteristics is provided along with a discussion of recommended sources of information. Section 3 describes both the unsaturated and saturated zone methodologies and the assessment nomograph. A detailed description of the assessment methodology. Section 3 also discusses linkage of unsaturated and saturated zone assessments and the assumptions and limitations of the assessment procedures - these should be carefully reviewed and understood by the user.

Section 4 provides guidelines for estimating the input parameters for both the unsaturated and saturated zone assessments. Emphasis is placed on obtaining local site and compound specific data in order to obtain realistic parameter estimates. However, quantitative guidelines are provided for most parameters as a last resort if no other information is available.

Section 5 presents example applications for the assessment nomograph for both zones and demonstrates linkage procedures. Section 6 includes cited references, Appendix A provides a description of the SCS Curve Number procedure for estimating surface runoff; Appendix B is a glossary of terms; and Appendix C provides blank worksheets and copies of enlarged nomographs for ease of use during an application.

1.4 CAVEAT

Although all efforts have been made to insure the accuracy and reliability of the methods and data included in this manual, the ultimate responsibility for accuracy of the final predictions must rest with the user. Since parameter estimates can range within wide limits, especially under the resource and time constraints of an emergency response, the user should assess the effect of methodology assumptions and parameter variability on predicted concentrations for the specific site. The methodology predictions must be evaluated with common sense, engineering judgement and fundamental principles of soil science, hydrogeology, and chemistry. Accordingly, neither the authors nor Anderson-Nichols assume liability from use of the methods and/or data described in this manual.

2. Overview of Rapid Assessment Methodology

An emergency response to releases of hazardous substances is generally comprised of three steps - characterization, assessment, mitigation - defined as follows (Battelle PNL, 1982a):

o Characterization - the acquisition, compilation, and processing of data to describe the scene so that a valid assessment of alternative actions can be made.

o Assessment - an analysis of the severity of an incident; the evaluation of possible response actions for effectiveness and environmental impact.

o Mitigation - the implementation of the best response action and followup activities.

The assessment procedures for potential ground-water contamination in this manual draw upon data and information developed in the characterization phase in order to provide a tool for performing parts of the assessment phase when ground-water contamination is suspect. The EPA Field Guide for Scientific Support Activities Associated with Superfund Emergency Response (Battelle PNL, 1982a) provides an excellent framework within which to view these procedures as part of the arsenal of the emergency response team for assessments of hazardous substance releases. This field guide identifies the calculation of transport rates of hazardous materials as an important element in the assessment phase. When subsurface fate and movement of hazardous substances is important at an emergency response site, these calculations can be made with the procedures described herein based on the methodology assumptions and data expected to be available within the emergency response time frame.

2.1 APPLICATION SCENARIOS

Ground-water contamination by hazardous materials may result from surface spills; seepage from waste injection operations, waste storage/burial sites; and leaks from underground containers (i.e., waste or storage) or pipelines. The rapid assessment procedures are designed for application in two typical scenarios, or cases, based on the temporal nature of the release:

Case 1 Analysis - Typically a hazardous waste site or chemical/waste storage facility where the release is relatively continuous and constant over an extended period of time (e.g. years).

Case 2 Analysis - Typically a spill incident (or a short-term release
from a storage facility) where the release can be assumed constant
over a relatively short span of time (e.g. weeks, months) producing
a pulse-type release.

The assumption of a constant release either on a continuous or pulse basis
is necessary for the analytical solutions from which the nomographs have
been developed. Consequently, although actual releases will be
time-varying, the user will need to approximate the actual release by either
the Case 1 or Case 2 assumptions above in order to perform an assessment
within the emergency response time frame. (See Section 3 for further
discussion.)

Superimposed on the temporal nature of the release is the time period of
concern for the assessments and the associated quantities of the forces
driving the movement of the contaminant. In most cases, the driving force
will be water movement through the soil to ground water; for large volume
spills the mass of the material may be sufficient to move through the soil.

The time period can vary from an assessment of the historical movement and
current extent of the contaminant plume, to a projection of the plume at
some time in the future. For the discovery of an abandoned hazardous waste
site, the user may need to evaluate the current extent of contamination
based on the age of the site, the period of release, and ground-water
recharge estimates during the past; whereas, for a spill situation the user
may need to project the future movement of the plume based on precipitation
forecasts and resulting expected recharge. Thus, the time period of concern
and the temporal nature of the release jointly determine the appropriate
type of analysis (i.e., Case 1 vs. Case 2) and parameter estimates for the
driving force behind contaminant movement.

2.2 METHODOLOGY FLOWCHART

The overall flowchart for the rapid assessment methodology is shown in
Figure 2.1. Prior to initiating application of these procedures, the
On-Scene Coordinator (OSC) at the emergency response site must determine
that (1) the potential for ground-water contamination exists, or (2)
contaminants have reached ground water, and (3) an assessment of the
potential or current extent of contamination must be made within the 24-hour
emergency response time frame. These decisions will be based on the results
of the characterization phase of the emergency response effort and will
depend on current conditions (e.g., current contamination of wells or
streams, weather forecasts), compound characteristics (e.g., toxicity,
solubility, sorption, volatility), and site characteristics (e.g., depth to
ground-water, soil/aquifer characteristics, distance to drinking water wells
and streams). If no emergency assessment is deemed necessary, the
procedures in this manual should not be used, except as preliminary guidance
for subsequent detailed sampling, analysis, and investigations possibly
including numerical modeling techniques.

If an emergency assessment is deemed necessary, the steps in Figure 2.1
should be followed as discussed below:

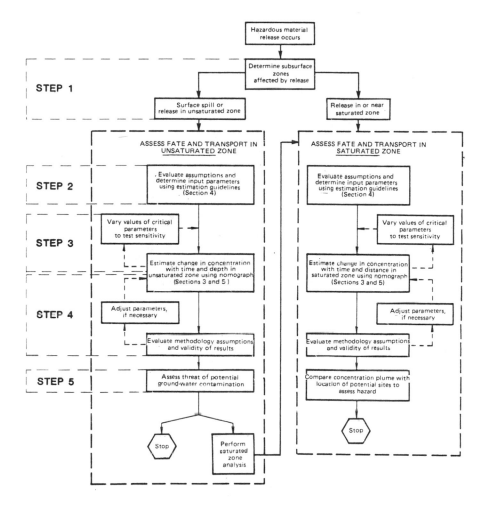

Figure 2.1   Flow Chart for Rapid Assessment of Potential Ground-Water
Contaminant Under Emergency Response Conditions.

STEP 1 involves the determination of which zone, unsaturated or saturated, will be affected by the contaminant release and which associated branch to follow in the flowchart. Most surface and near-surface releases will need to move through the unsaturated zone before reaching ground water; thus an unsaturated zone analysis (left branch in Figure 2.1) will be needed. For shallow ground-water depths, highly permeable soils, and/or highly fractured surface materials, the user may choose to ignore the unsaturated zone and assume direct release to the saturated, ground-water zone. This assumption ignores any attenuation or retardation in the unsaturated zone and, in many cases, will over-estimate actual concentrations reaching ground water.

STEP 2 involves an initial evaluation of the methodology assumptions (both unsaturated and saturated zones) for the specific site, and estimation of the nomograph input parameters based on the guidelines in Section 4. These two aspects are closely linked since parameter values can be adjusted to partially compensate for certain assumptions and limitations. However, significant parameter uncertainties should be identified early in the application so that associated impacts can be assessed.

STEP 3 includes calculation of concentrations with time and distance using the nomograph described in Section 3. For the unsaturated zone the depth to ground water will usually be the distance measurement of interest; for the saturated zone horizontal distances to nearby wells or streams may be needed. Sensitivity analyses should be performed on critical parameters (e.g., decay rate and retardation in the unsaturated zone, ground-water velocity in the saturated zone) in order to assess the effects of possible inaccuracies in parameter estimation.

STEP 4 requires the user to re-evaluate the methodology assumptions based on the predicted concentrations and results of sensitivity analyses. Further parameter adjustments and re-calculation of concentrations may be necessary. This is a critical step since the predictions will be used next to assess the potential or current extent of ground-water contamination.

STEP 5 provides the assessment results upon which to make decisions on needed emergency response actions. The need for an emergency response, and the possible alternative actions, are decisions to be made by the On-Scene Coordinator and other emergency personnel which are not addressed in this manual. For unsaturated zone analysis, concentration estimates for various depths will indicate if the contaminant will reach ground water at levels and within the time frame where emergency response actions may be needed. If ground-water contamination is predicted the user may need to perform a saturated zone analysis, using the results of the unsaturated zone analysis as input to estimate the contaminant plume migration in ground water. The results of the saturated zone analysis can provide the concentrations and associated travel times at potential impact sites (e.g., wells, streams) where human exposure or ecological impacts may occur.

Complete application of the assessment procedures may require a number of iterations of the steps discussed above, as new data becomes available or as additional questions arise from the emergency situation. Following the step-by-step procedures outlined above and shown in Figure 2.1 will allow the user to perform consistent assessments of potential ground-water contamination in a variety of circumstances.

2.3 CRITICAL COMPOUND AND SITE CHARACTERISTICS

The extent of contaminant fate and transport following releases to the land surface and subsurface depends upon a variety of critical compound and site characteristics. Table 2.1 lists the major characteristics of concern for determining potential ground-water contamination at a specific site. This section briefly describes the compound and site characteristics listed in Table 2.1 to provide the user with an understanding of the types of information needed to perform a valid assessment. Guidelines for translating these characteristics into specific parameter values required by the assessment procedures are provided in Section 4.

2.3.1 Critical Compound Characteristics

To assess the potential for ground-water contamination in an emergency response situation, several properties of the compound or waste must first be determined. Much of this information may be difficult to accurately quantify within a 24-hour time frame, but it is likely that an applicable range of values can be estimated. Some properties are used directly in the assessment or to estimate parameters, while others are needed to interpret the results. Those characteristics deemed crucial to an informed assessment and listed in Table 2.1 are discussed below:

1. Contaminant Identity

    The identities of the contaminants must be known to determine those physical/chemical properties necessary for assessing pollutant fate and migration. The physical state of the contaminant (gas, liquid, or solid) should be assessed as part of the identification process. Within the emergency response time frame, it may be possible to identify only general classes of chemicals rather than specific compounds. In such instances, parameter estimation will be especially difficult.

2. Extent of the Contamination

    The extent of the contamination must be defined to determine the source term used in estimating transport into the soil and ground water. This assessment should provide an estimate of the mass of the pollutant entering, or potentially entering, the subsurface environment by adjusting for volatilization into the air, runoff, and containment or removal measures on the land surface,

TABLE 2.1 CRITICAL COMPOUND AND SITE CHARACTERISTICS

Critical Compound Characteristics

    1.   Contaminant Identity and Physical State

    2.   Extent of the Contamination

    3.   Solubility

    4.   Adsorption

    5.   Degradation

    6.   Toxicity

    7.   Concentration and Loading

    8.   Density, Viscosity, and Temperature

Critical Site Characteristics (Applicable to Both the Unsaturated and Saturated Zones Unless Otherwise Indicated)

    1.   Identity of Subsurface Medium

    2.   Age of Site

    3.   Distances to Wells, Streams, Property Boundaries

    4.   Porosity

    5.   Infiltration, Net Recharge, and Volumetric Water Content (Unsaturated Zone Only)

    6.   Bulk Density

    7.   Hydraulic Conductivity (Saturated Zone Only)

    8.   Chemical Characteristics of Medium

    9.   Dispersion

  10.   Depth to Ground Water (Unsaturated Zone Only)

  11.   Hydraulic Gradient (Saturated Zone Only)

  12.   Effective Aquifer Thickness (Saturated Zone Only)

  13.   Structural and Geologic Features

if necessary. Information on the volatility and reactivity of the waste may be required in making this assessment. In addition, the cross-sectional area of the spill or the disposal site should be ascertained.

3.   Solubility

The solubility of a compound affects its mobility in the soil and ground water. The release of the contaminant from a landfill or surface spill is usually controlled by its tendency to dissolve in the water moving through the soil. A material's solubility may also affect the ease with which it can adsorb on soil particles, with less soluble wastes being more easily adsorbed. Solubility generally provides an upper limit on dissolved concentrations that can be found in the soil environment. The existence of solvents other than water should also be determined since it can affect the compound's miscibility with soil water and ground water.

4.   Adsorption

Adsorption can be a significant means of retarding contaminant movement through the soil or ground water. It is a property dependent upon both the nature of the compound and the soil. Adsorption capabilities for organic, nonionic compounds are often described in terms of adsorption (or partition) coefficients for a particular compound/soil combination. These coefficients are often estimated from the organic carbon (or organic matter) content of the soil and the organic carbon partition coefficient (which in turn can be estimated from compound characteristics such as the octanol/water partition coefficient). Adsorption of ionic compounds is also a function of ion exchange capacities and clay type and content. This is especially important for soils or media with low organic matter.

5.   Degradation

Degradation by both chemical and biological mechanisms is important because it can prevent contaminants from reaching ground water and can reduce levels of contaminants already present. Common degradation mechanisms in the environment are hydrolysis, photolysis, biodegradation, chemical oxidation, and radioactive decay. Hydrolysis and chemical oxidation are important primarily for contaminants in soils and saturated media. Photolysis can occur only in surface waters or on the surface of the soil. Biodegradation is most important in surface waters and in the top few feet of soil where bacterial concentrations are high; however, anaerobic decomposition in deep soils and ground water is possible. Radioactive decay occurs in all environments under all conditions.

6. Toxicity

    To assess the hazard of any predicted or observed ground-water
    contamination, the toxicity of the pollutants must be determined.
    Since nearly all chemicals are toxic at very high concentrations,
    the concern in this assessment is for materials that are moderately
    to severely toxic or are carcinogenic, mutagenic, or teratogenic to
    humans or aquatic organisms.

7. Concentration and Release/Loading Rate

    Compound concentrations and volume or release/loading rates from a
    spill or waste site are especially important because of the effects
    on other characteristics and the extent of contamination.
    Concentration will affect solubility, adsorption, degradation, and
    toxicity. Since many of these characteristics are usually measured
    at low concentrations and/or in aqueous solutions, changes at high
    concentrations can be significant, such as exceeding solubility
    limits or adsorption capacities, or reducing effective microbial
    populations. Low volume releases from spills may only contaminate
    a few feet of soil which could be removed by excavation; whereas
    large volume and/or continuous releases can result in much larger
    scale contamination.

8. Density, Viscosity, and Temperature

    These compound parameters are important in evaluating the mixing
    characteristics of the contaminant in soil water and ground water.
    Differences in these properties between the water and the
    contaminant can lead to density stratification, floating, or
    sinking of materials which will significantly impact transport
    behavior. Major differences in these characteristics may require
    an evaluation of the validity of the assessments which assume
    contaminant transport with the water movement.

2.3.2 Critical Site Characteristics

To assess potential ground-water contamination at a hazardous waste or spill
site, a number of site characteristics listed in Table 2.1 are important in
addition to the waste characteristics discussed above. Critical site
properties for both the unsaturated and saturated zones are identified and
briefly discussed below. Many of the parameters which define important site
characteristics are shared by both subsurface zones, although the values for
the parameters may be different for each zone. The discussions are intended
to provide an overview of the information needed to characterize an
emergency response site in appropriate detail to estimate contaminant
transport and fate in the subsurface environment; specific guidelines on
parameter estimation are presented in Section 4.

1. Identity of Subsurface Medium (Unsaturated and Saturated Zones)

    Perhaps the most critical site characteristics which must be
    determined is the dominant material types of the subsurface zones.

While the subsurface materials for either zone will rarely be homogeneous, it is necessary to identify the major soil or rock types in order to assign reasonable values to such parameters as porosity, bulk density, hydraulic conductivity, dispersion coefficients, and chemical characteristics.

2. Age of the Site (Unsaturated and Saturated Zones)

The age of the site will be most important in analyzing newly-discovered landfills, uncontrolled waste disposal sites, or leaking chemical storage facilities. The extent of pollutant migration at the emergency response site cannot be adequately assessed without knowledge of the length of time that contamination has been occurring, unless other data are available. Many surface chemical spills are investigated immediately after their occurrence and thus the age of the incident is known.

3. Distances (Unsaturated and Saturated Zones)

Distances to water wells, streams, and property boundaries from the hazardous waste or spill site are fundamental concerns in an emergency response. This information represents horizontal distances that the waste material must travel on the land surface or in the ground, before reaching potential receptor sites of concern.

4. Porosity

The total porosity, usually stated as a fraction or percent, is that portion of the total volume of the material that is made up of voids (air) and water. In determining the retardation coefficient, a measure of adsorptive capabilities, the total porosity of the aquifer is needed. Due to dead-end or unconnected pores, effective porosity is somewhat less than total porosity. Effective porosity is often estimated as the specific yield in unconfined aquifers which is the quantity of water that will drain from a unit volume of aquifer under the influence of gravity. Effective porosity is required for the calculation of the interstitial pore-water velocity in ground water based on Darcy's Equation.

5. Infiltration, Net Recharge, and Volumetric Water Content (Unsaturated Zone Only)

Infiltration and net recharge refer to water movement below the land surface to the unsaturated soil zone and ground water. Infiltration is generally greater than net recharge since it includes evaporation and transpiration quantities which are usually deducted to estimate net recharge to ground water. Both of these components are a function of climatic, topographic and soil properties, and are important in estimating contaminant movement into and through the unsaturated zone to ground water. Their relative importance depends on the time frame of the analysis (See Section 4).

The volumetric water content is the volume of water in a given volume of media, usually expressed as a fraction or percent. It depends on properties of the media and the water flux estimated by infiltration or net recharge. The volumetric water content is used in calculating the water movement through the unsaturated zone (pore water velocity) and the retardation coefficient. In saturated media, the volumetric water content equals total porosity.

6. Bulk Density (Unsaturated and Saturated Zones)

The bulk density of the medium is required in calculating the retardation factor, a measure of adsorption processes. The bulk density is the dry mass per unit volume of the medium (soil or aquifer), i.e., neglecting the mass of the water.

7. Hydraulic Conductivity (Saturated Zone)

The velocity of ground-water flow is essential to assessing the spread of contamination; it is an especially sensitive parameter for plume migration in the saturated zone. The hydraulic conductivity (or permeability) of the aquifer is needed to estimate flow velocity based on Darcy's Equation. It is a measure of the volume of liquid that can flow through a unit area of media with time; values can range over nine orders of magnitude depending on the nature of the media. Heterogeneous conditions produce large spatial variations in hydraulic conductivity, making estimation of a single, effective value extremely difficult.

In the unsaturated zone, conductivity is an extreme function of soil moisture, increasing by orders of magnitude as moisture content increases. This indicates the difficulty in assessing dynamic pollutant transport through the unsaturated zone as a function of dynamic soil moisture conditions.

8. Chemical Characteristics of Medium (Unsaturated and Saturated Zones)

The primary chemical characteristics of the medium include organic carbon content, ion exchange capacity, clay type and clay content. These properties are used in conjunction with the adsorption characteristics of the compound (as discussed in Section 2.3.1) to allow formulation of an appropriate partition coefficient for the specific compound/medium combination. The partition coefficient is used with bulk density, and either total porosity (saturated zone) or volumetric water content (unsaturated zone) to calculate a retardation factor, to represent the impact of adsorption on retarding contaminant movement through the medium.

9. Dispersion (Unsaturated and Saturated Zones)

    Hydrodynamic dispersion in subsurface media is a phenomenon that
    causes the spreading of a contaminant. The complicated system of
    interconnected passages comprising a porous media system causes a
    continuous division of the contaminant mass into finer offshoots.
    Variations in the local velocity (both magnitude and direction)
    along and between these tortuous flow paths gives rise to this ever
    increasing spreading on the microscopic scale. On a larger or
    macroscopic scale, inhomogeneity due to variations in permeability
    and porosity also gives rise to further spreading. On a megascopic
    scale, the effects of layering and the associated differences in
    permeabilities and porosities can give rise to further spreading
    (Pickens, et al, 1977).

    Dispersion is often considered together with molecular diffusion in
    determining a dispersion coefficient. Because the actual spread of
    a contaminant depends on inhomogeneity at various scales in
    addition to the tortuosity and local velocity variation on a
    microscopic scale, the selection and measurement of the dispersion
    parameter (i.e., dispersivity) should be related to the scale and
    detail of the modeling effort. This dependence on scale is
    demonstrated by the fact that dispersivity values measured in the
    laboratory can range from 10^{-2} to 1 cm, while field values can
    range from 10's to 100's of meters.

10. Depth to Ground Water (Unsaturated Zone)

    The depth to ground water must be estimated in order to evaluate
    the likelihood that contaminants moving through the unsaturated
    soil will reach the ground water. Seasonal fluctuations, if
    significant, should be identified, as well as the impacts of
    pumping and recharge sources, natural or man-made.

11. Hydraulic Gradient (Saturated Zone)

    To determine the magnitude and direction of ground-water flow, the
    hydraulic gradient must be known. It is the slope of the water
    table in an unconfined aquifer, or the piezometric surface for a
    confined aquifer. As for the ground-water depth, the effects of
    pumping and recharge should be considered in estimating the
    hydraulic gradient since these actions can reverse expected
    ground-water flow directions.

12. Effective Aquifer Thickness (Saturated Zone)

    The available zone of mixing in the aquifer is described using an
    effective aquifer thickness. For good mixing between the ground
    water and the contaminant, this effective thickness may equal the
    actual total thickness of the aquifer, but in many cases it will be
    considerably less. In cases where the pollutant is of a
    significantly different density than water, the extent of mixing

may be reduced and the contaminant plume will be concentrated over only a portion of the aquifer's thickness.

13.   Structural & Geologic Features (Unsaturated & Saturated Zones)

A general assessment of the soils, topographic and geologic environment of the study site is necessary to effectively evaluate the potential for ground-water contamination. Rapid assessments made within an emergency response time frame must assume homogeneous conditions due to time constraints, but heterogeneous properties will retard or increase contaminant migration and should be at least qualitatively assessed. Folds, faults, fractures, sinkholes, clay lenses, and soil variations are examples of features that should be considered when estimating appropriate ranges of parameters used in the rapid assessment methodology.

2.4 AUXILIARY SOURCES OF INFORMATION

To obtain the data necessary to evaluate critical compound and site characteristics during an emergency response, a variety of information sources should be consulted prior to and during the emergency. As noted in Section 1.2, the need to be familiar with the various sources of information that might be needed during an emergency response cannot be over-emphasized. The EPA Field Guide (Battelle PNL, 1982a) mentioned previously includes a useful check-list of activities for chemical characterization that should be performed before and between emergency responses, during the response, and following the response; an analogous checklist is provided for hydrologic assessments. In support of our recommendations, the EPA Field Guide also emphasizes the importance of pre-emergency planning and preparation especially in the collection and aggregation of data sources for compound and site characterization. This guide should be an important part of the library of an emergency response team.

This manual is not intended to be a stand-alone document since the supporting data that might be needed in an emergency response would fill multiple volumes many times the size of this report. The sections below describe various information sources for both compound and site characteristics; these sources will be further referenced in the specific parameter estimation guidelines in Section 4.

2.4.1   Sources of Compound Characterization Information

During an emergency response, data on waste characteristics are available from five major sources:

1.   Records

2.   Onsite Observations

3.   Analyses

4. Handbooks and Data Bases

5. Experts

These information sources must be applied jointly to determine the necessary input data for a ground-water contamination assessment. For example, transportation records may first be used to determine the chemical identity of a spilled cargo of waste before consulting a data base for a list of the physical/chemical properties of the waste. Much of the information in this section is published in the chemical characterization section of the EPA Field Guide, to which the user is referred for additional sources.

1. Records

Records can provide the most rapid, positive identification of the materials involved at an emergency, and, if available, should be the preferred means of identification. A variety of useful records (e.g., shipping papers and transportation labels) are now required when transporting hazardous materials. Transportation records contain information on the quantities of hazardous materials transported and may be used to estimate quantities involved in emergencies. A complete description of available records and how to use them in identifying spilled material is provided by Huibregtse, et al, (1977). Also, the Association of American Railroads is developing a computerized tracking system for rapidly identifying railcars containing hazardous materials (Guinan 1980). The use of records to identify chemicals present at uncontrolled waste sites is much more difficult. Waste manifests, which describe each shipment of waste received at a facility, are a possible source. In many cases, however, these have only recently been required.

2. Onsite Observations

Observable characteristics such as odor, color, density, and reaction may be useful in rapidly identifying an unknown material. An excellent method of quick identification of spilled materials based on easily observable characteristics is presented in the Field Detection and Damage Assessment Manual for Oil and Hazardous Materials Spills (EPA 1972). Over 300 hazardous materials are identified by odor, color, reaction, etc.

The U.S. Coast Guard Chemical Hazard Response Information System (CHRIS) Manual CG-446-1 and CG-446-2 (U.S.C.G. 1974a, 1974b) describes observable characteristics of approximately 900 hazardous chemicals. The OHM-TADS data system maintained by EPA can be used to identify chemical substances based on observable characteristics. Physical properties of the unknown material (physical state, odor, color turbidity, miscibility, reactions) are input to the computer system, which then performs a search to obtain possible identities.

It should be noted that the use of observable characteristics may be limited to identifying general classes of chemicals rather than specific compounds.

Onsite observations may also be important in establishing the extent of the contamination. Aerial photography and remote sensing may be needed to supplement ground observations in detecting the boundaries of a large spill or dump site, but such information may not be available within the 24-hour emergency response time frame.

3. Analyses

Analytical methods may be employed if other methods fail to identify the contaminants present. In emergency conditions where rapid response is required, the available techniques may be limited to qualitative field methods. Laboratory methods, while providing more definitive results, require considerably more time. Mobile laboratories have now made many complex instrumental methods available for use in the field, helping to reduce the time requirements of laboratory analysis.

The Field Detection and Damage Assessment Manual for Hazardous Materials Spills (EPA 1972) describes analytic tests that may be used in the field to identify chemicals. A variety of commercial products are currently available for infield detection and identification of hazardous materials. These products include portable spectrophotometers, ion-specific electrodes, gas chromatographs, and organic vapor analyzers. Information on such systems can be obtained from manufacturers and scientific supply houses.

Once the identity of the contaminant is known, analytical methods can be used in conjunction with a sampling program to determine the extent of the contamination. Under emergency response conditions, maximum use must be made of existing sampling sites such as wells, ponds, drainage ditches, runoff collection devices, and so on. Hand or gasoline powered augers provide a rapid means of quickly obtaining subsurface samples over a large area. Sampling techniques are described in EPA (1980).

4. Handbooks and Data Bases

Handbooks and data bases are an excellent source of physical/chemical data on hazardous wastes including toxicities, solubilities, densities, degradation rates, reactivities, volatilities, and adsorption data. As were previously discussed, data bases and handbooks also aid in identifying wastes based on observable characteristics. The data source descriptions provided below were taken largely from the EPA Field Guide:

Handbooks -

CHRIS, the Coast Guard Hazardous Chemical Data Manuals CG 446-1 and CG 446-2, are excellent sources of data on approximately 900 hazardous materials. The data contained in these and other CHRIS manuals are designed for use with the Coast Guard's Hazard Assessment Computer System (HACS), a computerized simulation system that models the physical behavior of chemical spills and provides information describing the extent of the hazard associated with these spills (Parnarouskis et al 1980).

Manual CG-446-1, A Condensed Guide to Chemical Hazards, contains a summary listing of physical/chemical properties of several hazardous materials. It is designed to be carried to the scene of an accident. Manual CG-446-2, Hazardous Chemical Data, contains detailed information on the properties of hazardous chemicals.

The EPA Field Detection and Damage Assessment Manual for Oil and Hazardous Materials Spills (EPA 1972) is useful for supplying data needed for identifying any of 329 hazardous materials in the field.

The Handbook of Environmental Data on Organic Chemicals (Verschueren 1977) is an excellent source of data describing the behavior of over 1,000 organic chemicals in the environment. This is perhaps the most complete collection of environmental chemical data that can be easily taken into the field.

Dangerous Properties of Industrial Materials (Sax 1979) is a collection of physical, chemical, and toxicological data on almost 13,000 common industrial and laboratory materials. The data deal primarily with the hazards posed by the materials and include acute and chronic toxic hazard ratings, toxicity figures, a description of toxicology, treatment of poisoning, and storage, handling, and shipping guidelines.

Physical Chemical Properties of Hazardous Waste Constituents (Dawson, English and Petty 1980) is a collection of data on 250 chemicals commonly found in hazardous waste streams. This collection is an excellent reference for predicting the behavior of chemicals following spills. For each chemical, quantitative estimates are included of the relative human health hazard posed by its release to the environment.

The Merck Index (1976) contains general chemical data on almost 10,000 chemical substances. This work contains descriptions of the preparation and chemistry of the various substances, with citations to the original published sources of the data.

Aquatic Fate Process Data for Organic Priority Pollutants (Mabey et al, 1982) this report includes physical transport, and transformation data for 114 organic priority pollutants in aqueous solutions, and provides methods of calculating partition coefficients and volatilization rates.

Handbook of Chemical Property Estimation Methods (Lyman et al, 1982) is a collection of estimation methods for several physical and chemical properties of organic chemicals with emphasis on environmental processes; it does not actually contain the data. The handbook includes definitions and principles of the properties, overviews of the available methods, and specific instructions for the use of each one including detailed examples. An appendix of the handbook also contains a listing of selected reference books which contain compilations of many physical/chemical properties of organic chemicals.

Data Bases -

OHM-TADS - The Oil and Hazardous Materials-Technical Assistance Data System contains chemical, physical, and biological data on over 850 hazardous chemicals and industrial materials. OHM-TADS contains data describing physical/chemical properties, toxicity, environmental fate and persistence, and emergency response methods. These data are maintained on computer by EPA and are accessible by remote terminal or by microfiche.

Octanol/Water Partition Coefficient Data Base, a data base containing octanol/water partition coefficients for several thousand chemicals, is maintained by Dr. Corlan Hansch at Pomona College, Pomona, California (714--621-8000 ext. 2225). This is perhaps the most complete source of K_{ow} values currently available. The material in this data base can be purchased in hard copy form, on microfiche, or on magnetic tape.

The Chemical Substances Information Network (CSIN) is a computerized data collection system currently being developed by EPA. Sources for this system will initially include the National Library of Medicine, the Chemical Information System, EPA's Chemicals in Commerce Information System, Bibliographical Retrieval Services, System Development Corporation, and Lockheed's Dialog System.

Table 2.2 summarizes the data available from the major handbooks and data bases notes above.

5. Experts

An additional source of information on compound characteristics lies with experts within the chemical industry, scientific community, and hazardous waste response teams.

The Chemical Manufacturers Association (CMA) Chemical Transportation Emergency Center (CHEMTREC) telephone hotline [(800) 424-9300 or 483-7616 in Washington, DC] maintains a directory of industry experts who can be contacted for information related to emergency response. CHEMTREC can rapidly provide information on approximately 18,000 chemicals and trade-name products.

TABLE 2.2 SUMMARY OF CHEMICAL/PHYSICAL DATA AVAILABLE FROM HANDBOOKS AND DATA BASES

	Handbook or Data Base							
	Chris Manual CG446-1,2 (U.S.C.G. 1974b)	EPA Field Detection Manual (1972)	Versch- ueren 1979	Sax 1979	Dawson English and Petty, 1980	Merck Index Wind- holz (1976)	OHM- TADS	Mabey et al SRI (1982)
Chemical Synonyms	X	X	X	X	X	X	X	X
Molecular Weight	X		X	X	X	X		X
Solubility in Water	X		X	X	X	X	X	X
Vapor Pressure	X		X	X	X		X	X
Boiling Point	X		X	X		X	X	X
Melting Point	X		X	X		X	X	X
Liquid Specific Gravity	X		X	X	X	X	X	
Vapor Specific Gravity	X		X	X	X	X	X	
Saturated Vapor Concentration			X					
Observable Characteris- tics	X	X	X	X		X	X	
Odor Threshold	X		X	X		X	X	
Sampling and Analysis Methods		X				X	X	
Chemical Reactivity	X	X	X		X	X	X	X
Reactions in Water	X	X	X		X		X	X
Reactions in Air			X		X			
Biodegradation Rate Constant			X		X			X
BOD	X		X		X		X	
Hydrolysis Rate Constant				X	X			
Photolysis Rate Constant					X			X
Bioconcentration Factor			X		X			X
Kow					X			X
Kd					X			X
Koc					X			X
Number of Chemicals	900	329	1,000	13,000	250	10,000	850	114

Source: after Battelle PNL, 1982a

Other contacts can be found within local universities, technical assistance teams (TAT) and regional response teams (RRT). Directories of possible contacts are also available through trade organizations and professional societies.

2.4.2   Sources of Site Characterization Information

Site characterization data by its very nature will be much more site and region specific than compound characteristics. Consequently, pre-emergency collection of relevant meteorologic, soils, geologic, and topographic data is especially important. Also, prior hydrologic and hydrogeologic studies of the region may provide a wealth of information. However, regional data must be examined to insure it is representative of site-specific conditions at the emergency response site.

In an emergency response situation, data on site characteristics should be sought from six major sources:

1) Prior Studies

2) Textbooks

3) Well Owners

4) Records

5) Experts

6) Onsite Observations

Textbooks, regional studies, and lists of consultants should be in the hands of the emergency response team before they reach the spill site. It will probably be necessary to refer to many of these data sources at each site, since the required information is seldom found in a single source.

1. Prior Studies

Federal, state, and local government agencies may have performed detailed soils, geologic, water supply, or water quality studies in the area of the site. These prior studies are valuable sources of data on site characteristics. An emergency response team should contact the U.S. Geological Survey, the state geological survey, the local health department and water district, and the local engineering department as a start in the search for prior technical reports. It is expected that many of the site properties might be available in detailed prior investigations. Appendix A of the EPA manual for ground-water/subsurface investigations at hazardous waste sites (EPA, 1981) summarizes an extensive list of contacts and information sources.

2. Textbooks

For some of the geologic and soils properties required in a rapid assessment, tables in geology or ground-water textbooks provide a readily available data source. Ranges of hydraulic conductivity, bulk density, and porosity should be correlated with types of materials in most texts.

3. Well Owners

Owners of nearby wells may be able to provide information on the aquifer thickness (based on perforated interval of well log), the depth to ground water, the hydraulic gradient in the area, and the nature of the water-bearing strata. Well locations and property boundaries should be sought in assessing the hazards of the spill, thus, conversations with well owners are recommended to search for possible data and sources, such as the drilling company or drillers familiar with the area.

4. Records

To determine the age of the site, records of waste disposal operations or property ownership should be consulted. Waste manifests, describing shipments to the site, may prove useful, but have only recently been required.

5. Experts

In describing the ground water and unsaturated zone of the site, local geologists, water resources engineers, county officials, and university professors will be of assistance. Without detailed prior studies, the estimation of many of the required parameters should be guided by as much expert advice as can be gathered. Local agencies can also aid in locating wells and property boundaries in the site area.

6. Onsite Observations

Wells, topography, property lines, and stream locations should be verified by field reconnaissance at the site.

The major factor which will determine the success and accuracy of the site characterization is the availability of soils/geologic data from previous investigations. Without existing knowledge of subsurface characteristics such as predominant composition and thickness of unsaturated and saturated layers, evaluation of many site parameters will be largely conjecture. It is not likely that field testing will be able to provide adequate geologic data within the time frame of an emergency response assessment. When subsurface material composition is known, many site characteristics including porosity, bulk density, hydraulic conductivity, dispersion, and chemical characteristics can be estimated with reasonable accuracy in some cases (see Section 4). Values for these media-related parameters can be combined with macrogeologic data from reports or regional experts to estimate contaminant transport rates. If available, additional localized structural and geologic data which identify nonhomogeneity of the subsurface materials can be used to adjust and/or interpret quantitative estimates of contaminant transport, which assume media homogeneity. Thus, the ultimate accuracy of any estimate of contaminant transport will be largely dependent on the amount of specific localized information available for the emergency response site.

3. Rapid Assessment Nomograph and Its Use

The quantitative, graphical procedures for contaminant fate and movement in both the unsaturated and saturated zones are presented in this section. A single nomograph was developed for predicting contaminant movement in both soil zones to provide a comprehensive, integrated methodology for use under emergency response conditions. Graphical procedures were selected so as not to require prior experience with computers or programmable calculators by emergency personnel. However, analogous techniques for both the unsaturated and saturated soil zones have been programmed on hand-held calculators providing greater flexibility for assessments (see Pettyjohn et al, (1982) for ground-water programs). With the rapid advances in personal computers and programmable calculators, as emergency response teams acquire the necessary capabilities the techniques described herein can be easily computerized for their use.

Section 3.1 describes in detail the development of the assessment nomograph and Section 3.2 describes its general use, while Section 3.3 describes procedures for linked unsaturated-saturated zone assessments. Finally, Section 3.4 discusses the assumptions and limitations of the technique so that the user can effectively assess the accuracy of predicted concentrations for the specific emergency response situation.

3.1 DEVELOPMENT OF THE ASSESSMENT NOMOGRAPH

This section describes the nomograph developed for assessment of potential ground-water contamination to predict contaminant movement based on input parameters for contaminant and site characteristics.

The background and basis for the methodology is presented, including a discussion of the convective-dispersive transport equation for porous media, the types of pollutant source inputs usually encountered in an emergency response situation (i.e. continuous and pulse inputs) and the corresponding analytical solutions for each input condition. The parameters required to perform an assessment are listed and discussed, followed by the description of the assessment nomograph and its usage. This nomograph is actually a graphical solution of the transport equation and is the heart of the rapid assessment methodology. The same nomograph is used for both zones assuming only vertical transport in the unsaturated zone and only horizontal (or longitudinal) transport in the aquifer (saturated zone). However, the input parameters are evaluated differently for each zone, as will be discussed in Section 4.

371

3.1.1 Contaminant Fate and Transport in Soils

Movement of contaminants in the soil (saturated or unsaturated) can be described by the following equation (Van Genuchten and Alves, 1982)

$$\frac{\partial c}{\partial t} = D^* \frac{\partial c^2}{\partial x^2} - V^* \frac{\partial c}{\partial x} - k^*C \tag{3.1}$$

where
$$C = \text{solution concentration (mg/l)}$$
$$D^* = D/R$$
$$V^* = V/R$$
$$k^* = k/R$$
$$R = 1 + \frac{B}{N} K_d = \text{retardation factor (dimensionless)}$$

$$D = \text{dispersion coefficient (cm}^2/\text{day)}$$

$$V = \text{average interstitial pore-water velocity (cm/day)}$$

$$k = \text{degradation rate coefficient (day}^{-1})$$

$$B = \text{bulk density (g/cm}^3)$$

$$N = \begin{cases} \theta, \text{ volumetric water content (dimensionless), for} \\ \quad \text{unsaturated zone} \\ n_e, \text{ effective porosity (dimensionless), for saturated} \\ \quad \text{zone} \end{cases}$$

$$K_d = \text{partition coefficient (ml/g)}$$

Equation 3.1 states that the change in contaminant concentration with time at any distance, (X) is equal to the algebraic sum of the dispersive transport (1st term to right of equal sign), the convective transport (2nd term) and the degradation or decay of the compound (3rd term). Van Genuchten and Alves (1982) note that various modified forms of this same basic equation have been used for a wide range of contaminant transport problems in soil science, chemical and environmental engineering, and water resources.

Equation 3.1 considers only one-dimensional transport of contaminants and is applicable under steady, uniform flow conditions, i.e. velocity, V, is constant with space and time. This equation considers dispersion, advection, equilibrium adsorption (linear isotherm), and degradation/decay (first-order kinetics). Analytical solutions to the transport equation have been developed for both continuous (step function) and pulsed inputs of contaminants as boundary conditions. A step function implies the input of a constant concentration contaminant for an <u>infinite</u> amount of time, while a pulse load is a constant concentration input for a <u>finite</u> amount of time. Clearly, the terms infinite and finite are relative to the time frame of the analysis.

When the pollutant source is applied as a step function (continuously) with the following boundary conditions:

$$C(x,o) = 0$$
$$C(o,t) = C_o$$

$$\frac{dc}{dx}(\infty, t) = 0$$

(3.2)

the analytical solution, as given by Cho (1971), Misra (1974), van Genuchten (1982) and Rao (1982), can be expressed as:

$$\frac{C(x,t)}{C_o} = \frac{1}{2}\left[\exp(A_1)\ \text{erfc}(A_2) + \exp(B_1)\ \text{efrc}(B_2)\right] = P(x,t)$$

(3.3)

where

$$A_1 = \frac{x}{2D^*}(V^* - \sqrt{V^{*2}+4D^*k^*})\qquad B_1 = \frac{x}{2D^*}(V^* + \sqrt{V^{*2}+4D^*k^*})$$

(3.4)

$$A_2 = \frac{x-t\sqrt{V^{*2}+4D^*k^*}}{\sqrt{4D^*t}}\qquad B_2 = \frac{x+t\sqrt{V^{*2}+4D^*k^*}}{\sqrt{4D^*t}}$$

(NOTE: Exp(A_1) denotes the exponential of A_1, i.e., e^{A_1}, while erfc (A_2) represents the "complementary error function" of A_2, a function commonly used in applied mathematics. Erfc(A_2) produces values between 0.0 and 2.0 (Abramowitz and Stegun, 1972)).

The boundary conditions shown in Equation 3.2 indicate that (1) no contaminant is present in the soil prior to input from the source, (2) the input concentration at the surface is constant at C_o, and (3) a semi-infinite column is assumed with a zero concentration gradient at the bottom. This last boundary condition is often assumed to allow development of the analytical solution; van Genuchten and Alves (1982) indicate that this assumption has a relatively small influence on the accuracy of the solution in most circumstances when applied to well-defined finite systems.

Note that for large values of x and/or t, the second term within the brackets in Equation 3.3 can be neglected (i.e., erfc(B_2) approaches zero) and produces the following:

$$\frac{C(x,t)}{C_0} = \frac{1}{2} [\exp(A_1) \; erfc(A_2)] \tag{3.5}$$

The validity of Equation 3.5 depends on the values of the parameters and variables that define A_1 and A_2. Moreover, Equation 3.5 is comprised of two terms: $\exp(A_1)$ is time-independent and represents the eventual steady-state concentration at x, while $erfc(A_2)$ is time-dependent and corrects for the moving pollutant front (Rao, 1982). Thus, the steady-state condition, where C/C_0 is constant and $erfc(A_2) = 2$, simplifies Equation 3.5 to

$$\frac{C(x)}{C_0} = \exp(A_1) \tag{3.6}$$

Under the appropriate conditions stated above, these equations can greatly simplify calculations of contaminant concentrations.

When the pollutant source is applied as a pulse with a pulse duration, t_0, and boundary conditions as shown below:

$$\begin{aligned} C\,(x,o) &= 0 \\ C\,(o,t) &= \begin{cases} C_0, & o \leq t \leq t_0 \\ 0 & t > t_0 \end{cases} \\ C\,(\infty,t) &= 0 \end{aligned} \tag{3.7}$$

the analytical solution, as given by van Genuchten and Alves (1982), and Rao, (1982), can be expressed as:

$$\frac{C(x,t)}{C_0} = P(x,t) \qquad 0 \leq t \leq t_0$$

$$\tag{3.8}$$

$$\frac{C(x,t)}{C_0} = P(x,t) - P(x,t-t_0) \qquad t > t_0$$

where $P(x,t)$ is as defined in Equation 3.3.

Comparing Equations 3.8 and 3.3 shows that the analytical solution to the pulse boundary condition is the result of subtracting the solutions to two continuous inputs lagged by the pulse duration, t_0. This is further explained below.

3.1.2 Continuous and Pulse Contaminant Inputs and Associated Responses

The rapid assessment procedures discussed in this section are directed to two types of contaminant releases found in most emergency situations: continuous and pulse. As noted above, continuous release (or continuous input to the zone) implies the input of a constant source concentration of contaminant to the soil profile for an extended amount of time. This pollutant source could be an uncontrolled hazardous waste site, an abandoned dump site, a waste lagoon, a leaking chemical/waste container, etc. A pulse input is the application of a constant source concentration for a short time period relative to the time frame of the analysis. In this case, the pollutant source could be a surface spill or a short-term leak from a storage tank. The assessment methodology can be used to predict movement of contaminants in the subsurface resulting from either one of these release situations under emergency response conditions.

Movement of contaminants in the subsurface zones can be expressed by either profile responses or time responses resulting from continuous or pulse inputs. Profile responses are plots of pollutant concentration with distance, x, at various defined times, t. Time responses are plots of concentration changes with time, t, at certain specific locations x. For the unsaturated zone, the distance measure will be the vertical soil depth or depth to ground water; for the saturated zone, the down-gradient horizontal distance to a specific point (e.g., well, stream) will be of interest.

Figure 3.1 graphically illustrates time responses (i.e. C/Co vs t) at a chosen soil depth or distance (x=L) resulting from both continuous and pulse contaminant inputs from the source (x=O). Note that the figure is designed to show that the superposition of two continuous input functions and their associated responses (Figure 3.1a and 3.1b), produces a pulse input and its response (Figure 3.1c). In effect, the continuous input starting at time t_2 is subtracted from the continuous input starting at time t_1; the result is an input pulse of duration t_o (i.e. $t_2 - t_1$). Similarly, at the point x=L, superposition of the two continuous response functions results in the response function produced by the pulse input. This concept is the basis for the analytical solution for the pulse boundary condition given in Equation 3.8.

Figure 3.2 shows profile and time responses for both the continuous and pulse type releases expected in emergency situations. Specific assessments may involve evaluation of concentrations at many different x and t values.

When profiles are desirable, concentrations must be evaluated at specific times for different values of x; when time responses are needed, concentrations will be estimated for different values of t, for defined soil depths or down-gradient locations. As noted above, for most unsaturated zone assessments, users will be concerned with the concentration and time of arrival of contaminants at the ground-water table. Thus, time responses for an x value equal to the depth to ground water will be commonly calculated. For ground-water (saturated zone) assessments, the horizontal distance in the direction of ground-water flow to a potential impact point is often used.

3.1.3 Required Parameters

In order to predict contaminant movement in soils and ground water, parameters regarding transport and pollutant fate, and boundary or source conditions of an emergency situation must be evaluated. These parameters are listed in Table 3.1, along with their symbols and recommended units.

TABLE 3.1   REQUIRED PARAMETERS FOR RAPID ASSESSMENT PROCEDURES

Parameter/Boundary Condition	Symbol	Recommended Unit
Source Concentration	C_o	mg/l
Interstitial Pore Water Velocity	V	cm/day
Dispersion Coefficient	D	cm^2/day
Degradation/Decay Rate Parameter	k	day^{-1}
Retardation Factor (function of following characteristics)	$R= 1 + \dfrac{B\ K_d}{\theta}$	dimensionless
Partition (Adsorption) Coefficient	K_d	ml/g
Soil Bulk Density	B	g/cm^3
Volumetric Water Content*	θ	dimensionless
Pulse Duration (Pulse input only)	t_o	day

* - For saturated zone assessments, the volumetric water content is equal to the effective porosity, n_e.

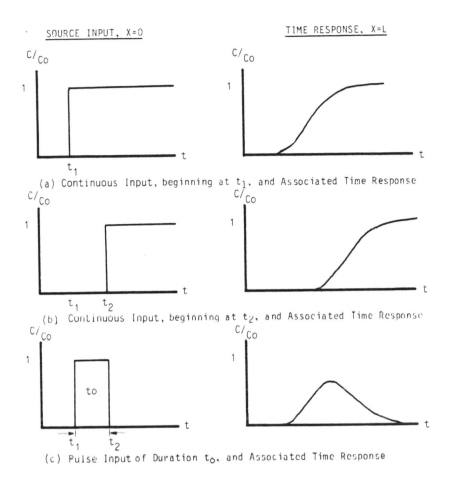

SOURCE INPUT, X=0   TIME RESPONSE, X=L

(a) Continuous Input, beginning at t_1, and Associated Time Response

(b) Continuous Input, beginning at t_2, and Associated Time Response

(c) Pulse Input of Duration t_o, and Associated Time Response

Figure 3.1   Continuous vs. Pulse Contaminant Inputs
and Associated Responses

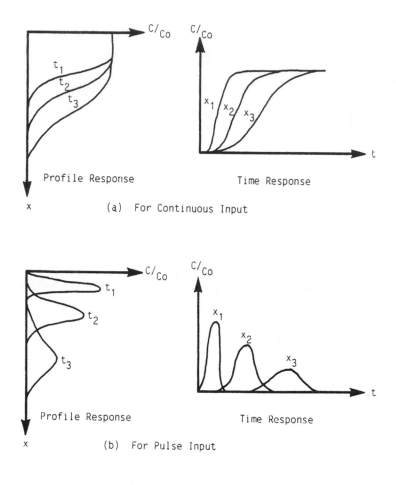

Profile Response   Time Response

(a) For Continuous Input

Profile Response   Time Response

(b) For Pulse Input

Figure 3.2   Contaminant Movement Expressed by Profile and
Time Response to Continuous and Pulse Inputs

Transport parameters include the interstitial pore water velocity (V) and dispersion coefficient, (D). Pollutant fate parameters include the degradation/decay coefficient (k) and retardation factor, (R). Retardation is primarily a function of the adsorption process which is characterized by a linear, equilibrium partition coefficient (K_d) representing the ratio of adsorbed and solution contaminant concentrations. This partition coefficient, along with soil bulk density (B) and volumetric water content (θ), are used to calculate the retardation factor. Retardation is important in contaminant transport in the unsaturated zone because it affects pollutant movement by modifying the convective, dispersive and degradation terms in the transport equation (Equation 3.1) as follows:

$$V^* = V/R$$
$$D^* = D/R \qquad\qquad (3.9)$$
$$k^* = k/R$$

Boundary conditions of a waste or spill situation are characterized by the contaminant concentration, Co, of the pollutant source. For a release situation characterized as a pulse input, the pulse duration, (t_o) must also be specified.

Section 4.2 includes further discussion of the parameters listed in Table 3.1 and provides guidelines for estimating their values.

3.2 THE NOMOGRAPH AND HOW TO USE IT

The assessment nomograph was developed to facilitate computation of the analytical solution to the transport equation for emergency situations which can be characterized as continuous (step function) input. However, through superposition (as discussed above) the same nomograph can be used for waste/spill conditions characterized as pulse input. The nomograph (Figure 3.3) predicts contaminant concentration as functions of both time and location in either the unsaturated or saturated zone. Separate computations, parameter estimates, and use of the nomograph is required for each zone. The prediction requires evaluation of four dimensionless input values - A_1, A_2, B_1, and B_2 - and subsequent evaluation of the result, C/Co, according to Equation 3.1 through use of the nomograph.

Direct computation of C/Co is quite cumbersome; in addition to parameter calculations, it involves evaluation of both the exponential and complementary error functions, and subsequent arithmetic operations. The nomograph facilitates these computations.

As shown in Figure 3.3; the nomograph consists of two groups of curves joined in the center by three vertical axes. Both curve groups have two axes, vertical and horizontal. The horizontal axis to the left is for entry of A_1 and to the right entry of B_1. Both axes are scaled to provide evaluation of their corresponding exponential functional values (exp [A_1] and exp [B_1], respectively). The vertical axis to the left is for entry of A_2 and to the right, entry of B_2. Both axes are scaled to provide

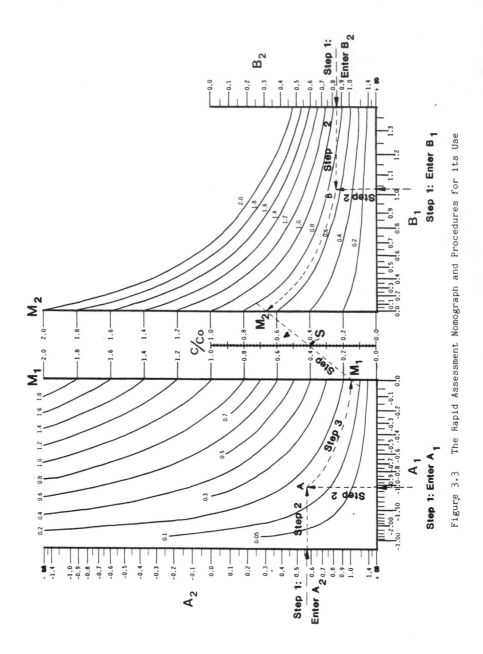

Figure 3.3   The Rapid Assessment Nomograph and Procedures for its Use

evaluation of their corresponding complementary error functional values (erfc [A_2] and erfc [B_2], respectively): the intersection of A_1 and A_2, and B_1 and B_2 (points A and B in Figure 3.3) represents the product of the axes, i.e. exp(A_1) times erfc(A_2). The two groups of curves represent points of equal multiplicands. The solution, C/Co, is located in the middle as represented by the center axis. The remaining two axes on both sides of the solution are multiplicands of the exponential and the erfc values. The curves represent points of equal multiplicands.

Step-by-step procedures are outlined below to demonstrate use of the nomograph.

Step 1: A_1, A_2, B_1, and B_2, must first be calculated. This can be done by inputting selected parameter values into Equation 3.4.

Step 2:   Once A_1, A_2, B_1, and B_2 are calculated, C/Co can be obtained from the nomograph (Figure 3.3). Start by entering values of A_1 and A_2 to the left group curves and B_1 and B_2 to the right. As represented by the dotted lines labeled "step 2," the entering lines join at points "A" and "B" respectively.

Step 3:   Then draw curves "AM_1" and "BM_2" by following the patterns in each respective curve group. As shown in Figure 3.3 these curves intersect the center axes at points "M_1" and "M_2".

Step 4:   The solution, C/Co, can finally be obtained by drawing a straight line connecting points "M_1" and "M_2". The solution is found at the point where line "M_1M_2" intersects the solution line, C/Co. In this example, the solution is located at point "S".

The precision of a nomograph is determined both by its size and the divisions of the axes. Large nomographs with fine divisions, in general, will allow greater precision. The full scale nomograph provided in this section (Figure 3.3) is precise enough for use with a continuous input or a long pulse input situation. However, higher precision is needed for conditions with a short contaminant pulse, especially for low C/Co values. For this reason, a nomograph with an enlarged scale (Figure 3.4) is provided to magnify the lower portion of the full nomograph, for C/Co values less than 0.4. For ease of use, enlarged foldout versions of both the full-scale and expanded-scale nomographs are provided in Appendix C.

To organize application procedures and provide a record of calculations and predicted concentrations, worksheets are provided to complement the nomograph for predicting contaminant concentrations for different values of x and t. Step-by-step procedures in applying these worksheets in emergency assessments are discussed below separately for the two contaminant input situations.

Worksheet Procedures for Continuous Input Assessment

Step 1:   Evaluate "required parameters" and enter values in Table 3.2.

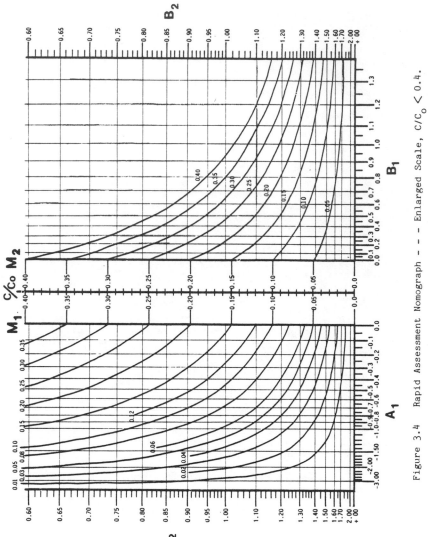

Figure 3.4   Rapid Assessment Nomograph - - - Enlarged Scale, $C/C_o < 0.4$.

Sheet _____ of _____
Calculated by _____ Date _____
Checked by _____ Date _____

Table 3.2

WORKSHEET FOR RAPID ASSESSMENT NOMOGRAPH

ZONE: **UNSATURATED** _____

SATURATED _____

Site: _____   Date of Incident: _____

Location: _____

On Site Coordinator: _____   Agency: _____

Scientific Support
Coordinator: _____   Agency: _____

Compound Name: _____

Compound Characteristics: _____

REQUIRED PARAMETERS: **DATA SOURCES / COMMENTS**

C_o = _____   _____

V = _____   _____

D = _____   _____

k = _____   _____

$R = 1 + \dfrac{B}{\theta} K_d$ = _____   _____

K_d = _____   _____

B = _____   _____

θ = _____   _____

PRELIMINARY CALCULATIONS:

1. $V^* = {}^V/_R$ = _____   3. $k^* = {}^k/_R$ = _____

2. $D^* = {}^D/_R$ = _____   4. $\sqrt{V^{*2} + 4D^*k^*}$ = _____

5	6	7	8	9				10	11	12
				See Footnote # 2				From Nomograph[3]		
x	t	${}^x/_{2D^*}$	$\sqrt{4D^*t}$	A_1	A_2	B_1	B_2	M_1	M_2	${}^C/_{Co}$	C

Table 3.2

Sheet _____ of _____
Calculated by _____ Date _____
Checked by _____ Date _____

NOMOGRAPH WORKSHEET (con't.)

ZONE: UNSATURATED _____
SATURATED _____

5	6	7	8	9				10		11	12
x	t	$x/2D*$	$\sqrt{4D*t}$	See Footnote # 2				From Nomograph[3]			c
				A_1	A_2	B_1	B_2	M_1	M_2	C/Co	

Footnotes: 1. Refer to Table 3.1 for definitions and units, and to Chapter 4 for estimation guidelines.

2. $A_1 = Col.7 \times (Item 1 - Item 4) = \frac{x}{2D*}(V* - \sqrt{V*^2 + 4D*k*})$

$A_2 = [Col.5 - Col.6 \times Item 4] / Col.8 = \frac{x - t\sqrt{V*^2 + 4D*k*}}{\sqrt{4D*t}}$

$B_1 = Col.7 \times (Item 1 + Item 4) = \frac{x}{2D*}(V* + \sqrt{V*^2 + 4D*k*})$

$B_2 = [Col.5 + (Col.6 \times Item 4)] / Col.8 = \frac{x + t\sqrt{V*^2 + 4D*k*}}{\sqrt{4D*t}}$

3. Figure 3.3 or Figure 3.4 (See Figure 3.3 for use of nomograph).

Step 2:   Perform preliminary calculations.

Step 3:   Enter values of x and t.
            o To obtain a profile response, enter different values of x
               for a selected time, t.

            o To obtain a time response, enter different values of t for a
               selected location, x.

Step 4:   Perform calculation and apply nomographs, to evaluate C/Co and
            C as instructed in the worksheet.

Step 5:   Go back to Step 3 for further evaluation, if necessary.

Worksheet Procedures for Pulse Input Assessment

As mentioned earlier in Section 3.1.2, the analytical solution for a pulse
contaminant input results from the superposition of solutions for two separate
continuous input functions lagged by the pulse duration. Since the assessment
requires substracting two continuous input response (i.e. C/Co) values, a
supplementary worksheet (Table 3.3) is provided.

Step-by-step procedures for the pulse input situation are provided below:

Step 1:   Evaluate "required parameters," enter pulse duration (t_o) and
            source concentration (Co) in Table 3.3 and other parameter
            values in Table 3.2.

Step 2:   Perform preliminary calculations in Table 3.2.

Step 3:   Enter values of x and t in Table 3.3, and in Table 3.2 for
            continuous input assessment.

            o To obtain a profile response, enter different values of x
               for a selected time, t, in both Tables

            o To obtain a time response, enter different values of t for a
               selected location, x, in both Tables

Step 4:   Perform continuous input assessment using work sheet Table 3.2
            and enter result C/Co, in column 11 of Table 3.2 and column 4
            of Table 3.3.

Step 5:   Evaluate ($t-t_o$) in Table 3.3. If $t > t_o$, go to Step 6.
            Otherwise, pulse concentration (Column 6) equals the continuous
            input concentration (Column 4). Go to step 8.

Step 6:   Evaluate C/Co at ($t - t_o$) using worksheet Table 3.2 and enter
            result in column 5 of Table 3.3.

Step 7:   Subtract column 5 from column 4 and enter result in column 6.

Table 3.3   Sheet _____ of _____

SUPPLEMENTARY WORKSHEET FOR PULSE INPUT ASSESSMENT

ZONE: **UNSATURATED** _____

to = _____ ,   Co = _____   **SATURATED** _____

			CONTINUOUS INPUT ASSESSMENT (From Worksheet)		PULSE ASSESSMENT	
					Col.4, t≤to Col.4-5, t > to	Co x Col. 6
1	2	3	4	5	6	7
x	t	t - to	$^C/_{Co}(t)$	$^C/_{Co}(t-to)$	$^C/_{Co}(t)$	C

Step 8:   Multiply column 6 by Co and enter result in column 7 of Table 3.3.

Step 9:   Go back to Step 3 for further evaluation if necessary.

Detailed examples demonstrating the use of the nomograph and the worksheets for both continuous and pulse inputs are provided in Section 5. The user is encouraged to work through these examples and procedures to become familiar with them prior to an emergency response situation.

3.3 LINKAGE OF UNSATURATED AND SATURATED ZONE ASSESSMENTS

Since the assessment nomograph can be applied to both the unsaturated and saturated zones individually, linkage procedures are required for situations where an assessment of contaminant movement through both saturated and unsaturated media is needed. The linkage procedures require the following two steps:

1.   Approximation of the time-varying concentrations leaving the unsaturated zone by either a continuous step function or pulse input.

2.   Estimation of C_0 (i.e., source concentration) for the saturated zone assessment based on Step 1 (above), recharge from the waste site, and ground-water flow.

Figure 3.5 shows typical time responses for concentrations reaching ground water as estimated by an unsaturated zone assessment for both continuous and step function inputs; the dashed lines show the approximations needed to convert the time responses into continuous or pulse inputs for applying the nomograph to the saturated zone. The approximations in Figure 3.5 are designed so that the area under the dashed line is approximately equal to the area under the associated time response curve. This ensures that the contaminant mass entering ground water is the same for both the time response and its approximation.

Since the arrival time of a contaminant at a particular point in the aquifer is often the primary reason for a saturated zone assessment, users should evaluate the sensitivity of these arrival times to the starting time of the input approximation. For example, in Figure 3.5 the starting dates for the step function and pulse input approximation are day 15 and day 10, respectively; varying these starting dates by 2 to 3 days would help to evaluate the impact of the approximation on the contaminant arrival time at the point of concern.

The second step in the linkage procedure is to determine the value of C_0, the source concentration, to use in the saturated zone assessment. Unless the waste/spill site is adjacent to a well and/or the ground-water table itself is the impact point of concern, dilution and mixing in the aquifer must be considered in estimating C_0 for the saturated zone assessment. The following equation should be used to estimate C_0 for the saturated zone:

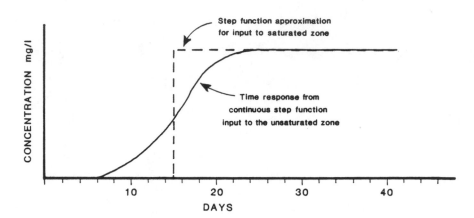

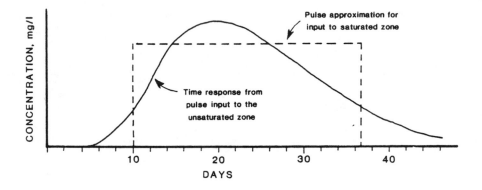

Figure 3.5   Time Responses From The Unsaturated Zone and Approximations
For Input To The Saturated Zone

$$C_o = \frac{C_u qL}{V_d m} \qquad\qquad (3.10)$$

where C_o = source concentration for saturated zone, mg/l

C_u = maximum step function or pulse concentration
from the unsaturated zone, mg/l

q = recharge rate from the site, cm/yr

L = width of leachate plume at the water table, m

V_d = ground-water (Darcy) velocity, cm/yr

m = effective aquifer thickness or zone of mixing, m

Figure 3.6 schematically illustrates the linkage and underlying assumptions in Equation 3.10, which considers dilution of the contaminant load by recharge from the site and ground-water flow. The dilution terms (i.e., qL and V_dm) in the equation are written as a velocity times a distance since the representation is a vertical plane with a unit width, which drops out of the calculation.

Users should note that the q and V_d terms in the equation are bulk or volumetric velocities, i.e., these are not pore-water velocities. Guidelines for estimating q, V_d and m are included in Section 4, L is determined from the dimensions of the waste/spill site, and C_u results from the approximation shown in Figure 3.5. With this information and nomograph parameter estimates for the saturated zone, the user can apply the nomograph to estimate contaminant concentrations in the aquifer.

3.4 ASSUMPTIONS, LIMITATIONS, AND PARAMETER SENSITIVITY

To effectively and intelligently use the rapid assessment procedures described in this manual, the user must understand and appreciate the impact of assumptions and limitations on which the procedures are based, and the relative sensitivity of the required parameters. These two aspects are interrelated; performing sensitivity analyses on certain parameters will allow the user to assess the impact of specific assumptions. Sensitivity analyses were noted in Section 2.2 as a key element in applying the assessment methodology.

3.4.1 Methodology Assumptions

The assumptions on which the assessment nomograph is based are as follows:

1.   All soil and aquifer properties are homogeneous and isotropic throughout each zone.

2.   Steady, uniform flow occurs only in the vertical direction throughout the unsaturated zone, and only in the horizontal

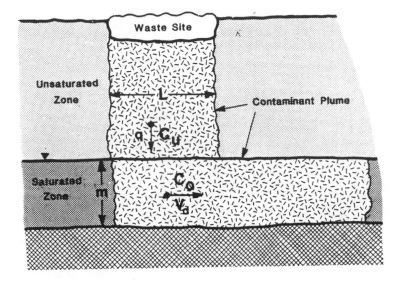

Figure 3.6   Schematic Linkage of Unsaturated and Saturated
             Zone Assessments

(longitudinal) plane in the saturated zone in the direction of ground-water velocity.

3.   Contaminant movement is considered only in the vertical direction for the unsaturated zone and horizontal (longitudinal) direction for the saturated zone.

4.   All contaminants are water soluble and exist in concentrations that do not significantly affect water movement.

5.   No contaminant exists in the soil profile or aquifer prior to release from the source.

6.   The contaminant source is applied at a constant concentration continuously; a pulse input can be handled by superposition (Section 3.1).

7.   There is no dilution of the plume by recharge outside the source area.

8.   The leachate is evenly distributed over the vertical dimension of the saturated zone.

The assumption of homogeneous and isotropic conditions is equally critical in both zones. In many cases, extensive heterogeneities will exist for both soil and aquifer properties, but the emergency response time frame precludes adequate consideration of variations even if they are known to exist. Adjustment of certain parameters may be possible to estimate an "effective" parameter value that partially accounts for property variations. However, conditions involving soil cracks, fractured media, impermeable layers, clay lenses, etc. will require the user to make a qualitative assessment on their potential impact on predicted concentrations.

The assumption of steady, uniform flow is much more critical in the unsaturated zone than in the saturated zone. Pore water velocities are significantly more dynamic and variable in the unsaturated zone since they depend on the percolation flux from rainfall and variable soil moisture conditions, which in turn affect other soil properties. Under-estimation of travel times in the unsaturated zone can occur if mean annual percolation rates are used to estimate movement of the contaminant front during shorter time periods (e.g., months). Ground-water flow velocities are also difficult to estimate, but they are less dynamic than in the unsaturated zone. Consequently, great care is needed in estimating velocities in both zones.

The assumption that contaminants are water-soluble and exist in concentrations that do not impact water movement is relevant to both zones but may be more critical for the unsaturated zone. Surface and unsaturated soils will likely experience higher concentrations than ground water due to accidental spills or releases. Also, the majority of contaminants that reach ground water after traveling through a reasonable depth of unsaturated soil will likely be water soluble. Although water solubility is assumed since the contaminant is moving with the water, the same basic form of the transport equation has also been used to assess movement of the soluble portion of oil spills in ground water (Duffy et al, 1980).

The accommodation of contaminant pulse inputs was discussed in Section 3.1 using the principle of superposition. The same principle and procedures are used for both zones to assess plume migration from a pulse input. However, for short pulses and low concentrations the precision of numbers read from the nomograph may be the primary limitation (see discussion below).

3.4.2 Limitations and Parameter Sensitivity

In addition to the assumptions noted above the major limitations of the procedures described herein include the precision with which numbers can be read from the nomograph (an inherent limitation of graphical procedures) and the reliability and accuracy of parameter estimates. As shown on the nomograph, the C/Co values can be read from Figure 3.3 (full-scale) to two decimal digits (i.e. 0.01) and from Figure 3.4 (expanded scales, C/Co < 0.4) to three decimal digits (i.e. 0.001). If greater precision is required, direct calculation of the solution by Equation 3.3 may be needed.

The greatest limitation on predictions will be the accuracy and reliability of the data for estimating parameters. In most emergency situations, specific compound and site data will be difficult to obtain; however, all efforts should be made to acquire the most reliable and site-specific data as possible through the sources and guidelines provided in Section 4. Even with relevant data for parameter estimation, users should perform sensitivity analyses as recommended in Section 2.2 in order to assess the impact of possible parameter variations and methodology assumptions on predicted concentrations.

Pettyjohn et al (1982) have performed sensitivity analyses on the major parameters for an analogous nomograph for the saturated zone only; the user is referred to that source for complete details. Depending on the specific data available for individual parameters, the user should consider assessing the sensitivity of the following parameters which are generally the most sensitive:

    Degradation/decay rate
    Retardation factor
    Pore Water Velocity
    Source Contaminant Concentration
    Effective aquifer thickness (saturated zone only)
    Dispersion coefficient

Degradation/decay and retardation are interrelated since retarding the movement of the contaminant will allow greater time for degradation to occur. Velocity is a sensitive parameter for both zones. Since it is highly variable and can range over orders of magnitude, assessments of its sensitivity in site-specific situations is highly recommended.

For the saturated zone, the effective aquifer thickness or zone of mixing represents the degree to which the contaminant is uniformly mixed in the vertical direction. For very shallow aquifers, using the entire thickness may be appropriate. For deep aquifers, mixing zones considerably less than the total may be required. Consequently, the effect of varying mixing depths should be assessed by the user.

Dispersion in ground water can be significant especially at low ground-water velocities. Since the coefficient can vary over a wide range, accurate estimates of expected subsurface conditions can be extremely difficult. Sensitivity should be analyzed.

Source contaminant concentrations may be the most difficult of all data to obtain and/or characterize, especially for landfill, lagoon, or other waste site situations. If a range of possible or probable values can be estimated, the user should definitely evaluate the concentration predictions that would result from the full potential range of source values.

4. Parameter Estimation Guidelines

The most important part of the rapid assessment methodology is estimation of reasonable and valid parameter values for a specific emergency response situation. Section 2.3 described and discussed the critical compound and site characteristics that determine the potential for ground-water contamination at a particular hazardous waste or spill site. This section provides specific guidelines for estimation of the parameter values needed for use of the rapid assessment nomograph described in Section 3. The format of this section is as follows:

> Section 4.1 General Parameter Estimation
>
> Section 4.2 Unsaturated Zone Parameter Estimation
>
> Section 4.3 Saturated Zone Parameter Estimation

For each parameter, guidelines are provided, to the extent possible, for calculating the parameter value and estimating the relevant compound and site characteristics on which it depends. Thus, discussions of characteristics are grouped according to the affected parameters. For example, since the retardation factor for organic compounds depends upon organic carbon content, organic carbon partition coefficient, bulk density, and porosity (saturated zone), these characteristics are discussed under the section on estimating the retardation factor. For parameters needed in <u>both</u> the unsaturated and saturated zone assessments, the primary discussion is Section 4.2 (unsaturated) with any adjustments required for the saturated zone in Section 4.3.

Some repetition of information in Section 2.3 (characteristics) and 2.4 (data sources) is included in this section to preclude the need to continuously turn back to the earlier sections and to clarify the presentation. Once the user is familiar with the content of this manual, this section will likely receive the most usage on a continuing basis especially during an emergency response.

The user will note that the following statement is repeated numerous times in this section:

> <u>Local site-specific information should be used whenever possible; significant errors can result from using general or regional data.</u>

394

This emphasizes the need to search for and use local site-specific information for the parameter being discussed. The statement is repeated to insure that the user is aware of possible errors that can result from the use of general or regional data whenever the parameter must be estimated.

4.1 GENERAL PARAMETER ESTIMATION

This section discusses general characteristics important to assessment procedures in both zones, including identity and concentration of contaminants, nature of the soils and geologic strata, age of the waste/spill site, and depth to ground water. The contaminant concentration is the only characteristic discussed that results in a specific parameter value used in the nomographs. However, the other characteristics are important in applying the assessment procedures, evaluating assumptions, and determining compound/site characteristics.

4.1.1   Identity of Contaminants

Obviously the identity of the contaminants present at the waste/spill site is necessary to evaluate the relevant physical/chemical properties needed for predicting fate and migration. In many cases the identity will have been established by emergency personnel (e.g. at a spill site) or prior analyses (e.g. drinking water problems) in order to determine the need for an emergency ground-water assessment. Identification can be established quickly, on the order of several hours, through the use of records and observable characteristics. Chemical analyses can be used, if necessary, but they require considerably more time, and may need to be limited to qualitative field methods in order to give results within the emergency response 24-hour time frame.

Records provide the most rapid, positive identification of contaminants involved in a hazardous waste accident and should therefore be the focus of the initial efforts at contaminant identification. Shipping papers and transportation labels are now required when transporting hazardous materials. In addition, the Association of American Railroads is developing a computerized tracking system for rapid identification of railcars containing hazardous materials (Guinan, 1980).

The use of records to identify chemicals present at uncontrolled waste sites is much more difficult. Waste manifests, listing each waste shipment received at the facility, are a possible source of data, but these manifests have only recently been required in many cases. Waste site owners and/or companies who have disposed materials at the site may be able to provide some information on the types of contaminants present.

If records are unavailable or incomplete, observable characteristics such as odor, color, density, and reaction should be investigated as clues to the identity of the waste. The following handbooks and data bases (described in section 2.4) provide information to aid in waste identification based on observable characteristics:

1)   Field Detection and Damage Assessment Manual for Oil and
     Hazardous Materials Spills, U.S. EPA, Washington, DC, 1972.

2)   U.S. Coast Guard Chemical Hazard Response Information System,
     Manual CG-446-1, A Condensed Guide to Chemical Hazards, and
     Manual CG-446-2, Hazardous Chemical Data, Washington, DC, 1974.

3)   OHM-TADS Data System, U.S. EPA.

In addition to printed and computerized information on observable
characteristics, experts within the chemical industry (Chemical Transportation Emergency Center (CHEMTREC) at (800) 424-9300 or 483-7616), at local
universities, and at regional response teams (RRT) can be contacted for
assistance.

Field analytical methods will be difficult to apply within an emergency
response time frame and should therefore be called upon for compound
identification only after first considering records and observable
characteristics.   The Field Detection and Damage Assessment Manual for
Hazardous Spills (EPA 1972), the EPA Field Guide for Scientific Support
(Battelle PNL 1982a) and the EPA's OHM-TADS system describe the use of
several analytical methods for identifying hazardous chemicals. A variety
of chemical products are available for in-field analysis.   The application
of these analytical methods will require the presence of a skilled
technican, experienced in the operation of these instruments.

4.1.2 Contaminant Concentration

The source concentration of the specific contaminant(s) to be analyzed is a
required input parameter for both the unsaturated and saturated zone
assessments.   For the unsaturated zone, the user must specify the
concentration of the contaminant available to the soil after deducting
potential losses due to volatilization, decay processes, clean-up/removal
operations, retention by liners and/or non-leaking drums, etc.   In many
emergency response situations, the initial contaminant concentration may be
the most difficult of all parameters to estimate. A variety of sources of
information should be consulted to uncover data specific to the waste site
or spill under investigation.

Records and industry experts should be the primary sources contacted
initially to uncover concentration data.   Although waste disposal site
records (if available) and transport manifests do not often contain
concentration data, they may identify the general category of the
waste/contaminant, the industry or companies that generated the compound,
and possible contacts for further information. Also, the procedures and
sources used to identify the contaminant (e.g. CHEMTREC, AAR) may also be
useful in estimating concentrations. Industry contacts and experts may be
able to provide estimates of concentrations at which the chemical is
normally transported (i.e. for spills) or resulting from a particular
industry or industrial process (i.e. for waste sites).

Chemical analyses by mobile laboratories or other emergency or field procedures can provide the needed information, and will usually be ordered by the on-scene coordinator once it is determined that a toxic or hazardous compound is involved. However, the results of chemical analyses may not be available within the emergency response time frame, especially if subsurface sampling is required. The above sources should be contacted concurrently while samples are being taken and analyzed in order to expedite obtaining concentration information and performing the assessment procedures.

Lacking any information on the contaminant concentration, we recommend that the user assume the source concentration equal to the water solubility of the contaminant. In most situations this is an appropriate assumption for an initial assessment since movement of the contaminant through the unsaturated zone will occur primarily by the infiltrating water carrying the water soluble portion of the compound. Although retardation and decay processes will subsequently reduce unsaturated zone concentrations, the water solubility is a reasonable estimate of the source contaminant concentration. This assumption has been used by Falco et al (1980) in a screening procedure for assessing potential transport of major solid waste constituents in releases from landfills and lagoons.

Water solubility data for specific compounds and hazardous waste constituents is available in the following data sources:

1)   CHRIS Manual CG 446-1,2 U.S. Coast Guard, 1974

2)   OHM-TADS, U.S. EPA Data Base

3)   Physical Chemical Properties of Hazardous Waste Constituents, U.S. EPA, 1980

4)   The Merck Index, Merck and Company, Inc. (Windholz, 1976)

5)   Handbook of Environmental Data on Organic Chemicals (Verschueren, 1977)

6)   Aquatic Fate Process Data for Organic Priority Pollutants, (Mabey et al 1982)

In addition, Lyman et al (1982) describe a variety of methods of estimating solubility in water and other solvents from data on melting point, structure, octanol-water partition coefficients, activity coefficients, and other compound characteristics.

Alternatively, for compounds that are considered to be a small fraction of the total waste volume at a site, Falco et al (1980) assumed the concentration in the leachate to be the equilibrium concentration resulting from partitioning between the solid and dissolved phases of the waste compound. Thus the source solution concentration could be estimated as follows:

$$C_s = \frac{F}{K_{om}} \tag{4.1}$$

where C_s = source solution concentration, mg/l

K_{om} = partition coefficient between organic matter and solution, 1/mg

F = fraction of solid waste that is the contaminant of interest

F must be evaluated from records and other information available for the specific waste site, and K_{om} is discussed in Section 4.2.

The user should be aware that for water insoluble or slightly soluble compounds from waste sites or spills, the assumption of using water solubility values could lead to significant errors. This will be especially important for large volume spills of such chemicals where gravity and the mass of the spill are the driving forces for moving the contaminant through the unsaturated zone. (See Section 4.2 for further discussion). The above assumptions and methods of estimating the source concentration should be used only as a last resort when no other data or information is available.

Chemical Loss Mechanisms

In addition to leaching to ground water and chemical decay processes in the soil, chemical losses from the spill site may occur via photochemical decay and volatilization. These processes will help to reduce the contaminant concentration available to move through the soil, and should be considered when estimating this concentration value.

Photolysis rates depend on numerous chemical and environmental factors including the light absorption properties of the chemical, the light transmission characteristics of the chemical (if pure) or its environment (water, soil, etc.), and the available solar radiation of appropriate wave length and intensity. Estimation of the chemical's general photolytic reactivity and the light transmission properties of its environment or solvent will usually be very difficult. Also, most existing models and data (e.g. Smith et al, 1977; Callahan et al, 1979) for predicting photolytic decay in the environment are applicable to atmospheric and aquatic systems. Consequently, the quantitative estimation of attenuation of a chemical concentration by photolysis at a spill site during a 24-hour emergency response period would be impossible. The best we can do is to assess the probability of photolysis being an important loss mechanism and then adjust the assessment results accordingly when photolysis is ignored. The following steps are recommended:

1) Determine whether the chemical is exposed to direct solar radiation. If most of the chemical has percolated into the soil,

photolysis can be neglected; if the chemical is directly exposed, go to step 2.

2) Determine whether the chemical is susceptible to photolysis by consulting with a) industry officials familiar with the specific chemical or b) environmental photolysis reference books and literature (e.g. Callahan et al, 1979; U.V. Atlas of Organic Compounds 1966-1971). This step may be subject to considerable error, however, since the photochemical reactivity of a chemical is determined by its physical state (dissolved, solid, liquid, adsorbed) and environment (solvent, etc.) as well as its molecular structure.

3) If both 1 and 2 above are positive, tne user may conclude that photolysis is a possible or significant depletion mechanism for the chemical. However, further analysis to quantitatively estimate this depletion would require laboratory studies not possible within an emergency response time frame.

Volatilization may provide a significant attenuation mechanism for chemical spills on land. The rate of loss of chemicals from soil or surface pools due to volatilization is affected by many factors, such as the nature of the spill, soil properties, chemical properties, and environmental conditions. The mechanisms for chemical loss from the land are direct evaporation from a pool or saturated soil surface, vapor and liquid phase diffusion from chemicals incorporated into dry soil, and advection with vapor and liquid water due to capillary action (i.e. the wick effect). Thus, a comprehensive model of the volatilization process would be extremely complex; however, a number of relatively simple methods exist to estimate these losses, and three of them are presented here. (Thibodeaux, 1979; Hamaker, 1972; Swann et at, 1979). The reader is referred to the original literature or the text Handbook of Chemical Property Estimation Methods (Lyman et al, 1982) where a number of models are described along with conditions of use and parameter estimation methods.

Volatilization - Method 1

This method (Thibodeaux, 1979) is primarily applicable for a liquid pool of pure chemical. However, it can be used for a mixture of chemicals to estimate the reduction in the source concentration of one specific chemical due to the volatilization flux, assuming a constant volume mixture. It requires estimation of the area of the pool, wind speed at the spill site, pool temperature, and the Schmidt Number (Sc) for the chemical vapor.

The flux of chemical is given by:

$$N = 0.468 \ U^{.78} \ L^{-.11} \ Sc^{-.67} \ P_{vp} \ M/T \qquad (4.2)$$

where N = flux of chemical from pool, $\mu g/m^2/hr$

U = wind speed at 10m height, m/hr

$$L \;\; = \;\; \text{length of pool, m}$$

$$Sc \;\; = \;\; \text{Schmidt Number for chemical vapor (see below)}$$

$$P_{vp} \;\; = \;\; \text{chemical vapor pressure, mmHg}$$

$$T \;\; = \;\; \text{pool temperature, } ^{\circ}K$$

$$M \;\; = \;\; \text{chemical molecular weight, g/mole}$$

The Schmidt number for a gas is defined by:

$$Sc \; = \; v/D \tag{4.3}$$

where v =   kinematic viscosity, cm^2/s

         D =   gas diffusion coefficient, cm/s

Schmidt numbers for many chemicals are tabulated by Thibodeaux (1979) and may be estimated for similar chemicals by the following equation:

$$\frac{S_{c1}}{S_{c2}} \; = \; \left(\frac{M_2}{M_1}\right)^{\frac{1}{2}} \tag{4.4}$$

     where   M   =   molecular weight

Method 2

This method is applicable to situations in which the chemical has been applied to or spilled on the soil surface. Researchers at Dow Chemical Company (Swann et at, 1979) correlated volatilization rate with a number of chemical properties. The first-order rate constant for volatilization of chemicals spilled or applied to the soil was found to be approximated by the following correlation equation:

$$k_v \; = \; 4.4 \times 10^7 \, \frac{Pvp}{K_{oc}S} \tag{4.5}$$

where k_v =   volatilization rate constant. day^{-1}

        K_{oc} =   soil adsorption coefficient based on organic carbon

                  content, ml/g

P_{vp} = vapor pressure of chemical, mmHg

S   = water solubility of chemical, $\mu g/ml$

Since k_v is a first-order rate constant, the concentration loss function due to volatilization is represented as:

$$C = C_o \exp(-k_v t) \tag{4.6}$$

where Co = initial concentration of the chemical, $\mu g/l$

C   = concentration of the chemical after time t, $\mu g/l$

t   = time, day

Method 3

This method (Hamaker, 1972) allows estimation of volatilization rates from chemicals distributed in a soil column such as after initial infiltration of a spill. It assumes a semi-infinite impregnated soil layer and no upward water flux. The loss of chemical is given by

$$Q_t = 2Co \, (Dt/\pi)^{\frac{1}{2}}$$

where

Q_t = total loss of chemical per unit area over time t, $\mu g/cm^2$

Co   = initial concentration of chemical in the soil, $\mu g/cm^3$

D   = diffusion coefficient of chemical vapor in the soil-air, cm^2/sec

t   = time, sec

π = 3.14159...

For the situation where chemical is incorporated in moist soil, the upward flux of water due to evaporation and capillary action will greatly enhance the movement of chemical to the surface and its subsequent volatilization. Estimation of this flux requires use of a more complex model (e.g. Hamaker, 1972) which necessitates the determination of water fluxes in the soil. The user should recognize that Method 3 will significantly under-estimate volatilization under moist soil conditions.

Generally, the preceding methods require knowledge of vapor pressure, solubility and diffusion coefficients, all of which are available from sources previously identified (See Table 2.2). Additional sources of data for these methods, including the Schmidt number (Method 1), can be found in the following:

1. Chemical Engineers Handbook (Perry and Chilton, eds., 1973)

2. Chemodynamics, (Thibodeaux, 1979)

3. Gaseous Diffusion Coefficients (Marrero and Mason, 1972)

4. Handbook of Chemical Property Estimation Methods (Lyman et al, 1982)

4.1.3 Nature of Soils and Geological Strata

A reliable study of contaminant movement through the unsaturated and/or saturated zones requires a careful assessment of the types of soils and/or geological formations present. The methodologies incorporated in this manual accept only homogeneous descriptions of the transport media being modeled, but users can choose parameter values that can partially account for any heterogeneities that are known to be present. For this reason, a thorough knowledge of the soils and geology at the site is important in the sound application of the relatively simple methods in this manual.

Data on the types of soils, presence of cracks or sinkholes, and occurrences of lenses of heterogeneous materials in the unsaturated zone can be found in soil surveys (performed by the U.S. Soil Conservation Service), well drillers' logs (usually kept by well owners or local departments of health or water), and construction design reports (on file with local engineering department or building inspector). If a soils expert is on the spill site, a quick evaluation of the general character of the surface material may be possible. The first aim of the soils assessment is the establishment of the predominant nature of the unsaturated zone so that the bulk density, porosity, organic content, and volumetric water content can be estimated. The presence of heterogeneities (cracks, clay lenses, sinkholes, etc.) can be used as a basis for adjusting the parameter values chosen under the assumption of homogeneity, or for interpreting the final model results.

An evaluation of the nature of the ground-water formations present at the site should include searches for prior hydrogeological investigations (by the U.S. Geological Survey, State Geological Survey and Department of Water Resources, and local and regional health and water agencies). A second major source of geological data lies with experts in universities, consulting firms, and government agencies. Drillers' well logs represent a third significant record of the composition of the saturated zone. Among the data being sought are the type of aquifers present (confined or water table), the predominant composition of the strata, the presence of fractures, and the existence of clay lenses. The assessment nomograph incorporated in this methodology is designed to simulate a single water table (unconfined) aquifer, but users should be aware of the existence of other types of aquifers and/or multiple water-bearing formations to assess the reliability of the predicted results and to perform qualitative assessments beyond the focus of the nomograph. Fractures can greatly increase the spread of contamination, while clay lenses retard this movement. Knowledge of their presence will govern the choice of parameter values and the interpretation of the predictions.

4.1.4 Age of the Waste Site or Spill

The age of the waste site or spill is essential in estimating the time duration of leaching of the contaminant into the unsaturated and saturated zones. For many surface spills, the investigation will occur immediately after the accident and the age of the site is therefore known. The analysis of newly-discovered uncontrolled disposal sites with the methodologies contained in this manual will require knowledge of the age of the site. Predictions of contaminant transport can then be related to real time and the extent of the plume at the time of the analysis can be estimated. To establish the age of an uncontrolled waste site, records of waste shipments should first be consulted. Any information found in the site records can be supplemented by tracing ownership of the property to determine the length of time the area was used as a landfill.

4.1.5 Depth to Ground Water

In evaluating transport in the unsaturated zone, the depth to ground water must be estimated in order to assess the likelihood that contaminants will reach the ground water. Since ground-water levels are often within 10 to 20 meters of the land surface, and can be 3 meters or less, the potential for ground-water contamination from waste sites and chemical spills is a significant problem. Seasonal fluctuations, if significant, should also be considered since these fluctuations can range from 1 to 5 meters or more in many parts of the country. Also, the effects of pumping and recharge areas should be evaluated.

Local site-specific information should be used whenever possible; significant errors can result from using general or regional data.

Prior hydrogeologic and water supply studies in the general region of the site are valuable sources of data on site characteristics, including depth to ground water. Contact should be made with the U.S. Geological Survey, the State Geological Survey, the State Department of Water Resources, and the local and county water, health, and engineering departments as a start in the search for existing technical reports and information.

The depth to ground water can be determined by talking to the owners of nearby wells or by making depth measurements at these wells as long as the wells are not being actively pumped, and therefore accurately represent the water-table level. Also, water-surface elevations in nearby perennial streams, lakes, marshes, and other waterbodies (e.g., mines, gravel pits, flooded excavations) can be used to estimate the depth to ground water since these are areas where the ground-water surface intersects the land surface.

If prior studies, observations, or information from nearby wells are not available or do not provide the required data on depth to ground water, then local experts in hydrogeology (at universities, consulting firms, and governments agencies) should be contacted for guidance.

4.2   UNSATURATED ZONE PARAMETER ESTIMATION

Table 4.1 lists the nomograph parameters for an unsaturated zone assessment and the various types of information needed and/or useful for their evaluation. Except for contaminant concentration, which was discussed in Section 4.2, estimation guidelines for each parameter are provided below in the order shown in Table 4.1.

4.2.1 Pore Water Velocity

Estimation of pore water velocity is a necessary and important element in analyzing transport of contaminants through the unsaturated soil zone. In essence, the water (or other fluid) moving through the pore spaces in the soil is the driving mechanism for contaminant movement through the soil and to ground water. Although the term conventionally implies water movement, pore velocity could also refer to the movement of other solvents or fluids as might occur in a large volume chemical spill infiltrating through the soil.

Pore water velocity is a function of the volumetric flux per unit surface area and the volumetric water content, as follows:

$$V = \frac{q}{\theta} \qquad\qquad (4.8)$$

Where V = pore water velocity, cm/day

   q = volumetric flux per unit area, cm/day

   θ = volumetric water content, dimensionless

In reality, the velocity of water movement through the unsaturated zone is a highly dynamic process resulting from the combined effects of stochastic rainfall inputs and soils, topographic, and vegetation characteristics of the site. However, under the steady flow assumption of our transport equation, the pore water velocity is assumed constant for the time period of interest. The specific time interval of concern also determines the appropriate method of estimating the volumetric flux, q, for the two types of problems addressed in this manual:

   Case 1 - Waste Sites: To assess the extent of the contaminant plume emanating from a leaking waste site, long-term or annual values of water infiltrating or percolating through the unsaturated zone of the site represents the volumetric flux, q. For the saturated zone, this value is also called the recharge rate representing the moisture actually reaching ground water.

   Case 2 - Spills: To estimate contaminant movement from a spill site, the volumetric flux is based on the volume of the spill (for large spills) and/or expected percolation/recharge volumes derived from short-term (5-day, 10-day, monthly) precipitation forecasts.

TABLE 4.1 UNSATURATED ZONE PARAMETERS AND ASSOCIATED INFORMATION
NEEDED/USEFUL FOR EVALUATION

Parameter	Name	Information Needed/Useful For Evaluation
Co	Initial contaminant concentration	Contaminant identity, solubility, waste/site records, organic carbon partition coefficient, decay rates and processes
V	Pore water velocity	Meteorologic and soil characteristics, infiltration, percolation, volumetric water content, spill volume/waste quantity, soil porosity
k	Degradation/decay rate	Contaminant identity, relevant attenuation processes, environmental conditions
R	Retardation factor	Contaminant identity, adsorption characteristics, soil organic carbon, bulk density, ion exchange capacity, clay content/type, volumetric water content
D	Dispersion Coefficient	Subsurface/soil characteristics, pore water velocity, dispersivity

Guidelines and recommendations for estimating the volumetric flux for each case and the volumetric water content are presented below:

Percolation/Recharge

To estimate percolation and recharge values for a specific site, the conventional water balance equation can be written in the following form.

$$PER = P - ET - DR \qquad\qquad (4.9)$$

where PER = Percolation and Recharge, cm/yr
      P   = Precipitation, cm/yr
      ET = Evapotranspiration, cm/yr
      DR = Direct Surface Runoff, cm/yr

As a simplification for use within the emergency response time frame, the equation ignores any man-made water additions (e.g. irrigation, which could be added to P if known) and any change in soil moisture storage. PER includes both percolation and recharge to the ground-water systems of concern. For sites where the ground-water table is close to the land surface, percolation and recharge will be equal. However, for most sites where ground water is considerably below the surface, some of the percolating water will move laterally within the soil or upon reaching the ground-water surface, and subsequently discharge to a surface stream. Thus, PER should be used to assess contaminant movement through the unsaturated zone, but this value may need to be reduced to estimate recharge to deep aquifers or where impermeable strata exist.

A variety of local meteorologic and hydrologic data sources should be contacted to estimate percolation and recharge values for the specific site based on the water balance components of Equation 4.9. As discussed above, the appropriate time frame for the needed data and associated data sources will be different for Case 1 and Case 2 analyses.

Case 1 Analyses will require an estimate of the age of the waste site, or the time when hazardous waste releases may have begun, in order to determine the time period for the needed data. In most cases, this time period will be a number of years. In order of preference, the following methods of obtaining site-specific estimates of percolation and recharge are recommended:

1)   Obtain annual estimates of PER from local sources and calculate an average value for the time period

2)   Obtain annual estimates of P, ET, and DR from local sources, calculate annual values of PER from Equation 4.9 and calculate an average value for the time period.

3)   Obtain mean annual values for PER from local sources, or obtain mean annual values of P, ET, and DR from which a mean annual value of PER can be calculated.

Local site-specific information should be used whenever possible;
significant errors can result from using general or regional data.

Local sources of historical data needed for estimating PER for Case 1
analyses include:

    o   Local or regional water agencies

    o   Local or regional offices of State and Federal water agencies
          (e.g. U.S. Geological Survey, National Weather Service;
          Forest Service, Department of Agriculture, EPA)

    o   University libraries and departments of engineering,
          agriculture, soils, etc.

    o   First order weather stations - usually found at airports

Lacking any local data, the user can obtain a preliminary estimate of mean
annual percolation for areas in the eastern half of the U.S. from Figures
4.1 and 4.2, based on the U.S. Soil Conservation Service hydrologic soil
classifications defined in Table 4.2. The isopleths of mean annual
percolation in these figures were derived from application of the U.S. Soil
Conservation Service Curve Number procedure (U.S. SCS, 1964) for estimating
potential direct runoff at more than fifty sites in the Eastern U.S.
(Stewart et al, 1976).

The Western U.S. was not included due to irrigation applications and the
highly variable rainfall patterns and steep gradients (due to orographic
effects) which preclude interpolation of percolation estimates between
widely separated meteorologic stations.

To use these figures, the user must determine or estimate the hydrologic
soil group for the soil at the waste site and then choose the appropriate
figure for that class i.e.

               A or B: Figure 4.1

               C or D: Figure 4.2

Hydrologic soil groups for a variety of soils have been determined by the
U.S. SCS (U.S. SCS, 1971); local offices and/or the state conservationist
should be contacted for this information for the site. Alternately, Figure
4.3 provides an approximate mapping of hydrologic soil groups based on
generalized soils information. Due to the extreme spatial variability of
soil characteristics, Figure 4.3 should be used only as a last resort when
site-specific information is not available.

Figure 4.4 provides an overview of the spatial variability of the three
independent variables of the water balance equation - precipitation,
evapotranspiration, surface runoff - on a national scale. This information
is provided to supply the user with general background with which to assess
possible major errors in locally supplied information. The national maps
should not be used to estimate PER for a number of reasons; significant

Figure 4.1a   Mean annual percolation below a 4-foot root zone in inches. Hydrologic Soil
Group A. Four inches available water-holding capacity. Straight-row corn.
(Stewart et al., 1976)

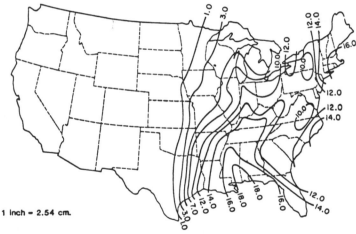

1 inch = 2.54 cm.

Figure 4.1b   Mean annual percolation below a 4-foot root zone in inches. Hydrologic Soil
Group B. Eight inches available water-holding capacity. Straight-row corn.
(Stewart et al., 1976)

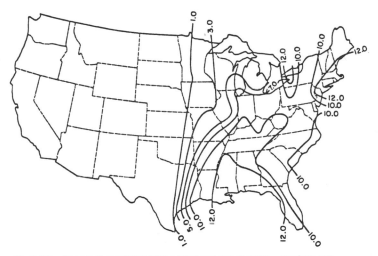

Figure 4.2a   Mean annual percolation below a 4-foot root zone in inches. Hydrologic Soil
Group C. Eight inches available water-holding capacity. Straight-row corn.
(Stewart et al., 1976)

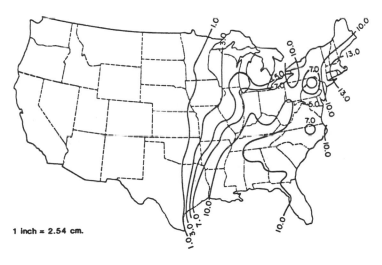

1 inch = 2.54 cm.

Figure 4.2b   Mean annual percolation below a 4-foot root zone in inches. Hydrologic Soil
Group D. Six inches available water-holding capacity. Straight-row corn.
(Stewart et al., 1976)

TABLE 4.2 HYDROLOGIC SOIL CLASSIFICATIONS (U.S. SCS, 1964)

Group/Runoff Potential	Description
Group A. Low Runoff Potential	Soils having high infiltration rates even when thoroughly wetted and consisting chiefly of deep, well-to excessively-drained sands or gravels. These soils have a high rate of water transmission.
Group B. Moderately Low Runoff Potential	Soils having moderate infiltration rates when thoroughly wetted and consisting chiefly of moderately deep to deep, moderately well to well-drained soils with moderately fine to moderately coarse textures. These soils have a moderate rate of water transmission.
Group C. Moderately High Runoff Potential	Soils having slow infiltration rates when thoroughly wetted and consisting chiefly of soils with a layer that impedes downward movement of water, or soils with moderately fine to fine texture. These soils have a slow rate of water transmission.
Group D. High Runoff Potential	Soils having very slow infiltration rates when thoroughly wetted and consisting chiefly of clay soils with a high swelling potential, soils with a permanent high water table, soils with a claypan or clay layer at or near the surface, and shallow soils over nearly impervious material. These soils have a very slow rate of water transmission.

Figure 4.3   Generalized Hydrologic Soil Groups For The U.S. (Battelle, 1982)

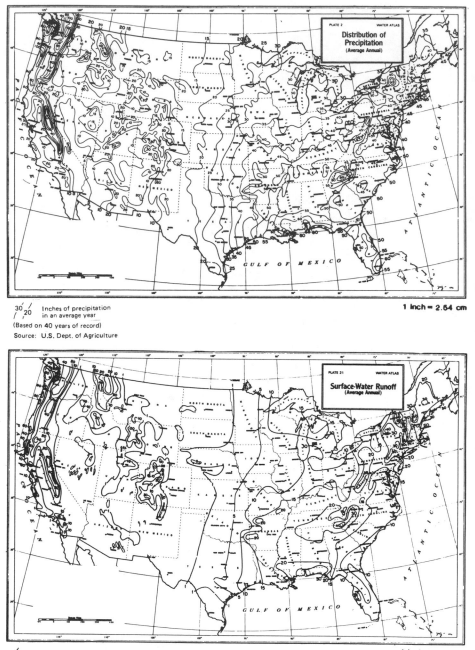

30 / 20 Inches of precipitation
in an average year
(Based on 40 years of record)
Source: U.S. Dept. of Agriculture

1 inch = 2.54 cm

10 / Average annual runoff in inches
Source: U.S. Geological Society

1 inch = 2.54 cm

Figure 4.4 Average Annual Precipitation, Potential Evapotranspiration,
and Surface Water Runoff for the U.S. (Geraghty et. al., 1973)

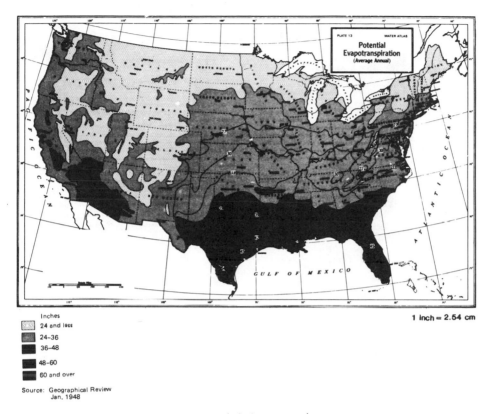

Inches
24 and less
24–36
36–48
48–60
60 and over

1 inch = 2.54 cm

Source: Geographical Review
        Jan. 1948

Figure 4.4 (continued)

local and regional variations are masked by the national isopleths; actual evapotranspiration is usually less than the potential evapotranspiration, especially in the arid west; surface runoff isopleths are derived from U.S.G.S. gaging station data which includes significant contributions of baseflow derived from ground water. In many areas of the country and especially the Western U.S., ignoring the runoff component in the calculation of annual PER values will not lead to significant errors, and may actually result in better percolation values since most runoff data include significant ground-water contributions.

Case 2 Analyses will require forecasts of expected future conditions, primarily rainfall and associated runoff, in order to assess the potential for ground-water contamination from a spill. The same water balance equation (Equation 4.9) is used to estimate PER but the P, ET, and DR terms must be evaluated differently from the Case 1 analyses. The primary differences result from the much shorter time frame of concern; spill situations will require assessment of the contaminant plume from a few days to a few months in the future in order to determine the appropriate emergency response actions. Because of this shorter time frame, recommendations for evaluating the water balance components are as follows:

1. P should be estimated from the quantitative precipitation forecasts (QPF) for the local region available from the local or regional office of the National Weather Service. Generally 5-day to 30-day forecasts are available; longer forecasts may often be qualitative in terms of "above-normal" or "below-normal" expected rainfall.

2. ET can be effectively ignored for the short time frames of 5 to 10 days without significant inaccuracy, especially during heavy rainfall periods. For time periods of one month or longer, ET estimates should be included in the water balance calculation.

3. Soil moisture storage and resulting effects on direct surface runoff become significant during the short time frame of a Case 2 analysis. Also, contaminated runoff although removed from the immediate spill site can reinfiltrate further downslope.

The same sources of local data noted under the Case 1 analysis are also important for a Case 2 analysis. One critical addition is the local or regional National Weather Service office; local precipitation forecasts are absolutely essential for a Case 2 analysis (except possibly in large-volume spills, discussed below). Other local information sources include TV/radio stations, local meteorologists, and other agencies either making or needing weather forecasts. Experience with local conditions in water agencies, universities, hydrologists, and other water experts is especially important in estimating ET and DR values for the short time frame analysis.

Local site-specific information should be used whenever possible; significant errors can result from using general or regional data.

Lacking any local data, the user should consult the following publications which contain meteorologic data on a national scale:

1. **Climatic Atlas of the U.S.** (Environmental Data Service, 1968)

2. **Water Atlas of the U.S.** (Geraghty et al, 1973)

Both publications include various types of evaporation and evapotranspiration data. The Climatic Atlas also includes normal monthly precipitation on a national basis from which forecasts of above or below normal rainfall might be estimated. In addition, Thomas and Whiting (1977) have published annual and seasonal precipitation probabilities for 93 weather stations across the U.S. which could be used to further quantify qualitative forecasts.

Short term estimates of direct runoff from forecasted storm events are difficult to make, and are highly dependent on local site-specific conditions and existing soil moisture conditions. Lacking any local information or guidance from local hydrologists or water agencies, the user can choose runoff coefficients from Table 4.3 to estimate the portion of the rainfall that will result in runoff. The values in this table were derived by applying the SCS curve number procedure (U.S. SCS, 1964) for one-inch and four-inch storm events for each hydrologic soil group, and under each of the three antecedent soil conditions. Thus, the user should choose the low values in Table 4.3 for a one-inch forecast, the high value for a four-inch forecast, and prorate other forecasted amounts between the extremes. The values were developed for pasture land in good condition with an average slope of 2-5 percent; for more accuracy and/or significantly different land conditions the user should apply the SCS procedures directly as described in Appendix A.

The methodology in this manual is not directly applicable for large volume chemical spills where gravitational forces and the hydraulic pressure head (due to ponding) are the driving forces behind the chemical movement through the unsaturated zone. The primary reason for this is because the equations and parameters from which the nomograph was developed are based on _water_ movement through porous media.

TABLE 4.3 RUNOFF COEFFICIENTS FOR HYDROLOGIC SOIL GROUPS*

5-Day Antecedent Rainfall (inches)

Dormant Season 10/1-3/31	Growing Season 4/1-9/30	Hydrologic Soil Groups			
		A	B	C	D
<0.5	<1.4	0.0	0.0 -0.2	0.0 -0.14	0.0 -2.3
0.5-1.1	1.4-2.1	0.0-0.04	0.0 -0.20	0.02-0.40	0.08-0.51
>1.1	>2.1	0.0-0.18	0.07-0.49	0.22-0.66	0.36-0.75

*Derived from 1" to 4" rainfall events on pasture land kept in good condition with average slopes of 2-5 percent.

However, if a qualitative or semi-quantitative assessment is required in an emergency situation, the methodology could be used for a gross or relative evaluation. The pore water velocity could be estimated by using the volume of the spill (i.e. area x average depth) and its infiltration rate to calculate the volumetric flux term in Equation 4.8; the infiltration rate might be grossly estimated by timing the drop in the surface of the ponded chemical, or estimating the total time for the ponded chemical to infiltrate or disappear. This estimated pore velocity could then be used, along with the other parameters (adjusted accordingly) in the methodology to estimate the concentration and time to enter the ground water. Clearly, the results will need to be analyzed and used with extreme caution, and only as a gross approximation. Chemicals with viscosities greater than water can be expected to move slower, while chemicals with a lower viscosity would likely move faster than water. The methodology predictions should be analyzed and adjusted in this manner.

For longer time frames, such as a few months, where infiltration from rainfall would be significantly greater than the spill volume, the pore water velocity should be derived from the water balance equation as described under the Case 2 analyses above.

Volumetric Water Content

The volumetric water content is the percent of the total soil volume which is filled with water. Under saturated conditions, the volumetric water content equals the total porosity of soil and is considerably less than porosity under unsaturated conditions. Conceptually, under steady flow conditions water (i.e. volumetric flux or percolation) is flowing through the pore spaces occupied by the volumetric water content. Thus, the flux and moisture content are directly related with higher flux values requiring higher moisture content, and vice versa.

Volumetric water content values will range from 5% to 10% at the low end to less than the total porosity (discussed below) at the higher end. For most soils, this results in a range of 5% to 50%. If no other local information is available, we recommend that the user select a value within this range (with the upper value modified to reflect total porosity of the site-specific soils) corresponding to the relative value of the flux as estimated by the percolation rate. Thus, for high percolation values (see Figures 4.1 and 4.2) water content values of 30% to 50% should be used, and for low percolation values 10% to 20% would be recommended. Alternately, the user may assume that the volumetric water content is equal to the field capacity for the particular soil type. Field capacity is the moisture retained by the soil after free drainage. Although this is not a rigorous definition, it is usually equal to the 1/3 bar soil moisture value by volume. Representative value ranges of field capacity by soil type are as follows:

	Field Capacity
Sandy soils	0.05 - 0.15
Silt/loam soils	0.13 - 0.30
Clay soils	0.26 - 0.45

Total and Effective Porosity

The total porosity, usually stated as a fraction or percent, is that portion of the total volume of the material that is made up of voids. Effective porosity is less than total porosity, being reduced by the amount of space occupied by dead-end pores. In unconfined aquifers, the term specific yield, a measure of the quantity of water that will drain from a unit volume of aquifer under the influence of gravity, can be used as an estimate of effective porosity. The terms total and effective porosity are applicable to both the saturated and unsaturated zones.

Total porosity is required in the saturated and unsaturated zone methodologies in determining the retardation effects of adsorption. The effective porosity of the aquifer is necessary for calculating the velocity of flow within the voids using Darcy's Law. Effective porosity is not needed for the unsaturated zone analysis contained within this manual. Tables 4.4 and 4.5 provide representative values of total porosity and specific yield (an estimate of effective porosity) for several different soils and geologic materials.

4.2.2 Degradation Rate

The unsaturated zone can serve as an effective medium for reducing contaminant concentration through a variety of chemical and biological decay mechanisms which transform or attenuate the contaminant. Depending on chemical and soil characteristics, processes such as volatilization, biodegradation, hydrolysis, oxidation, and radioactive decay may be important in reducing concentrations prior to reaching the ground-water table. Also, both volatilization and photolysis may be important in reducing the concentrations of surface spills (see Section 4.1.2) and thus reduce the amount and concentration of contaminants available to move through the unsaturated zone.

The equations and nomograph for contaminant migration allow the use of a degradation or decay rate to represent disappearance of the pollutant by the attenuation mechanisms listed above. A first-order rate process is assumed with the degradation rate representing the aggregate disappearance rate of the compound by all significant decay or transformation processes. The input degradation rate is in units of inverse time (i.e. per day) and is related to the half-life of a compound as follows:

$$k = \frac{0.693}{t_{1/2}} \qquad\qquad (4.10)$$

where k = degradation rate, day^{-1}
         $t_{1/2}$ = half-life, days

In evaluating an appropriate degradation rate, the following steps are recommended:

1. Determine if degradation can be significant for the specific time frame, compound, and situation being analyzed. For most instances

TABLE 4.4 REPRESENTATIVE VALUES OF POROSITY

Material	Porosity, Percent	Material	Porosity, Percent
Gravel, coarse	28[a]	Loess	49
Gravel, medium	32[a]	Peat	92
Gravel, fine	34[a]	Schist	38
Sand, coarse	39	Siltstone	35
Sand, medium	39	Claystone	43
Sand, fine	43	Shale	6
Silt	46	Till, predominantly silt	34
Clay	42	Till, predominantly sand	31
Sandstone, fine-grained	33	Tuff	41
Sandstone, medium-grained	37	Basalt	17
Limestone	30	Gabbro, weathered	43
Dolomite	26	Granite, weathered	45
Dune Sand	45	Granite, weathered	45

[a]These values are for repacked samples; all others are undisturbed.

Source: Pettyjohn, W.A., et al, 1982

TABLE 4.5 SPECIFIC YIELDS, IN PERCENT, OF VARIOUS MATERIALS
(Rounded to nearest whole percent)

Material	# of Determinations	Specific Yield Max.	Min.	Ave.
Clay	15	5	0	2
Silt	16	19	3	8
Sandy clay	12	12	3	7
Fine sand	17	28	10	21
Medium sand	17	32	15	26
Coarse sand	17	35	20	27
Gravelly sand	15	35	20	25
Fine gravel	17	35	21	25
Medium gravel	14	26	13	23
Coarse gravel	14	26	12	22

Source: Pettyjohn, W.A. et al, 1982

involving the fate and movement of non-persistent compounds, degradation should be considered. However, high concentrations of toxic chemicals may effectively sterilize the soil and reduce or eliminate microorganisms that biologically degrade the compound.

Also, if decay of the compound is extremely slow relative to the time frame of interest, or if daughter products produced by transformation are also toxic, the user may decide to ignore degradation in order to estimate maximum potential concentrations.

2. Assess the major decay mechanisms for the specific compound of concern.

3. Evaluate compound-specific rates for each major decay mechanism.

4. Use the sum of the decay rate or the maximum if one decay mechanism is predominant, as the value of the decay rate for the assessment nomograph calculations.

The same information sources used in identifying the compound may be helpful in determining major decay or loss mechanisms and associated rate values. Companies associated with the waste/spill incident, or companies within the same industry, can be an extremely valuable source of this information.

Table 4.6 provides a summary of the relative importance of different chemical fate processes for a wide variety of compounds in various classifications. If the specific compound is not included in Table 4.6, industry sources may be able to provide the classification or names of other compounds with similar degradation mechanisms. (For example, volatilization is a major process for most halogenated aliphatic hydrocarbons). Although Table 4.6 was developed primarily for the aquatic environment, it may be appropriate for many spill situations and appears to be the best summary of the relative importance of different chemical processes for a variety of compounds. The user should confirm the validity of the compound-specific information in Table 4.6 with any other available data.

Degradation rates for specific mechanisms have been compiled for numerous chemicals and hazardous compounds in the following publications:

1. Physical/Chemical Properties of Hazardous Waste Constituents, Dawson et al (1980).

2. Aquatic Fate Process Data for Organic Priority Pollutants, Mabey et al (1982) (Note: This publication includes available data for all compounds listed in Table 4.6, except metals and inorganics).

3. Handbook of Environmental Data on Organic Chemicals, Verschueren, K. (1977).

Also, degradation rates for pesticides in both field and laboratory conditions have been collected and published by Rao and Davidson (1980), Nash (1980), and Wauchope and Leonard (1980). This information is based on

TABLE 4.6   RELATIVE IMPORTANCE OF PROCESSES INFLUENCING AQUATIC FATE OF PRIORITY POLLUTANTS   (After Mills et al., 1982;   Callahan et al., 1979)

Compound	Sorption	Volatilization	Biodegradation	Photolysis-Direct	Hydrolysis	Bioaccumulation
PESTICIDES						
Acrolein	?	+	+	+	-	-
Aldrin	+	+	?	-	-	+
Chlordane	+	+	?	-	-	+
DDD	+	+	-	-	-	+
DDE	+	+	-	+	-	+
DDT	+	+	-	-	+	+
Dieldrin	+	+	-	+	-	+
Endosulfan and Endosulfan Sulfate	+	+	+	?	+	-
Endrin and Endrin Aldehyde	?	?	?	+	-	+
Heptachlor	+	+	-	?	++	+
Heptachlor Epoxide	+	-	?	?	-	+
Hexachlorocyclohexane (α, β, δ isomers)	+	?	+	-	-	-
-Hexachlorocyclohexane (Lindane)	+	-	+	-	-	-
Isophorone	?	-	?	+	-	-
TCDD	+	-	-	?	-	+
Toxaphene	+	+	+	-	-	+
PCBs and RELATED COMPOUNDS						
Polychlorinated Biphenyls	+	+	+[a]	?	-	+
2-Chloronaphthalene	-	?	+	+	-	-
HALOGENATED ALIPHATIC HYDROCARBONS						
Chloromethane (methyl chloride)	-	+	-	-	-	-
Dichloroethane (methylene chloride)	-	+	?	-	-	-
Trichloromethane (chloroform)	-	+	?	-	-	-
Tetrachloromethane (carbon tetrachloride)	?	+	-	-	-	?
Chloroethane (ethyl chloride)	-	+	?	-	+	-
1,1-Dichloroethane (ethylidene chloride)	-	+	?	-	-	-
1,2-Dichloroethane (ethylene dichloride)	-	+	?	-	-	-
1,1,1-Trichloroethane (methyl chloroform)	-	+	-	-	-	-
1,1,2-Trichloroethane	?	+	-	-	-	?
1,1,2,2-Tetrachloroethane	?	+	-	-	-	?

Key to Symbols:

++ Predominant fate determining process   - Not likely to be an important process
 + Could be an important fate process   ? Importance of process uncertain or not known

TABLE 4.6 continued

Compound	Sorption	Volatilization	Biodegradation	Photolysis-Direct	Hydrolysis	Bioaccumulation
Hexachloroethane	?	?	?	?	?	+
Chloroethene (vinyl chloride)	+	-	-	-	-	..
1,1-Dichloroethene (vinylidene chloride)	?	+	?	-	-	?
1,2-trans-Dichloroethene	-	+	?	-	-	-
Trichloroethene	-	+	?	-	-	-
Tetrachloroethene (perchloroethylene)	-	+	+	-	-	-
1,2-Dichloropropane	?	+	-	?	+	?
1,3-Dichloropropene	?	+	-	?	+	-
Hexachlorobutadiene	+	+	?	-	?	+
Hexachlorocyclopentadiene	+	+	-	+	+	+
Bromomethane (methyl bromide)	-	+	-	-	+	-
Bromodichloromethane	?	?	?	?	-	+
Dibromochloromethane	?	+	?	?	-	+
Tribromomethane (bromoform)	?	+	?	?	-	+
Dichlorodifluoromethane	?	+	-	?	-	?
Trichlorofluoromethane	?	+	-	-	-	?
HALOGENATED ETHERS						
Bis(choromethyl) ether	-	-	?	-	++	-
Bis(2-chloroethyl) ether	-	+	-	-	-	?
Bis(2-chloroisopropyl) ether	-	+	-	-	-	?
2-Chloroethyl vinyl ether	-	+	?	-	+	-
4-Chlorophenyl phenyl ether	+	?	?	+	-	+
4-Bromophenyl phenyl ether	+	?	?	+	-	+
Bis(2-chloroethoxy) methane	-	-	?	-	+	?
MONOCYCLIC AROMATICS						
Benzene	+	+	-	-	-	-
Chlorobenzene	+	+	-	?	-	+
1,2-Dichlorobenzene (o-dichlorobenzene)	+	+	-	?	-	+
1,3-Dichlorobenzene (m-dichlorobenzene)	+	+	?	?	?	+
1,4-Dichlorobenzene (p-dichlorobenzene)	+	+	-	?	-	+
1,2,4-Trichlorobenzene	+	+	-	?	-	+
Hexachlorobenzene	+	-	-	-	-	-

Key to Symbols:
++ Predominant fate determining process   - Not likely to be an important process
 + Could be an important fate process   ? Importance of process uncertain or not known

TABLE 4.6   continued

Compound	Sorption	Volatilization	Biodegradation	Photolysis-Direct	Hydrolysis	Bioaccumulation
Ethylbenzene	?	+	?	-	-	-
Nitrobenzene	+	-	-	+	-	-
Toluene	+	+	?	-	-	-
2,4-Dinitrotoluene	+	-	-	+	-	?
2,6-Dinitrotoluene	+	-	-	+	?	?
Phenol	-	+	+	+	-	-
2-Chlorophenol	-	-	?	+	-	-
2,4-Dichlorophenol	-	-	++	-	-	-
2,4,6-Trichlorophenol	?	-	?	?	-	-
Pentachlorophenol	+	-	+	++b	-	+
2-Nitrophenol	-	-	-	++b	-	-
4-Nitrophenol	+	-	-	++b	-	-
2,4-Dinitrophenol	+	-	-	++	-	-
2,4-Dimethyl phenol (2,4-xylenol)	-	-	?	+	-	-
p-chloro-m-cresol	-	-	?	++	-	-
4,6-Dinitro-o-cresol	+	-	-	++	?	?
PHTHALATE ESTERS						
Dimethyl phthalate	+	-	+	-	-	+
Diethyl phthalate	+	-	+	-	-	+
Di-n-butyl phthalate	+	-	+	-	-	+
Di-n-octyl phthalate	+	-	+	-	-	+
Bis(2-ethylhexyl) phthalate	+	-	+	-	-	+
Butyl benzyl phthalate	+	-	+	-	-	+
POLYCYCLIC AROMATIC HYDROCARBONS						
Acenaphthene[c]	+	-	+	+	-	-
Acenaphthylene[c]	+	-	+	+	-	-
Fluorene[c]	+	-	+	+	-	-
Naphthalene	+	-	+	+	-	-
Anthracene	+	+	+	+	-	-
Fluoranthene[c]	+	+	+	+	-	-
Phenanthrene[c]	+	+	+	+	-	-
Benzo(a)anthracene	+	-	+	+	-	-
Benzo(b)fluoranthene[c]	+	-	+	+	-	-
Benzo(k)fluoranthene[c]	+	-	+	+	-	-
Chrysene[c]	+	-	+	+	-	-

Key to Symbols:
++ Predominant fate determining process
 + Could be an important fate process
 - Not likely to be an important process
 ? Importance of process uncertain or not known

TABLE 4.6 continued

Compound	Sorption	Volatilization	Biodegradation	Photolysis-Direct	Hydrolysis	Bioaccumulation
Pyrene[c]	+	-	+	+	-	-
Benzo(ghi)perylene[c]	+	-	+	+	-	-
Benzo(a)pyrene	+	+	+	+	-	-
Dibenzo(a,h)anthracene[c]	+	-	+	+	-	-
Indeno(1,2,3-cd)pyrene	+	-	+	+	-	-
NITROSAMINES AND MISC. COMPOUNDS						
Dimethylnitrosamine	-	-	-	++	-	-
Diphenylnitrosamine	+	-	?	+	-	?
Di-n-porpyl nitrosamine	-	-	-	++	-	-
Benzidine	+	-	?	+	-	-
3,3'-Dichlorobenzidine	++	-	-	+	-	-
1,2-Diphenylhydrazine (Hydrazobenzene)	+	-	?	+	-	+
Acrilonitrile	-	+	?	-	-	+
METALS AND INORGANICS						
Asbestos	+	-	-	-	-	-
Antimony	+	-	-	-	+	+
Arsenic	+	+	+	-	+	+
Berylumm	+	-	?	-	+	-
Cadmium	+	-	-	-	+	+
Copper	+	-	-	-	+	+
Chromium	+	-	-	-	+	+
Cyanides	-	+	+	+	-	-
Lead	+	-	+	+	-	+
Mercury	+	+	+	+	-	+
Nickel	+	-	-	-	+	-
Selenium	+	+	+	-	+	+
Silver	+	-	-	-	-	-
Thallium	+	-	-	-	-	+
Zinc	+	-	-	-	+	+

Key to Symbols:
++ Predominate fate determining process   − Not likely to be an important process
+ Could be an important fate process   ? Importance of process uncertain or not known

Notes

[a]Biodegradation is the only process knoen to transform polychlorinated biphenyls under environmental conditions, and only the lighter compounds are measurably biodegraded. There is experimental evidence that the heavier polychlorinated biphenyls (five chlorine atoms or more per molecule) can be photolyzed by ultraviolet light, but there are no data to indicate that this process is operative in the environment.

[b]Based on information for 4-nitrophenol

[c]Based on information for PAH's as a group. Little or no information for these compounds exists.

agricultural pesticide applications and should be used with caution for the waste/spill situation only if other data is lacking.

4.2.3 Retardation Factor

The process of adsorption of contaminants onto soil particles and associated organic matter retards the movement of the contaminant through both unsaturated and saturated media. As discussed in Section 3, the adsorption process is included in the assessment nomograph by the retardation factor which is defined as follows:

$$R = 1 + \frac{B}{N} \; K_d \tag{4.11}$$

where R = Retardation factor (dimensionless)

B = Bulk density, g/cc

N = Effective porosity (saturated conditions), or θ, volumetric water content (unsaturated conditions), dimensionless

K_d = Partition coefficient, ml/g

Thus, the major determinant of the retardation factor, R, is the partition coefficient, K_d, which represents the ratio of the adsorbed pollutant concentration to the dissolved (or solution) concentration. Under the linear, equilibrium isotherm assumption employed in this manual, the form and units of K_d are as follows:

$$K_d = \frac{Cs}{Cw} \tag{4.12}$$

Where K_d = Partition coefficient (ml/g)

C_s = Pollutant concentration on soil (ppm)

C_w = Pollutant concentration in water (mg/l)

Since B and N usually vary within a small range of values and K_d can vary by many orders of magnitude, the resulting value of R is primarily determined by K_d, which in turn is a function of the specific compound and soil combination.

Guidelines for estimating N, either as total porosity under saturated conditions or volumetric water content under unsaturated conditions, are presented in Section 4.2.1. Guidelines for evaluating K_d and B are discussed below.

Partition Coefficient

Since K_d can have a different value for each compound and soil combination, values of K_d from previous studies (or other sources) at the waste/spill site should be used whenever possible. For most spill sites and many waste sites, this information will not be available and K_d will need to be estimated by other means.

For neutral organic compounds, a body of knowledge has been developed over the past decade (see Lyman et al, 1982) whereby K_d values can be estimated from the soil organic carbon content and the organic carbon partition coefficient for the compound as follows:

$$K_d = K_{oc} \frac{OC}{100}$$ (4.13)

where K_{oc} = Organic carbon partition coefficient (ml/g)

OC = Percent organic carbon content of soil or sediment (dimensionless)

Equation 4.13 assumes that the organic carbon in the soil or sediment is the primary means of adsorbing organic compounds onto soils and sediments. This concept has served to reduce much of the variation in K_d values for different soil types.

K_{oc} values for a number of chemicals and hazardous compounds have been tabulated by Rao and Davidson (1980), Dawson et al (1980) and Mabey et al (1982). Also, a variety of regression equations relating K_{oc} to solubility, octanol-water partition coefficients (K_{ow}), and other compound characteristics have been developed; Table 4.7 from Lyman et al (1982) presents the major regression equations available, the chemical classes represented, the number of compounds investigated, and the associated correlation coefficient. Users should review the discussion by Lyman et al (1982) to comprehend the limitations, assumptions, and parameter ranges underlying the equations in Table 4.7. K_{oc} estimates from more than one equation should be evaluated in order to assess the variability in the estimates.

Data on the compound characteristics needed for the regression equations can be obtained from Dawson et at (1980), Mabey et al (1982), and other sources listed in Table 2.2. Also, a very complete data base of K_{ow} values is maintained by Dr. Corlan Hansch at Pomona College, Pomona, California (714-621-8000 ext. 2225). This data base is available in microfiche form for easy use in the field. K_{oc} values should be used directly whenever available; otherwise estimation of K_{oc} from K_{ow} is appropriate.

Organic Carbon/Organic Matter Content

Organic content of soils is described in terms of either the percent organic carbon, which is required in our estimation of K_d, or the percent organic matter. These two values are conventionally related as %OC = %OM/1.724.

Typical values of percent organic matter range from 0.4% to 10.0% (Brady, 1974). Table 4.8 lists the range and average organic matter content for mineral surface soils in various parts of the U.S; organic soils, such as

TABLE 4.7   REGRESSION EQUATIONS FOR THE ESTIMATION OF K_{oc}.   Lyman et al., 1982.
(Reference numbers keyed to Lyman et al., 1982, Chapter 4)

Regression Equations for the Estimation of K_{oc}

Eq. No.	Equation[a]	No.[b]	r^2[c]	Chemical Classes Represented	Ref.
4-5	$\log K_{oc} = -0.55 \log S + 3.64$ (S in mg/L)	106	0.71	Wide variety, mostly pesticides	[26]
4-6	$\log K_{oc} = -0.54 \log S + 0.44$ (S in mole fraction)	10	0.94	Mostly aromatic or polynuclear aromatics; two chlorinated	[25]
4-7[d]	$\log K_{oc} = -0.557 \log S + 4.277$ (S in μ moles/L)	15	0.99	Chlorinated hydrocarbons	[11]
4-8	$\log K_{oc} = 0.544 \log K_{ow} + 1.377$	45	0.74	Wide variety, mostly pesticides	[26]
4-9	$\log K_{oc} = 0.937 \log K_{ow} - 0.006$	19	0.95	Aromatics, polynuclear aromatics, triazines and dinitroaniline herbicides	[9]
4-10	$\log K_{oc} = 1.00 \log K_{ow} - 0.21$	10	1.00	Mostly aromatic or polynuclear aromatics; two chlorinated	[25]
4-11	$\log K_{oc} = 0.94 \log K_{ow} + 0.02$	9	e	s-Triazines and dinitroaniline herbicides	[7]
4-12	$\log K_{oc} = 1.029 \log K_{ow} - 0.18$	13	0.91	Variety of insecticides, herbicides and fungicides	[36]
4-13[d]	$\log K_{oc} = 0.524 \log K_{ow} + 0.855$	30	0.84	Substituted phenylureas and alkyl-N-phenylcarbamates	[5]
4-14[d,f]	$\log K_{oc} = 0.0067 (P - 45N) + 0.237$	29	0.69	Aromatic compounds: ureas, 1,3,5-triazines, carbamates, and uracils	[18]
4-15	$\log K_{oc} = 0.681 \log BCF(f) + 1.963$	13	0.76	Wide variety, mostly pesticides	[26]
4-16	$\log K_{oc} = 0.681 \log BCF(t) + 1.886$	22	0.83	Wide variety, mostly pesticides	[26]

a. K_{oc} = soil (or sediment) adsorption coefficient; S = water solubility; K_{ow} = octanol-water partition coefficient; BCF(f) = bioconcentration factor from flowing-water tests; BCF(t) = bioconcentration factor from model ecosystems; P = parachor; N = number of sites in molecule which can participate in the formation of a hydrogen bond.

b. No. = number of chemicals used to obtain regression equation.

c. r^2 = correlation coefficient for regression equation.

d. Equation originally given in terms of K_{om}. The relationship $K_{om} = K_{oc}/1.724$ was used to rewrite the equation in terms of K_{oc}.

e. Not available.

f. Specific chemicals used to obtain regression equation not specified.

peat or muck soils, can have values in the range of 15% to 20% or greater. Agricultural soils are commonly in the range of 1% to 5% organic matter. Figure 4.5 shows a national distribution of % Nitrogen in the surface foot of soil; % Nitrogen and %OC are generally related as %OC = 11 x % N. This information can be used to estimate %OM and %OC as a basis for determining K_d.

The values in Table 4.8, and those mentioned above are primarily for the top 15cm of the soil profile. Organic content normally decreases sharply with depth, as shown in Figure 4.6 which compares the relative change in percent organic matter with depth for a prairie soil and a forest soil. Below 60 cm in depth, percent organic matter values of less than 2% are common. Users must evaluate appropriate %OM values for the specific region or regions of the soil profile through which the contaminant will be moving. Thus for surface spills, a weighted value of surface and subsurface %OM for the unsaturated zone should be used; whereas subsurface releases from waste sites will require the subsurface %OM at the appropriate depth. For many subsurface or saturated zone releases, a %OM value of less than 1% may be reasonable.

Local site-specific information should be used whenever possible; significant errors can result from using general or regional data.

TABLE 4.8   AVERAGE ORGANIC MATTER CONTENTS AND RANGES OF
MINERAL SURFACE SOILS IN SEVERAL AREAS OF THE
UNITED STATES
(Lyon et al, 1952)

Soils	Organic Matter (%)	
	Range	Av.
240 West Va. soils	0.74-15.1	2.88
15 Pa. soils	1.70- 9.9	3.60
117 Kansas soils	0.11-3.62	3.38
30 Nebraska soils	2.43-5.29	3.83
9 Minn. prairie soils	3.45-7.41	5.15
21 Southern Great Plains soils	1.16-2.16	1.55
21 Utah soils	1.54-4.93	2.69

Retardation Factors for Ionic Species

The processes which govern the adsorption of substances which ionize are very different from those for substances that are nonionic. Most soils have

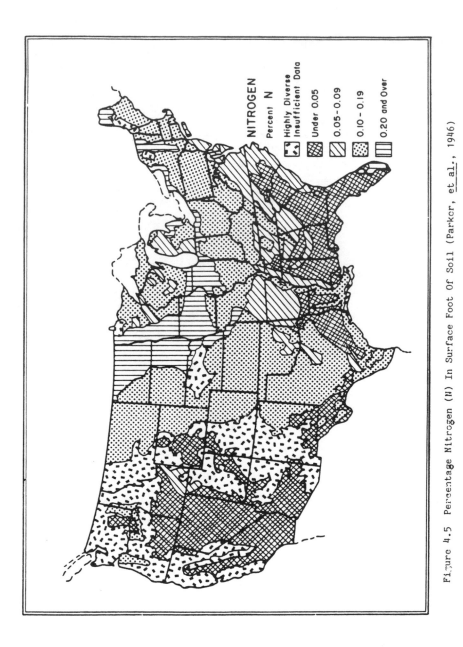

Figure 4.5 Percentage Nitrogen (N) In Surface Foot Of Soil (Parker, et al., 1946)

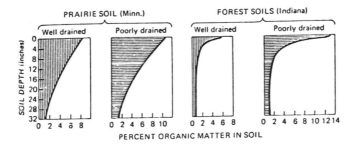

Figure 4.6 - Distribution of organic matter in four soil
profiles. (Brady, 1974)

a net negative charge, and therefore ions that are positively charged are attracted to them. Some positively charged ions are preferentially adsorbed to soil materials and will displace other positively charged ions already on the exchange sites. This process is referred to as <u>cation exchange</u> or base exchange. Anions (negatively charged particles) can either be attracted or repelled by soil particles depending upon the net charge of the soil. The anion exchange capacity of soils is usually less than cation exchange capacity, unless extremely low pH's are encountered or high amounts of Fe or Al oxides, or hydroxides are present. A rule of thumb is that anion repulsion (negative adsorption) is roughly 1 to 5% of the cation exchange capacity (CEC) in non-alkaline soils, and up to 15% in alkaline soils (pH 8.5) (Bolt, 1976).

<u>Acids and Bases in Solution</u> - By definition acids are substances which give-up (donate) protons (hydrogen atoms) in solution. Bases, on the other hand, take-on (accept) protons from solution. A typical reaction for a monoprotic (one hydrogen) acid dissolved in water is

$$HA \quad = \quad H^+ \quad + \quad A^- \tag{4.14}$$

where the double arrow indicates an equilibrium dissociation reaction. The ratio between the products and reactants in this reaction is always a constant known as the dissociation constant, K_a, where

$$K_a = \frac{[H^+] \, [A^-]}{[HA]} \tag{4.15}$$

K_a is usually expressed as a logarithm, pK_a,

$$pKa = -\log (K_a) \tag{4.16}$$

The reaction for a monoprotic base in solution is

$$B + H_2O \quad = \quad HB^+ \quad + \quad OH^- \tag{4.17}$$

and the reaction constant is

$$K_b = \frac{[HB^+] \, [OH^-]}{[B]} = \frac{[HB^+] \, Kw}{[B] \, [H^+]} \tag{4.18}$$

The constants pK_a, pK_b, and pK_w are related by

$$14 = pK_a + pK_b = pK_w \qquad (4.19)$$

for any given compound. K_w, is the dissociation constant for water, equal to 10^{-14}.

Retardation Factor for Acids and Bases - If the constants pK_a or pK_b are known for an acid or base and the pH of the solution is known, the fraction of unionized acid, or base can be determined. For the acid the fraction unionized acid is

$$\alpha = \left(\frac{1 + K_a}{[H^+]} \right)^{-1} \qquad (4.20)$$

For a base, the fraction unionized is

$$\beta = \left(\frac{1 + K_b[H^+]}{K_w} \right)^{-1} \qquad (4.21)$$

If we assume that the ionized portion of the acid is unattracted to soil materials, then the retardation factor for the acid is

$$R_{acid} = 1 + \alpha \frac{K_d B}{n} \qquad (4.22)$$

For the base we will assume that the ionized portion is exchanged similarly to any monovalent ion ($Kd \approx 100$) and that the unionized portion is adsorbed hydrophobically. Thus, the retardation factor for the base is

$$R_{base} = \frac{1 + \beta K_d B + 100 (1 - \beta)}{n} \qquad (4.23)$$

The value of K_d in either of these cases is determined exactly as described above for the neutral (nonionic) species.

The number 100 in Equation 4.23 is an estimate of $K_d^+ B/n$ for a model monovalent cation. This number can range from less than 1 up to 10^5 for various species. A substitute for this number for a particular soil can be estimated by

$$B \frac{K_d^+}{n} = \frac{CEC}{100 \sum z^+} \quad \frac{B}{n} \qquad (4.24)$$

when K_d^+ = adsorption partition coefficient for the charged cation, cc/g

   CEC   = cation exchange capacity of the soil, milliequivalents/100g

   $\sum z^+$ = sum of all positively charged species in the soil location, milliequivalents/cc

The quantity $\sum z^+$ is about 0.001 for most agricultural soils.

In reality, the exchange of the cation or anion is governed by a selectivity coefficient which varies for different soils, and competing ion pairs. When the ionized substance is adsorbed, it reduces the concentration of ionized substances in solution which causes more of the unionized substance to accept or donate protons. Thus, for strongly adsorbed ions, the concentration of the substance in solution could approach very small values. On the other hand if the selectivity for the ion is low or repulsion occurs, virtually all the substance could remain in solution. Thus, the above approach will give an answer between these two extremes; a conservative assumption would be a retardation factor of 1.

To use this methodology, the user should first decide whether the substance is an organic acid or base. This may not be easy to determine. If the substance is not listed in the tables in this section or one does not have prior knowledge about the compound, a retardation factor of 1 should be used. Some pK_a and pK_b values for specific compounds are found in Tables 4.9 and 4.10. Values of $pK_a > 14$ indicate fully protonated forms, while values < 0 indicate fully deprotonated forms of the acid; for bases, pK_b values less than 0 indicate complete protonation while values greater than 14 indicate a completely deprotonated form.

Harris and Hayes (1982) give references which contain pK_a and pK_b values for various organic acids and bases. These are listed below:

   Dissociation Constants of Organic Acids in Aqueous Solution, (Kortum et al 1961)

   Dissociation Constants for Organic Bases in Aqueous Solution, (Perrin, 1965)

   Ionization Constants of Organic Acids in Aqueous Solution (Sergeant and Dempsey, 1979)

TABLE 4.9 pK_a VALUES FOR SELECTED ORGANIC ACIDS

Compound	pKa	Ref.
Aliphatic Acids	3.8 - 5.0	A
Acetic Acid (Substituted)	0.2 - 4.3	A
Aliphatic Acids (Diabasic)		A
1st Carboxyl	1.3 - 4.3	
2nd Carboxyl	4.3 - 6.2	
p-Aminobenzoic Acid		B
K_1 (NH_3 group)	2.29	
K_2 (COOH group)	4.89	
m-Aminobenzoic Acid		B
K_1 (NH_3 group)	3.07	
K_2 (COOH group)	4.73	
m-Aminophenol		B
K_1 (NH_3 group)	4.17	
K_2 (OH group)	9.87	
Aminocyanomethane	5.34	B
Aniline	27.	B
Benzoic Acid	4.2	A
Benzoic Acid (Halogenated)		A
Ortho	2.8 - 3.3	
Meta	3.8	
Para	3.9 - 4.1	
3-Bromo-4-methoxy anilinium ion	4.08	B
Bromoacetic Acid	2.90	B
But-3-enoic Acid	4.34	B
t-Butane	19.	B
Benzoic Acid (Dicamba)	1.93	C
Benzoic Acid (Amiben)	3.40	C
p-Cyanophenol	7.95	B
4-Chloro-3-nitroanilinium ion	4.08	B
Cyanoacetic Acid	2.47	B
Chloromethylphosphonic Acid	1.40	B
Carboxylic Acids	4.5 ± 0.5	B
$CH_3OH_2^+$	-2	B
$C_6H_5OH_2^+$	-6.7	B
2-Chlorophenol	8.52	D
Dichloroacetic Acid	2.90	B
2, 4 Dichlorophenol	7.85	D
2, 4 Dinitrophenol	4.04	D
2, 4 Dimethylphenol	10.6	D
4, 6-Dinitro-o-cresol	4.35	D
Glycine		B
K_1	2.35	
K_2	9.78	
Hydroxymethylphosphonic Acid	1.91	B
p-Methoxybenzoic Acid	4.47	B

(Continued)

TABLE 4.9 (Cont.)

Compound	pKa	Ref.
m-Methylsulfonybenzoic Acid	3.52	B
Methane	40.	B
p-Nitrophenylarsenic Acid	2.90	B
p-Nitrophenol	7.2	B
p-Nitroanilinium	1.0	C
2-Nitrophenol	7.21	D
4-Nitrophenol	7.15	D
Phenol	10.	A, B
m-Phenoxybenzoic Acid	4.47	B
Pyridinium Ion	5.2	B
Phenoxy Acid (2, 4D)	2.8	C
Picolinic Acid (Picloram)	1.90	C
Phenol (Dinoseb)	4.4	C
Pentachlorophenol	4.74	D
RNH_3^+	10.	B
p-Tolyacetic Acid	4.37	B
Tetralol-2	10.48	B
1, 3, 5 - Trihydroxybenzene (K_1)	8.45	B
Trifluoroacetic Acid	0.23	B
Toluene	35	R
2,4,6 - Trichlorophenol	5.99	D

References:

A. Stevenson, 1982

B. Harris and Hayes, 1982

C. Weed and Weber, 1974

D. Mills et al, 1982

TABLE 4.10 pK_b VALUES FOR SELECTED ORGANIC BASES

Compound	pKa	Ref.
Aliphatic Amine Homologues	3.10-4.20	A
Anilines (substituted)	6.90-9.40	A
Acetanilide	13.6	A
Acetamide	14.5	A
Atrazine	12.32	C
Amitrole	9.83	C
Benzidine	9.34, 10.43	D
$CH_3:^-$	-26	B
$C_6H_5CH_2:^-$	-21	B
$C_6H_5NH:^-$	-13	B
$C_4H_9O:^-$	-5	B
$C_6H_5O:^-$	4.0	B
Carboxylate Anions	9.5 ± 0.5	B
Methanol	16	B
$p-NO_2-C_6H_4O:^-$	6.8	B
p-Nitroaniline	13.0	B
Pyridine	8.8	A,B
Pyrimidine	12.7	A
Phenol	20.7	B
Propazine	12.15	C
Prometryne	9.95	C
Prometone	9.72	C
RHN_2	4.00	B
Simazine	12.35	C

References:
- A. Stevenson, 1982
- B. Harris and Hayes, 1982
- C. Weed and Weber, 1974
- D. Mills et al, 1982

Bulk Density

Bulk density is the mass of a unit volume of dry soil, as measured in the field, usually expressed in g/cc or lb/ft^3. The entire volume is taken into consideration including both solids and pore spaces. Thus, loose porous soils will have low values of bulk density and more compact soils will have higher values. Bulk density values normally range from 1.0 to 2.0 g/cc, and soils with high organic matter content will generally have low bulk density values.

Brady (1974) has presented the following ranges of bulk density for selected surface soil types commonly found in agricultural areas:

	Bulk Density (g/cc)
well-decomposed organic soil	0.2 - 0.3
cultivated surface mineral soils	1.25 - 1.45
clay, clay loam, silt loam	1.00 - 1.60
sands and sandy loams	1.20 - 1.80

Ritter and Paquette (1967) have listed the following bulk density ranges for material classes encountered in road and airfield construction:

	Bulk Density (g/cc)
silts and clays	1.3 - 2.0
sand and sandy soils	1.6 - 2.2
gravel and gravelly soils	1.8 - 2.3

Subsoils will generally be more compact than surface soils and thus have higher bulk densities. Very compact subsoils regardless of texture can have bulk densities of 2.0 g/cc or greater; values of 2.3 to 2.5 g/cc should be considered as upper limits. Because of this relatively small range of values, users can choose bulk density values for the waste/spill site from the above information if local site-specific data are not available. Mean or average values for a soil type can be used, and if no data are available a value of 1.5 g/cc can be used with reasonable accuracy for many soils.

4.2.4 Dispersion Coefficient

The dispersion process is exceedingly complex and difficult to quantify, especially for the unsaturated zone. It is sometimes ignored in the unsaturated zone, with the reasoning that pore water velocities are usually large enough so that pollutant transport by convection i.e. (water movement) is paramount. Consequently, unless site specific information or studies are available to establish that dispersion is or is not significant, and data is available to estimate the dispersion coefficient, we recommend that the user perform at least two separate assessments. The first assessment would ignore dispersion and the second assessment should include a reasonable value of a dispersion coefficient to evaluate the importance of dispersion in the unsaturated zone for the specific site. A dispersion coefficient, D, of 0.01 will effectively ignore dispersion and subsequently simplify calculations. However, dispersion should not be ignored for saturated zone analyses. Since most available information on dispersion is for ground-water systems, discussion and parameter guidelines for the dispersion coefficient are provided in Section 4.3.3. Users should consult that section to estimate a coefficient for the unsaturated zone.

4.3 SATURATED ZONE PARAMETER ESTIMATION

Table 4.11 lists the parameters required for a saturated zone assessment and the types of information needed or useful in their estimation. The following sections provide guidelines for estimating each of these input parameters in the order shown in Table 4.11.

4.3.1 Effective Aquifer Thickness (or Zone of Mixing)

The extent of the aquifer subject to contamination is described using an effective aquifer thickness which represents a zone of mixing. For good mixing between the ground water and the contaminant, this effective thickness may equal the total thickness of the aquifer. However, in most cases it will be less than the total thickness, especially for deep aquifers. In cases where the pollutant has a significantly different density and/or viscosity than water, the extent of mixing may be reduced and the contaminant plume will be concentrated over only a portion of the aquifer's thickness. The saturated zone methodology in this manual assumes that the chemical pollutant mixes with the ground water to the effective thickness or mixing zone. The model does not consider immiscible wastes or portions of wastes that either entirely float on top of the water table or sink to the bottom of the aquifer and remain there. For example, the major portion of gasoline is immiscible in water and its total movement in the subsurface cannot be studied effectively with this manual. However, that portion of gasoline that is soluble in water can be analyzed using the assessment methodology.

Local site-specific information should be used whenever possible; significant errors can result from using general or regional data.

The user should search for prior hydrogeological investigations in the offices of Federal, State, County, and Municipal agencies as the initial step in gathering estimates of the total thickness of the aquifer being studied. Hydrogeologists in neighboring universities, consulting firms, and government agencies are another possible source of data on the structural thickness of water-bearing strata and may be able to provide recommendations for an effective mixing depth. If these reports and contacts are not helpful, nearby well owners can be consulted. The perforated intervals of their water supply wells provide a lower limit estimate of the thickness of underlying aquifers since most wells are not perforated for the entire thickness. This information, contained on their drilling logs, should be used carefully and only in the absence of other data.

An estimate of the minimum thickness to use for the mixing zone can be obtained as follows:

Table 4.11 SATURATED ZONE PARAMETERS AND ASSOCIATED INFORMATION
NEEDED/USEFUL FOR EVALUATION

Parameter	Name	Information Needed/Useful for evaluation
m	Effective aquifer thickness (or zone of mixing)	Aquifer characteristics, total aquifer thickness, contaminant density, ground-water density
V	Ground-water (interstitial pore water) velocity	Hydraulic conductivity, hydraulic gradient, effective porosity, specific yield
D	Dispersion coefficient	Aquifer characteristics, dispersivity, molecular diffusion, ground-water velocity
R	Retardation factor	Partition coefficient, bulk density, total porosity, K_{oc}, ionic characteristics
k	Degradation/decay rate	Contaminant identity, relevant attenuation processes, environmental conditions
Co	Source contaminant concentration	Contaminant identity, solubility, waste/site records, organic carbon partition coefficient, decay rates and processes, unsaturated zone assessment.

$$m = \frac{q\,L}{V_d} \qquad\qquad (4.25)$$

where

m	=	effective (minimum) aquifer thickness, m
q	=	recharge from the site, cm/day
V_d	=	Darcy flow velocity, cm/day
L	=	Width of leachate plume at water table, m

Equation 4.25 calculates the minimum aquifer thickness that will accept recharge from the site based on the aquifer properties; calculation of V_d is discussed in the next section. This calculated minimum value should be used only as a guide to the lower limit of a reasonable mixing zone depth.

In assessing which portion of the total aquifer thickness is subject to mixing with the contaminant, knowledge of the density and viscosity of the ground water and the pollutant is needed. Major differences in these characteristics indicate a tendency toward reduced mixing and therefore a smaller effective thickness.

In the temperature range normally expected in ground water, the water density can be assumed as 1 g/cc or 62.4 lb/ft^3. Viscosity is generally reported in units of centipoise (.01 g/sec-cm) and common values for organic liquids are in the range of 0.3 to 20 centipoise at ambient temperatures; water has a viscosity of 1 centerpoise at 20°C (Grain, 1982). To establish these characteristics of the contaminant, first measure or estimate its temperature, and then determine its density from one of the following sources:

1. OHM-TADS - U.S. EPA Data Base.

2. CHRIS Manuals - U.S. Coast Guard, 1974.

3. Dangerous Properties of Industrial Materials by N.I. Sax, 1979.

4. Handbook of Environmental Data on Organic Chemicals, by Verschueren, 1977.

5. The Merck Index, Merck and Co., (Windholz, 1976)

6. Physical/Chemical Properties of Hazardous Waste Constituents, U.S. EPA, 1980.

Information on viscosity is less wide-spread. Data can be found in the Handbook of Chemistry and Physics (Weast 1973) and in Grain (1982) which is contained in the Handbook of Chemical Property Estimation Methods (Lyman et al, 1982); Grain (1982) also provides methods of estimating viscosity from other chemical data.

Information on the compound may also be available from experts in the chemical industry or at universities.

The estimation of effective aquifer thickness will be quite difficult because of uncertainties concerning the mixing properties of the contaminant. It is recommended that a range of thickness values be used in the computation to evaluate the effect of errors in estimating this parameter on the predicted pollutant concentrations.

4.3.2 Ground-Water (Interstitial Pore Water) Velocity

The velocity of ground-water flow within the voids (i.e., interstitial pore water velocity) is required as an input to the saturated zone methodology in this manual. If the value of this parameter has not been established in previous investigations it can be calculated by using Darcy's Law.

In the Darcy equation, the Darcy flow velocity, V_d is equal to the product of the saturated hydraulic conductivity, K, and the hydraulic gradient, $\dfrac{dh}{dl}$ as follows:

$$V_d = -K \frac{dh}{dl} \tag{4.26}$$

To determine the flow velocity in the voids, the Darcy velocity is adjusted to account for the area actually available for flow:

$$V = \frac{K}{n_e} \frac{dh}{dl} \tag{4.27}$$

where n_e is the effective porosity.

Total and effective porosity are discussed in Section 4.2, while hydraulic conductivity and gradient are discussed below.

Hydraulic Conductivity

Saturated hydraulic conductivity is a measure of the quantity of liquid that can flow through a unit cross-sectional area of a medium with time, while under the influence of a hydraulic gradient. It is a function of the permeability of the medium as well as the density and viscosity of the flowing liquid. In ground-water systems, the fluid is water, existing within a fairly narrow range of temperatures, and hydraulic conductivity and permeability can therefore be directly related. Hydraulic conductivity is sometimes referred to as the coefficient of permeability, a term that should not be confused with permeability as discussed above.

In this manual, hydraulic conductivity is a required input for the saturated zone analysis if the ground-water velocity is unknown and must be determined using Darcy's Law as discussed above. The units of hydraulic conductivity are those of velocity (L/T) or alternatively, flow per unit area. In the ground-water (saturated zone) methodology used in this manual, flow and therefore hydraulic conductivity are considered only in the horizontal direction.

Local site-specific information should be used whenever possible; significant errors can result from using general or regional data.

Tables 4.12 and 4.13 contain values of horizontal hydraulic conductivity (K) and permeability (k) for a variety of geologic media. As can be seen in these tables, hydraulic conductivity increases with increasing particle size and with increasing occurrences of fractures. These values should be used only if site specific information is not available.

Ground-Water Flow Gradient

In order to determine the velocity of ground-water flow using Darcy's Law, the ground-water flow hydraulic gradient must be estimated. The flow gradient is influenced by both natural and man-made factors. In most cases, ground water moves in roughly the same direction as surface water drainage, and the ground-water flow gradient varies in magnitude in a direct relationship with surface topography (i.e. the gradient is steepest where the land slopes most steeply and vice versa). Man-made influences on the flow gradient include artificial recharge areas (disposal wells), areas of enhanced recharge (landfills) and pumping wells. Areas of increased recharge tend to cause the ground water to flow radially outward from the recharge point, while pumping wells tend to cause ground water to flow radially inward towards the well. These artificial influences on the ground-water flow patterns may be _very important_ in assessing the local _magnitude_ and _direction_ of the ground-water flow gradient.

Data on the ground-water flow gradient should be sought in existing hydrogeological reports from the U.S. Geological Survey, state water and geological agencies, and local health, water, and engineering departments. Experts in the engineering and geology departments of nearby universities, in consulting firms, and in government agencies may also be able to provide guidance. Also, water table elevations at several points in the area can be used to estimate the gradient and direction (See Section 4.1.5 for estimating depth to ground water).

Local site-specific information should be used whenever possible; significant errors can result from using general or regional data.

If these data sources are not helpful, the flow gradient can be roughly estimated as equivalent in magnitude and direction to the general slope of the land surface in the area of the waste/spill site. This estimate will suffer from substantial error if significant pumping or artificial recharge is occurring in the area. It is most appropriate as a regional estimate of ground-water flow, and becomes less valid when applied to smaller regions.

4.3.3 Dispersion Coefficient

Hydrodynamic dispersion in subsurface media is a process that causes the spreading of a contaminant beyond that which results from convection alone. Variations in local velocity (magnitude and direction) give rise to dispersive spreading on microscopic, macroscopic, and regional scales. The magnitude of dispersion varies significantly with the scale of the analysis,

RANGE OF VALUES OF HYDRAULIC CONDUCTIVITY AND PERMEABILITY

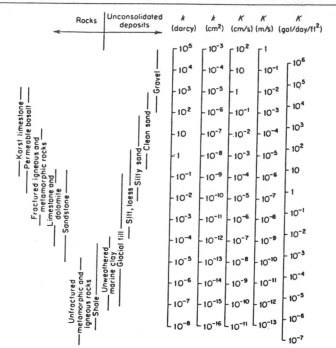

CONVERSION FACTORS FOR PERMEABILITY AND HYDRAULIC CONDUCTIVITY UNITS

	Permeability, k*			Hydraulic conductivity, K		
	cm^2	ft^2	darcy	m/s	ft/s	gal/day/ft²
cm^2	1	1.08×10^{-3}	1.01×10^{8}	9.80×10^{2}	3.22×10^{3}	1.85×10^{9}
ft^2	9.29×10^{2}	1	9.42×10^{10}	9.11×10^{5}	2.99×10^{6}	1.71×10^{12}
darcy	9.87×10^{-9}	1.06×10^{-11}	1	9.66×10^{-6}	3.17×10^{-5}	1.82×10^{1}
m/s	1.02×10^{-3}	1.10×10^{-6}	1.04×10^{5}	1	3.28	2.12×10^{6}
ft/s	3.11×10^{-4}	3.35×10^{-7}	3.15×10^{4}	3.05×10^{-1}	1	5.74×10^{5}
gal/day/ft²	5.42×10^{-10}	5.83×10^{-13}	5.49×10^{-2}	4.72×10^{-7}	1.74×10^{-6}	1

*To obtain k in ft^2, multiply k in cm^2 by 1.08×10^{-3}.

TABLE 4.12   RANGE OF VALUES OF HYDRAULIC CONDUCTIVITY AND
              PERMEABILITY. (After Freeze and Cherry, 1979)

TABLE 4.13 REPRESENTATIVE HORIZONTAL FIELD HYDRAULIC
CONDUCTIVITY RANGES FOR SELECTED ROCKS

Rock	Horizontal Field Hydraulic Conductivity (gpd/sq ft)
Gravel	$1x10^3$ $- 3x\ 10^4$
Basalt	$1x10^{-6}$ $- 2x10^4$
Limestone	$2x10^{-2}$ $- 2x10^4$
Sand and gravel	$2x10^2$ $- 5x10^3$
Sand	$1x10^2$ $- 3x10^3$
Sand, quick	50 $- 8x10^3$
Dune sand	$1x10^2$ $- 3x10^2$
Peat	4 $- 1x10^2$
Sandstone	$1x10^{-1}$ $- 50$
Loeses	$2x10-^{-3}$ $- 20$
Clay	$2x10^{-4}$ $- ?$
Till	$5x10^{-4}$ $- 1$
Shale	$1x10^5$ $- 1x10^1$

Source: Pettyjohn et at (1982)

Note: 1 gpd/sq. ft. = $4.72 x 10^{-5}$ cm/sec

and choosing appropriate coefficients often requires careful assessment of earlier studies. Applications of this manual for rapid assessment of potential ground-water contamination may be directed toward either local or regional evaluation of contaminant plume migration. Consequently, prior investigations may not provide data on an appropriate scale for the application.

For both the unsaturated and saturated zones, the effects of dispersion are based upon the input of dispersion coefficients with dimensions of L^2/T. These coefficients incorporate two forms of the dispersive process, dynamic dispersion (or dispersivity) and molecular diffusion. For typical flow velocities, molecular diffusion is a negligible part of total dispersion (Pettyjohn et al 1982), and thus it is often ignored.

In a saturated zone assessment, a longitudinal (horizontal) and transverse dispersion coefficient is required as an input parameter. As discussed earlier, the dispersion coefficient, D, is made up of a molecular diffusion component and a dynamic dispersion component as follows:

$$D = aV + D* \qquad\qquad (4.28)$$

where   D = Total dispersion coefficient, cm^2/day

a = Dispersivity, cm

V = Ground-water (interstitial) velocity, cm/day

D* = Molecular diffusion coefficient, cm^2/day

Dispersivity is far more significant than molecular diffusion except when ground-water flow velocities are very low (Freeze and Cherry, 1979). Table 4.14 provides regional dispersivities determined in earlier studies in a variety of aquifer types; dispersivities for local or small scale applications may be less than these values by an order of magnitude or more.

Evidence indicates that a general rule of thumb for dispersivity would be to set it equal to 10% of the distance measurement of the analysis (Gelhar and Axness, 1981). Thus, for a well or stream 100 meters down gradient from the source, a dispersivity of 10 meters would be appropriate. For the unsaturated zone, a 5-meter depth to ground water would require a 0.5 m dispersivity. This approximate rule of thumb, along with Table 4.14 and discussion above, should help the user to estimate a dispersion coefficient in the absence of other data. Sensitivity analyses on the dispersion coefficient are strongly recommended.

4.3.4   Retardation Factor

In using a nomograph for evaluating landfill permits, Pettyjohn et al (1982) recommend that a retardation factor of 1.0 be used unless the permit applicant can show that retardation is significant through field data and testing. In effect, this produces a "worst case" situation since retardation is ignored and the contaminant is routed straight through the aquifer. Since ion exchange is the major retardation mechanism for the saturated zone (since organic matter content is usually low), the clay

TABLE 4.14 REGIONAL DISPERSIVITIES (a)

Type of aquifer	Location	Longitudinal dispersivity (a_x) (ft)
Alluvial sediments	Rocky Mountain Arsenal, Co	100
	Colorado	100
	California	100
	Lyon, France	40
	Barstow, CA	200
	Sutter Basin, CA	260-6600
	Alsace, France	49
Glacial deposits	Long Island, N.Y.	70
	Alberta, Canada	10-20
Limestone	Brunswick, GA	200
Fractured basalt	Idaho	300
	Hanford site, Washington	100
Glacial till	Alberta, Canada	10-20
Limestone	Cutler area, Fla.	22
	Hypothetical	0.01-100
		70
		33
		1.6-330

Source: Pettyjohn, et al, 1982

Note: 1 ft. = .3048 m

content of the aquifer material tends to control retardation. In performing a saturated zone assessment, if the user feels that retardation is significant based on contaminant characteristics and aquifer composition, the guidelines for estimating the retardation factor in Section 4.2.3 may assist in evaluation; otherwise a value of R = 1 is recommended.

4.3.5 Degradation/Decay Rate

Degradation and decay mechanisms are generally more significant in surface and unsaturated soils than in ground water. However, hydrolysis and chemical oxidation can occur in saturated media, and anaerobic decomposition is possible even in deep aquifers. Considering the long travel times that occur in most ground-water systems, even decay rates that correspond to half-lives of 2 years or more can substantially reduce ground-water concentrations prior to discharge to a well or surface waters. Consequently, the user should carefully consider the use of a non-zero decay rate in the saturated zone assessment and analyze the impacts of a reasonable range of decay rates for the compound of concern. Section 4.2.2 discusses estimation of decay rates and sources of information.

4.3.6 Source Contaminant Concentration

As described previously, the assessment nomograph requires source contaminant concentration in ground water as a critical input parameter for a saturated zone assessment. If contaminant movement through the unsaturated zone is important, the Co value for the saturated zone is based upon the concentration predicted by the unsaturated zone and the linkage procedures described in Section 3.3. If the water table is sufficiently high so that the leachate directly enters ground water, the Co value is the estimate of the leachate concentration (see Section 4.1.2).

5. Example Applications and Result Interpretation

Two examples are given to demonstrate how the assessment nomograph and accompanying worksheets can be used for assessments of emergency response situations involving continuous input and pulse input of contaminants to the unsaturated zone with subsequent linkage to the saturated zone.

The nomographs and worksheets are computational tools for evaluation of contaminant concentrations at different values of x and t. As mentioned in Section 3, the user must determine from the potential hazards of the emergency situation which C(x,t) values need to be evaluated. If a time response is desirable C(x) should be evaluated at different values of t; or if a profile response is needed, C(t) should be evaluated at different x values.

In most emergency response situations especially involving chemical or waste spills, time responses which provide expected contaminant concentrations and time of arrival at the ground-water table and/or at a point in the aquifer are usually needed. This is the type of information an On-Scene Coordinator may need, to assess the potential for ground-water contamination and associated emergency actions. Profile responses are not often evaluated in a rapid assessment situation, but they are helpful in showing the movement of a contaminant through the unsaturated or saturated zones.

For both examples below, time responses for both the unsaturated and saturated zones are calculated and plotted as they are commonly needed for emergency assessments. In the first example, a profile response is also evaluated in order to familiarize the user with the associated calculation steps and some fundamental concepts of fate and transport phenomena. Since time responses are most often needed for subsequent saturated zone assessments, the unsaturated zone results are further interpreted and analyzed to demonstrate how time responses are used as input to the subsequent saturated zone assessment.

5.1 EXAMPLE #1: ASSESSMENT OF A CONTINUOUS CONTAMINANT SOURCE

Consider a recently discovered (continuous) leak of an industrial solvent from a surface storage tank. The following data are developed from past investigations conducted by the company and chemical characteristics of the solvent:

$$V = 0.55 \text{ cm/day} \qquad B = 1.5 \text{ gm/cm}^3$$
$$D = 13.75 \text{ cm}^2/\text{day} \qquad \Theta = 0.15$$

$k \quad = 0.004 \text{ day}^{-1} \qquad\qquad Co = 1500 \text{ mg/l}$

$K_d \quad = 0.07 \text{ ml/gm}$

Depth to water table = 250 cm

The worksheet in Table 5.1 describes the development of the above parameter values under the 'Data Sources/Comments' heading.

5.1.1 Evaluation of Profile Responses

Concentration profiles expected to result from this chemical leak are evaluated at different times to show the potential movement of the compound. As shown in Table 5.1, three profiles are evaluated at 50, 200, and 1000 days after the leak began. The calculations are performed according to the procedures discussed in Section 3.1.4. Results of these profile responses -- concentrations (C/Co) vs depth (x) at specific times (t) -- are plotted in Figure 5.1.

The plot (Figure 5.1) indicates that most of the compound remains in the top 20 cm of the soil for 50 days. In 200 days, the compound has leached below 150 cm, and in 1000 days, steady-state is attained. While moving downward, the chemical is being adsorbed and degraded. With a retardation factor of 1.7, very little retardation is occurring. Degradation is the major cause for the decrease in concentration values found at greater depths.

5.1.2 Evaluation of Time Response at the Ground-Water Table

The time response is evaluated at the ground-water table. In this example, the mean depth to ground water was 250 cm. Concentrations at different times are estimated as shown in the worksheet, Table 5.2, according to the procedures presented in Section 3.1.4. Results are plotted in Figure 5.2.

The plot (Figure 5.2) shows a steady state concentration of 300 mg/l (C/Co = 0.20) at the ground-water table. This concentration is then used to develop the source concentration (Co) for the saturated zone assessment.

5.1.3 Evaluation of Time Response in Ground Water

The spill site is located 100 m up gradient from a local stream that supplies a water supply reservoir. The goal of the assessment is to determine when and in what concentrations the contaminant plume will reach the stream. A recent hydrogeologic study of the area indicated the following parameter estimates:

$K_s = 10^{-3} \text{ cm/sec} \qquad\qquad \dfrac{dh}{dl} = 0.1\% \text{ (gradient)}$

$n_t = 0.33 \qquad\qquad a = 2.6 \text{ m (dispersivity)}$

$n_e = 0.26 \qquad\qquad B = 1.9 \text{ g/cc}$

TABLE 5.1   PROFILE RESPONSE FOR <u>CONTINUOUS</u>
INPUT TO UNSATURATED <u>ZONE</u>

Sheet __1__ of __2_____

Calculated by _____ Date _____

Checked by _____ Date _____

WORKSHEET FOR RAPID ASSESSMENT NOMOGRAPH

ZONE: UNSATURATED __X____

SATURATED _____

Site: Example No. 1 _____   Date of Incident: _____

Location: _____

On Site Coordinator: _____   Agency: _____

Scientific Support
Coordinator: _____   Agency: _____

Compound Name: _____

Compound Characteristics: _____

REQUIRED PARAMETERS:

DATA SOURCES / COMMENTS

C_o = __1500 mg/l__   Company records

V = __0.55 cm/day__   Based on 30 cm/yr recharge rate

D = __13.75 cm^2/day__   Dispersivity = 25 cm, i.e., 10% depth

k = __0.004 day^{-1}__   Company data on compound

$R = 1 + \frac{B}{\theta}K_d =$ __1.700__

K_d = __0.07 ml/gm__   Company data on compound

B = __1.5 gm/cm^3__   Company soils data

θ = __0.15__   Field capacity for sandy loam

PRELIMINARY CALCULATIONS:

1. $V^* = {}^V/_R =$ __0.324 cm/day__

2. $D^* = {}^D/_R =$ __8.088 cm^2/day__

3. $k^* = {}^k/_R =$ __0.0024 day^{-1}__

4. $\sqrt{V^{*2} + 4D^*k^*}$ = __0.427__

5	6	7	8	9				10	11	12
				See Footnote # 2				From Nomograph[3]		
x	t	$^x/_{2D^*}$	$\sqrt{4D^*t}$	A_1	A_2	B_1	B_2	M_1	M_2	$^C/_{Co}$	C
10	50	0.62	40.22	−0.06	−0.28	0.46	0.78	1.23	0.43	0.83	1244
30	50	1.85	40.22	−0.19	0.21	1.39	1.28	0.63	0.28	0.46	686
75	50	4.64	40.22	−0.48	1.33	3.48	2.40	0.04	0.03	0.03	52
10	200	0.62	80.44	−0.06	−0.94	0.46	1.19	1.70	0.15	0.93	1388

TABLE 5.1 continued

Sheet __2__ of __2__
Calculated by _____ Date _____
Checked by _____ Date _____

NOMOGRAPH WORKSHEET (con't.)

ZONE: UNSATURATED ___X___
SATURATED _____

5	6	7	8	9				10		11	12
x	t	$x/2D*$	$\sqrt{4D*t}$	See Footnote # 2				From Nomograph[3]			C
				A_1	A_2	B_1	B_2	M_1	M_2	C/Co	
40	200	2.47	80.44	−0.26	−0.57	1.86	1.56	1.22	0.18	0.70	1049
100	200	6.18	80.44	−0.64	0.18	4.64	2.31	0.42	0.16	0.29	434
150	200	9.27	80.44	−0.96	0.80	6.97	2.93	0.10	0.0	0.05	75
50	1000	3.09	179.87	−0.32	−2.10	2.32	2.65	1.45	0.0	0.73	1088
150	1000	9.27	179.87	−0.96	−1.54	6.97	3.21	0.76	0.0	0.38	567
250	1000	15.45	179.87	−1.60	−0.99	11.61	3.77	0.37	0.0	0.19	279

Footnotes: 1.   Refer to Table 3.1 for definitions and units, and to Chapter 4 for estimation guidelines.

2. A_1 = Col.7 X (Item 1 − Item 4) = $\frac{x}{2D*} (V* - \sqrt{V*^2 + 4D*k*})$

A_2 = [Col.5 − Col.6 X Item 4] / Col.8 = $\frac{x - t\sqrt{V*^2 + 4D*k*}}{\sqrt{4D*t}}$

B_1 = Col.7 X (Item 1 + Item 4) = $\frac{x}{2D*} (V* + \sqrt{V*^2 + 4D*k*})$

B_2 = [Col.5 + (Col.6 X Item 4)] / Col.8 = $\frac{x + t\sqrt{V*^2 + 4D*k*}}{\sqrt{4D*t}}$

3.   Figure 3.3 or Figure 3.4 (See Figure 3.3 for use of nomograph).

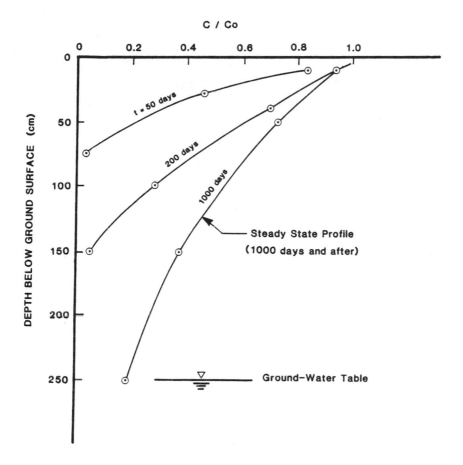

Figure 5.1   Soil Profile Response For Example #1:   Demonstrating Fate and
Movement of Pollutant.

TABLE 5.2   TIME RESPONSE FOR <u>CONTINUOUS</u>
            <u>INPUT</u> TO UNSATURATED ZONE

Sheet __1__ of __2_____
Calculated by _____ Date _____
Checked by _____ Date _____

WORKSHEET FOR RAPID ASSESSMENT NOMOGRAPH

ZONE: UNSATURATED __X____

SATURATED _____

Site: Example No. 1 _____   Date of Incident: _____

Location: _____

On Site Coordinator: _____   Agency: _____

Scientific Support
Coordinator: _____   Agency: _____

Compound Name: _____

Compound Characteristics: _____

REQUIRED PARAMETERS:

C_o = __1500 mg/l__

V = __0.55 cm/day__

D = __13.75 cm^2/day__

k = __0.004 day^{-1}__

R = $1 + \frac{B}{\theta} K_d$ = __1.700__

K_d = _____ 0.07 ml/gm

B = _____ 1.5 gm/cm^3

θ = _____ 0.15

DATA SOURCES / COMMENTS

Company records

Based on 30 cm/yr recharge rate

Dispersivity = 25 cm, i.e., 10% depth

Company data on compound

Company data on compound

Company soils data

Field capacity for sandy loam

PRELIMINARY CALCULATIONS:

1. $V^* = V/_R$ = __0.324 cm/day__

2. $D^* = D/_R$ = __8.088 cm^2/day__

3. $k^* = k/_R$ = __0.0024 day^{-1}__

4. $\sqrt{V^{*2} + 4D^*k^*}$ = __0.427__

5	6	7	8	9				10	11	12
				See Footnote # 2				From Nomograph[3]		
x	t	$x/_{2D^*}$	$\sqrt{4D^*t}$	A_1	A_2	B_1	B_2	M_1	M_2	$C/_{Co}$	C
250	25	15.45	28.44	−1.60	8.42	11.61	9.17	0.0	0.0	0.0	0.0
250	50	15.45	40.22	−1.60	5.69	11.61	6.75	0.0	0.0	0.0	0.0
250	75	15.45	49.26	−1.60	4.43	11.61	5.73	0.0	0.0	0.0	0.0
250	100	15.45	56.88	−1.60	3.64	11.61	5.15	0.0	0.0	0.0	0.0

TABLE 5.2 continued

NOMOGRAPH WORKSHEET (con't.)

ZONE: UNSATURATED ___X___
SATURATED _____

5	6	7	8	9				10		11	12
x	t	$x/2D*$	$\sqrt{4D*t}$	See Footnote # 2				From Nomograph[3]			C
				A_1	A_2	B_1	B_2	M_1	M_2	C/C_0	
250	125	15.45	63.59	-1.60	3.09	11.61	4.77	0.0	0.0	0.0	0.0
250	150	15.45	69.66	-1.60	2.67	11.61	4.51	0.0	0.0	0.0	0.0
250	175	15.45	75.24	-1.60	2.33	11.61	4.32	0.0	0.0	0.0	0.0
250	200	15.45	80.44	-1.60	2.05	11.61	4.17	0.0	0.0	0.0	0.0
250	300	15.45	98.52	-1.60	1.24	11.61	3.84	0.016	0.0	0.008	12.0
250	400	15.45	113.76	-1.60	0.70	11.61	3.70	0.065	0.0	0.03	45.0
250	500	15.45	127.18	-1.60	0.29	11.61	3.65	0.14	0.0	0.07	105.0
250	600	15.45	139.32	-1.60	-0.05	11.61	3.64	0.21	0.0	0.11	165.0
250	800	15.45	160.88	-1.60	-0.57	11.61	3.68	0.32	0.0	0.16	240.0
250	1000	15.45	179.87	-1.60	-0.99	11.61	3.77	0.37	0.0	0.19	285.0
250	1500	15.45	220.29	-1.60	-1.78	11.61	4.05	0.40	0.0	0.20	300.0

Footnotes: 1. Refer to Table 3.1 for definitions and units, and to Chapter 4 for estimation guidelines.

2. A_1 = Col.7 X (Item 1 - Item 4) = $\frac{x}{2D*}$ $(V* - \sqrt{V*^2 + 4D*k*})$

A_2 = [Col.5 - Col.6 X Item 4] / Col.8 = $\frac{x - t\sqrt{V*^2 + 4D*k*}}{\sqrt{4D*t}}$

B_1 = Col.7 X (Item 1 + Item 4) = $\frac{x}{2D*}$ $(V* + \sqrt{V*^2 + 4D*k*})$

B_2 = [Col.5 + (Col.6 X Item 4)] / Col.8 = $\frac{x + t\sqrt{V*^2 + 4D*k*}}{\sqrt{4D*t}}$

3. Figure 3.3 or Figure 3.4 (See Figure 3.3 for use of nomograph).

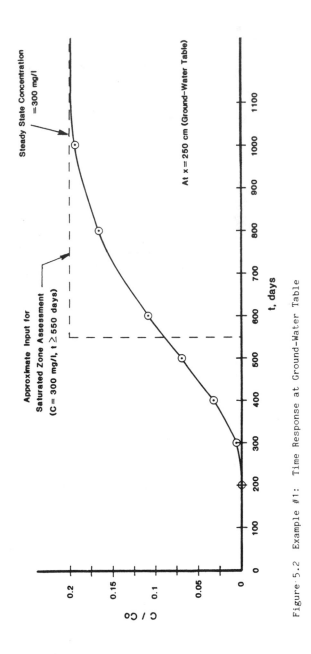

Figure 5.2   Example #1: Time Response at Ground-Water Table

Total aquifer thickness   = 18 m

Effective aquifer thickness = 6 m

Based on the linkage procedures and Equation 3.10 in Section 3.3, the source concentration was calculated as follows:

$$Co = \frac{(300 \text{ mg/l})(.0822 \text{ cm/day})(20 \text{ m})}{(.864 \text{ cm/day})(6m)}$$

$$Co = 95 \text{ mg/l}$$

This calculation assumes a plume width of 20 meters, an effective mixing depth of 6 meters, a recharge rate of 30 cm/yr (0.0822 cm/day), and a ground-water velocity of .864 cm/day (i.e., $V_d = K_s(dh/dl) = (10^{-3}$ cm/sec) (.001)).

Figure 5.2 shows the step function input to the saturated zone used to approximate the actual contaminant outflow from the unsaturated zone. The step function was assumed to begin at day 550 (i.e., 550 days after the beginning of the leak). As discussed in Section 3.3, this beginning time should be varied to evaluate the influence of the step function approximation on the arrival time and response at the discharge or impact point.

Table 5.3 is the worksheet with calculations for the ground-water time response at the stream and Figure 5.3 plots the calculated C/Co values. Figure 5.3 shows that the plume begins to reach the stream between 1000 and 2000 days and reaches a steady-state concentration of 27 mg/l at 6000 days. Note that the time scale is in days after entering the ground water under the site. The total travel time of the spill to the stream would be the above numbers plus 550 days, the beginning day of the step function input to ground water. Note that changing the beginning day of the step function approximation in Figure 5.2 by 200 to 300 days would not have a major impact on the relative arrival time of the plume at the stream. Thus the step function approximation is reasonable.

5.2 EXAMPLE #2: ASSESSMENT OF A PULSE CONTAMINANT SOURCE

Consider a chemical leak similar to Example No. 1 except that the leak is discovered and fixed after 200 days. The following data from Example #1 apply:

V	= 0.55 cm/day	B	= 1.5 gm/cm^3
D	= 13.75 cm^2/day	θ	= 0.22
k	= 0.004 day^{-1}	Co	= 1500 mg/l
K_d	= 0.07 ml/gm	t_o	= 200 days

Depth to water table = 250 cm

TABLE 5.3   TIME RESPONSE FOR <u>CONTINUOUS</u>
            <u>INPUT</u> TO SATURATED ZONE

Sheet __1__ of __2__
Calculated by _____ Date _____
Checked by _____ Date _____

WORKSHEET FOR RAPID ASSESSMENT NOMOGRAPH

ZONE: UNSATURATED _____

SATURATED _____X____

Site: _Example No. 1_   Date of Incident: _____

Location: _____

On Site Coordinator: _____   Agency: _____

Scientific Support
Coordinator: _____   Agency: _____

Compound Name: _____

Compound Characteristics: _____

REQUIRED PARAMETERS:

C_o = _95 (mg/l)_

V = _3.32 (cm/day)_

D = _860 (cm^2/day)_

k = _0.0004 (day^{-1})_

$R = 1 + \frac{B}{\theta} K_d$ = _1.06_

K_d = _0.01 (ml/gm)_

B = _1.9 (gm/cm^3)_

θ = _0.33_

DATA SOURCES / COMMENTS

Assumes L = 20m,   m = 6m

= $(10^{-3}$cm/sec) (.001) /.26

Dispersivity = 2.6m

Company data, 4.75 yr half-life in G.W.

_____

Company data

Recent G.W. Study

Recent G.W. Study

PRELIMINARY CALCULATIONS:

1. $V^* = V/R$ = _3.132 (cm/day)_   3. $k^* = k/R$ = _3.774 x 10^{-4} (day^{-1})_

2. $D^* = D/R$ = _811.32 (cm^2/day)_   4. $\sqrt{V^{*2} + 4D^*k^*}$ = _3.322_

5	6	7	8	9				10	11	12
				See Footnote # 2				From Nomograph[3]		
x	t	$x/2D^*$	$\sqrt{4D^*t}$	A_1	A_2	B_1	B_2	M_1	M_2	C/C_o	C
10000	1000	6.16	1801.5	-1.17	3.71	39.77	7.39	0.0	0.0	0.0	0.0
	2000	6.16	2547.7	-1.17	1.32	39.77	6.53	0.02	0.0	0.01	0.95
	2500	6.16	2848.4	-1.17	0.60	39.77	6.43	0.12	0.0	0.06	5.7
	2800	6.16	3014.4	-1.17	0.23	39.77	6.40	0.23	0.0	0.12	11.44

TABLE 5.3 continued

NOMOGRAPH WORKSHEET (con't.)

ZONE: UNSATURATED _____

SATURATED _____ X _____

5	6	7	8	9				10	11	12
				See Footnote # 2				From Nomograph[3]		C
x	t	$x/_{2D*}$	$\sqrt{4D*t}$	A_1	A_2	B_1	B_2	M_1	M_2 $C/_{Co}$	
10000	3000	6.16	3120.2	-1.17	0.01	39.77	6.40	0.31	0.0 0.15	14.25
	3200	6.16	3222.6	-1.17	-0.20	39.77	6.40	0.38	0.0 0.19	18.05
	3500	6.16	3370.2	-1.17	-0.48	39.77	6.42	0.47	0.0 0.23	21.85
	4000	6.16	3602.9	-1.17	-0.91	39.77	6.46	0.56	0.0 0.28	26.60
	6000	6.16	4412.7	-1.17	-2.25	39.77	6.78	0.62	0.0 0.31	29.45

Footnotes: 1.   Refer to Table 3.1 for definitions and units, and to Chapter 4 for estimation guidelines.

2. $A_1 = Col.7 \times (Item\ 1 - Item\ 4) = \frac{x}{2D*}(V* - \sqrt{V*^2 + 4D*k*})$

$A_2 = [Col.5 - Col.6 \times Item\ 4] / Col.8 = \frac{x - t\sqrt{V*^2 + 4D*k*}}{\sqrt{4D*t}}$

$B_1 = Col.7 \times (Item\ 1 + Item\ 4) = \frac{x}{2D*}(V* + \sqrt{V*^2 + 4D*k*})$

$B_2 = [Col.5 + (Col.6 \times Item\ 4)] / Col.8 = \frac{x + t\sqrt{V*^2 + 4D*k*}}{\sqrt{4D*t}}$

3.   Figure 3.3 or Figure 3.4 (See Figure 3.3 for use of nomograph).

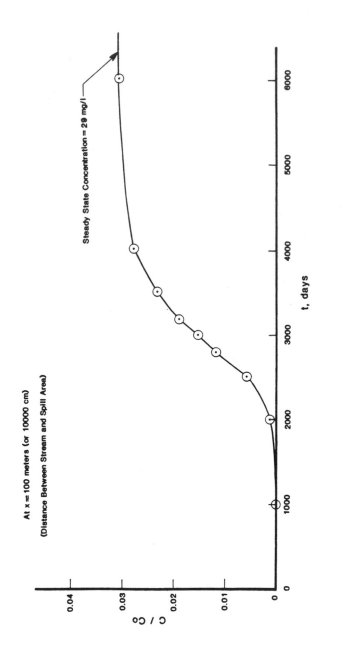

Figure 5.3 Example No. 1: Time Response At The Stream (x=100m)

A time response is evaluated to assess the chemical concentration as it reaches the ground-water table. The calculations are shown in worksheets, Tables 5.4 and 5.5, according to the procedures stated in Section 3.1.4. The response is evaluated at 250 cm, the estimated depth to ground water. The reader should note that concentrations are evaluated at times which differ by one pulse period (i.e. 200 days). This procedure can help to minimize the number of calculations required, since values of C/Co in column 5 of Table 5.5 can be directly entered by shifting the values in column 4 down to the appropriate row. The results are plotted in Figure 5.4 indicating a bell-shape time response curve. The plot indicates that the plume begins to arrive at the ground-water table in approximately 200 days, with a peak concentration of 120 mg/l (C/Co = 0.08) occurring in about 600 days. It also indicates that the plume would completely enter the ground water in about 1300 - 1400 days.

The primary purpose of the unsaturated zone analysis is to obtain a concentration-time response at the ground-water table, so that the bell-shape time response curve can be approximated by a pulse input and applied as a pollutant source for the saturated zone assessment. Following the same approximation procedures discussed in Section 3.3 and using the same parameter values from Example #1, the pulse concentration of 105 mg/l shown in Figure 5.4 produces a Co = 33.0 mg/l for the saturated zone assessment. Tables 5.6 and 5.7 show the worksheet calculations for the time response at the stream, which is plotted in Figure 5.5. Note that the arrival times for the pulse and step function inputs are similar, but that the maximum concentration for the pulse input is only 3.4 mg/l at 3400 days after the spill while the step function (Figure 5.3) produced 29 mg/l at 6550 days.

TABLE 5.4   TIME RESPONSE FOR <u>PULSE INPUT</u>
            TO UNSATURATED ZONE - STANDARD
            WORKSHEET

Sheet __1__ of __2__

Calculated by _____ Date _____

Checked by _____ Date _____

WORKSHEET FOR RAPID ASSESSMENT NOMOGRAPH

ZONE: UNSATURATED __X__

SATURATED _____

Site: _Example No. 2_____   Date of Incident: _____

Location: _____

On Site Coordinator: _____   Agency: _____

Scientific Support
Coordinator: _____   Agency: _____

Compound Name: _____

Compound Characteristics: _____

REQUIRED PARAMETERS:

Co = ___1500 mg/l___

V = ___0.55 cm/day___

D = ___13.75 cm^2/day___

k = ___0.004 day^{-1}___

$R = 1 + \dfrac{B}{\theta} K_d$ = ___1.700___

K_d = ___0.07 ml/gm___

B = ___1.5 gm/cm^3___

θ = ___0.15___

DATA SOURCES / COMMENTS

Company records

Based on 30 cm/yr recharge rate

Dispersivity = 25 cm, i.e., 10% depth

Company data on compound

Company data on compound

Company soils data

Field capacity for sandy loam

PRELIMINARY CALCULATIONS:

1. $V^* = {}^V/_R$ = ___0.324 cm/day___

2. $D^* = {}^D/_R$ = ___8.088 cm^{-2}/day___

3. $k^* = {}^k/_R$ = ___0.0024 day^{-1}___

4. $\sqrt{V^{*2} + 4D^*k^*}$ = ___0.427___

5	6	7	8	9				10	11	12
				See Footnote # 2				From Nomograph[3]		
x	t	${}^x/_{2D^*}$	$\sqrt{4D^*t}$	A_1	A_2	B_1	B_2	M_1	M_2	${}^C/_{Co}$	C
250	100	15.45	56.88	−1.60	3.64	11.61	5.15	0.0	0.0	0.0	0.0
250	200	15.45	80.44	−1.60	2.05	11.61	4.17	0.0	0.0	0.0	0.0
250	300	15.45	98.52	−1.60	1.24	11.61	3.84	0.016	0.0	0.008	12.0
250	400	15.45	113.76	−1.60	0.70	11.61	3.70	0.065	0.0	0.03	45.0

TABLE 5.4 continued

NOMOGRAPH WORKSHEET (con't.)

ZONE: UNSATURATED ___X_____

SATURATED _____

5	6	7	8	9				10	11	12
x	t	$x/_{2D*}$	$\sqrt{4D*t}$	See Footnote # 2				From Nomograph[3]		C
				A_1	A_2	B_1	B_2	M_1	M_2	$C/_{Co}$	
250	500	15.45	127.18	-1.60	0.29	11.61	3.65	0.14	0.0	0.07	105.0
250	600	15.45	139.32	-1.60	-0.05	11.61	3.64	0.21	0.0	0.11	165.0
250	800	15.45	160.88	-1.60	-0.57	11.61	3.68	0.32	0.0	0.16	240.0
250	1000	15.45	179.87	-1.60	-0.99	11.61	3.77	0.37	0.0	0.19	285.0
250	1200	15.45	197.03	-1.60	-1.33	11.61	3.87	0.39	0.0	0.20	300.0

Footnotes: 1. Refer to Table 3.1 for definitions and units, and to Chapter 4 for estimation guidelines.

2. A_1 = Col.7 X (Item 1 - Item 4) = $\frac{x}{2D*} (V* - \sqrt{V*^2 + 4D*k*})$

A_2 = [Col.5 - Col.6 X Item 4] / Col.8 = $\frac{x - t\sqrt{V*^2 + 4D*k*}}{\sqrt{4D*t}}$

B_1 = Col.7 X (Item 1 + Item 4) = $\frac{x}{2D*} (V* + \sqrt{V*^2 + 4D*k*})$

B_2 = [Col.5 + (Col.6 X Item 4)] / Col.8 = $\frac{x + t\sqrt{V*^2 + 4D*k*}}{\sqrt{4D*t}}$

3. Figure 3.3 or Figure 3.4 (See Figure 3.3 for use of nomograph).

TABLE 5.5   TIME RESPONSE FOR <u>PULSE INPUT</u> TO   Sheet __1__ of __1__
            UNSATURATED ZONE – SUPPLEMENTARY WORKSHEET

SUPPLEMENTARY WORKSHEET FOR PULSE INPUT ASSESSMENT

ZONE: UNSATURATED __X__

to = __200 days__ ,   Co = __1500 mg/l__   **SATURATED _____**

			CONTINUOUS INPUT ASSESSMENT (From Worksheet)		PULSE ASSESSMENT	
					Col.4, t ≤ to Col.4-5, t > to	Co x Col. 6
1	2	3	4	5	6	7
x	t	t - to	$C/C_o(t)$	$C/C_o(t-t_o)$	$C/C_o(t)$	C
250	200	0	0.0	0.0	0.0	0.0
250	300	100	0.008	0.0	0.008	12.0
250	400	200	0.03	0.0	0.03	45.0
250	500	300	0.07	0.008	0.06	93.0
250	600	400	0.11	0.03	0.08	120.0
250	800	600	0.16	0.11	0.05	75.0
250	1000	800	0.19	0.16	0.03	45.0
250	1200	1000	0.20	0.19	0.01	15.0

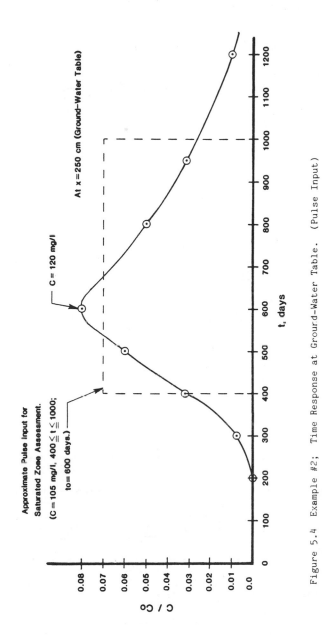

Figure 5.4 Example #2; Time Response at Ground-Water Table. (Pulse Input)

TABLE 5.6   TIME RESPONSE FOR <u>PULSE INPUT</u>
            TO SATURATED ZONE – STANDARD
            WORKSHEET

Sheet __1__ of __2__

Calculated by _____ Date _____

Checked by _____ Date _____

WORKSHEET FOR RAPID ASSESSMENT NOMOGRAPH

ZONE: UNSATURATED _____

SATURATED ___X___

Site: __Example No. 2__   Date of Incident: _____

Location: _____

On Site Coordinator: _____   Agency: _____

Scientific Support
Coordinator: _____   Agency: _____

Compound Name: _____

Compound Characteristics: _____

REQUIRED PARAMETERS:

C_o = __33__ (mg/l)

V = __3.32__ (cm/day)

D = __860__ (cm^2/day)

k = __0.0004__ (day^{-1})

$R = 1 + \frac{B}{\theta} K_d =$ __1.06__

K_d = _____ 0.01 (ml/gm)

B = _____ 1.9 (gm/cm^3)

θ = _____ 0.26

DATA SOURCES / COMMENTS

Results from pulse input linkage

= (10^{-3}cm/sec) (.001) /.26

Dispersivity = 2.6m

Company data, 4.75 yr half-life in G.W.

_____

Company data

Recent G.W. Study

Recent G.W. Study

PRELIMINARY CALCULATIONS:

1. $V^* = V/_R =$ __3.132__ (cm/day)

2. $D^* = D/_R =$ __811.32__ (cm^2/day)

3. $k^* = k/_R =$ __3.774 x 10^{-4}__ (day^{-1})

4. $\sqrt{V^{*2} + 4D^*k^*} =$ __3.322__

5	6	7	8	9				10	11	12
		$x/_{2D^*}$	$\sqrt{4D^*t}$	See Footnote # 2				From Nomograph[3]		
x	t			A_1	A_2	B_1	B_2	M_1	M_2	$C/_{Co}$	C
10000	1400	6.16	2131.5	−1.17	2.51	39.77	6.87	0.0	0.0	0.0	0.0
10000	2000	6.16	2547.7	−1.17	1.32	39.77	6.53	0.02	0.0	0.01	.33
10000	2600	6.16	2904.8	−1.17	0.47	39.77	6.42	0.16	0.0	0.08	2.64
10000	3200	6.16	3222.6	−1.17	−0.20	39.77	6.40	0.38	0.0	0.19	6.27

TABLE 5.6 continued

Sheet __2__ of __2__

Calculated by _____ Date _____

Checked by _____ Date _____

NOMOGRAPH WORKSHEET (con't.)

ZONE: UNSATURATED _____

SATURATED ___X___

5	6	7	8	9				10	11	12
x	t	$x/2D*$	$\sqrt{4D*t}$	See Footnote # 2				From Nomograph[3]		c
				A_1	A_2	B_1	B_2	M_1	M_2	c/c_o	
10000	3800	6.16	3511.7	−1.17	−0.75	39.77	6.44	0.53	0.0	0.27	8.91
10000	4400	6.16	3778.8	−1.17	−1.22	39.77	6.51	0.59	0.0	0.30	9.90
10000	5000	6.16	4028.2	−1.17	−1.64	39.77	6.61	0.61	0.0	0.31	10.23
10000	5600	6.16	4263.1	−1.17	−2.02	39.77	6.71	0.62	0.0	0.31	10.23
10000	1700	6.16	2348.8	−1.17	1.85	39.77	6.66	0.0	0.0	0.0	0.0
	2300	6.16	2732.1	−1.17	0.86	39.77	6.46	0.07	0.0	0.03	.99
	2900	6.16	3067.8	−1.17	0.12	39.77	6.40	0.27	0.0	0.13	4.29
	3500	6.16	3370.2	−1.17	−0.48	39.77	6.42	0.47	0.0	0.23	7.59
	4100	6.16	3647.7	−1.17	−0.99	39.77	6.48	0.57	0.0	0.29	9.57

Footnotes:
1. Refer to Table 3.1 for definitions and units, and to Chapter 4 for estimation guidelines.

2. A_1 = Col.7 X (Item 1 - Item 4) = $\frac{x}{2D*}$ $(V* - \sqrt{V*^2 + 4D*k*})$

   A_2 = [Col.5 - Col.6 X Item 4] / Col.8 = $\frac{x - t\sqrt{V*^2 + 4D*k*}}{\sqrt{4D*t}}$

   B_1 = Col.7 X (Item 1 + Item 4) = $\frac{x}{2D*}$ $(V* + \sqrt{V*^2 + 4D*k*})$

   B_2 = [Col.5 + (Col.6 X Item 4)] / Col.8 = $\frac{x + t\sqrt{V*^2 + 4D*k*}}{\sqrt{4D*t}}$

3. Figure 3.3 or Figure 3.4 (See Figure 3.3 for use of nomograph).

TABLE 5.7   TIME RESPONSE FOR <u>PULSE INPUT</u> TO   Sheet __1__ of __1__
            SATURATED ZONE - SUPPLEMENTARY WORKSHEET

SUPPLEMENTARY WORKSHEET FOR PULSE INPUT ASSESSMENT

ZONE: UNSATURATED _____

to = __600 days__ ,   Co = ____33 mg/1____   **SATURATED** ___X___

			CONTINUOUS INPUT ASSESSMENT (From Worksheet)		PULSE ASSESSMENT	
					Col.4,t≤to Col.4-5, t >to	Co x Col. 6
1	2	3	4	5	6	7
x	t	t - to	C/C_o (t)	C/C_o (t - to)	C/C_o (t)	C
10000	2000	1400	0.01	0.0	0.01	.33
	2600	2000	0.08	0.01	0.07	2.31
	3200	2600	0.19	0.08	0.11	3.63
	3800	3200	0.27	0.19	0.08	2.64
	4400	3800	0.30	0.27	0.03	0.99
	5000	4400	0.31	0.30	0.01	0.33
	5600	5000	0.31	0.31	0.0	0.0
10000	2300	1700	0.03	0.0	0.03	0.99
	2900	2300	0.13	0.03	0.10	3.30
	3500	2900	0.23	0.13	0.10	3.30
	4100	3500	0.29	0.23	0.06	1.98

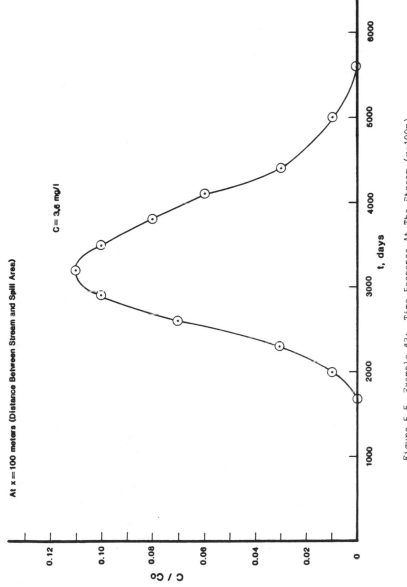

Figure 5.5 Example #2: Time Response At The Stream (x=100m)

References

Abramowitz, M. and I. A. Stegun. 1972. Handbook of Mathematical Functions.
Dover Publications. New York 1045p.

Battelle PNL, 1982a. EPA Field Guide for Scientific Support Activities
Associated with Superfund Emergency Response. U.S. EPA Office of
Emergency and Remedial Response, Washington, D.C. and Environmental
Research Laboratory. Corvallis, OR. EPA-600/8-82-025.

Battelle PNL. 1982b. Section 3.3, Module Four -- Failure Prediction, in:
Post-Closure Liability Trust Fund Simulation Model, Draft Report by ICF
Inc., Washington, D.C. Report to U.S. EPA Office of Solid Waste.

Bolt, G.H. 1976. Chapter 5 In: Soil Chemistry, A. Basic Elements. G. H.
Bolt and G.M. Bruggenwert, eds., Elsevier Scientific Publishing Co.,
New York.

Brady, N.C. 1974. The Nature and Properties of Soils. 8th ed., Macmillan
Publishing Co., New York.

Callahan, M.A., M.W. Slimak, N.W. Gable, I.P. May, C.F. Fowler, J.R. Freed,
P. Jennings, R.L. Durfee, F.C. Whitmore, B. Maestri, W.R. Mabey,
B.R. Holt, and C. Gould. 1979. Water-Related Environmental Fate of 129
Priority Pollutants. Volumes I and II. Prepared for EPA Office of
Water Planning and Standards, Washington, D.C. by Versar, Inc.,
Springfield, Virginia. Available from NTIS. PB80-204373.

Cho, D.M. 1971. Convective Transport of Ammonium with Nitrification in
Soil. Canadian Jour. Soil Sci. 51:339-350.

Clapp, R.B. and G.M. Hornburger. 1978. Empirical Equations for Some Soil
Hydraulic Properties, Water Resources Research, 14:601-604.

Dawson, G.W., C.J. English, and S.E. Petty. 1980. Physical Chemical
Properties of Hazardous Waste Constituents, U.S. EPA Environmental
Research Laboratory, Athens, GA.

Duffy, J.J., E. Peake, and M.F. Mohtadi. 1980. Oil Spills on Land as
Potential Sources of Groundwater Contamination, Environment
International Vol. 3: 107-120.

Enfield, C.G., R.F. Carsel, S.Z. Cohen, T. Phan, and D.M. Walters. 1982. Approximating Pollutant Transport to Ground Water, U.S. EPA Environmental Research Laboratory, Ada, OK (unpublished draft paper).

Environmental Data Service. 1968. Climatic Atlas of the United States. U.S. Department of Commerce, Washington, D.C.

EPA. 1972. Field Detection and Damage Assessment Manual for Oil and Hazardous Materials Spills, U.S. EPA, Washington, D.C.

EPA. 1980. Test Methods for Evaluating Solid Waste, Physical/Chemical Methods U.S. EPA, Washington, D.C. EPA/SW-846.

EPA. 1981. NEIC Manual for Groundwater/Subsurface Investigations at Hazardous Waste Sites, National Enforcement Investigations Center, Denver CO. EPA-330/9-81-002.

Falco, J.W., L.A. Mulkey, R.R. Swank, R.E. Lipcsei, and S.M. Brown. 1980. A Screening Procedure for Assessing the Transport and Degradation of Solid Waste Constituents in Subsurface and Surface Waters. In: Proceedings of the First Annual Meeting of the Society of Environmental Toxicology and Chemistry, Washington, DC (in press).

Freeze, R.A. and J.A. Cherry. 1979. Groundwater, Prentice-Hall, Inc., Englewood Cliffs, NJ.

Gelhar, L. W. and C. J. Axness. 1981. Stochastic Analysis of Macro-Dispersion in 3-Dimensionally Heterogeneous Aquifers. Report No. H-8. Hydrologic Research Program. New Mexico Institute of Mining and Technology, Soccorro, NM.

Geraghty, J.J. D.W. Miller, F. VanDer Leeden, and F.L. Troise. 1973. Water Atlas of the U.S., Water Information Center, Port Washington, NY.

Grain, C.F. 1982. Liquid Viscosity, Chapter 22. In: Lyman et al 1982.

Guinan, D.K. 1980. The Railroad Industry Hazard Information and Response System. In: Control of Hazardous Material Spills, Proceedings of 1980 National Conference, Louisville, KY. 350-357.

Hamaker, J.W. 1972. Diffusion and Volatilization, Chapter 5 in Organic Chemicals in the Soil Environment. Vol 1. ed. by G.A. Goring and J.W. Hamaker, Marcel Dekker, New York.

Harris, J.C. and M.J. Hayes. 1982. Acid Dissociation Constants, Chapter 6 In: Lyman et al 1982.

Huibregtse, K.R. et al. 1977. Manual for the Control of Hazardous Material Spills. U.S. EPA Cincinnati, OH. EPA-600/2-77-227.

Kortum, G., W. Voyel, and K. Andrussow. 1961. Dissociation Constants of Organic Acids in Aqueous Solution, Butterworths, London.

Lyman, W.J., W.F. Reehl, and D.H. Rosenblatt. 1982. Handbook of Chemical Property Estimation Methods. McGraw Hill Co., New York.

Marrero, T.R. and E.A. Mason. 1972. Gaseous Diffusion Coefficients. J. Phys. Chem. Ref. Data 1, 3–118.

Lyon, T.L., H.O. Buckman, and N.C. Brady. 1952. The Nature and Properties of Soils. Macmillan, Inc. New York.

Mabey, W.R., J.H. Smith, R.T. Podoll, H.L. Johnson, T. Mill, T.W. Chou, J. Gates, I. Waight Partridge, and D. Vandenberg. 1982. Aquatic Fate Process Data for Organic Priority Pollutants. Prepared by SRI International, Menlo Park, CA for U.S. EPA Office of Water Regulations and Standards. Washington, D.C.

McBride, G.B. 1982. Nomographs for Rapid Solutions for the Streeter-Phelps Equations. Journal W.P.C.F. 54 (4).

Mills, W.B., J.D. Dean, D.B. Porcella, S.A. Gherini, R.J.M. Hudson, W.E. Frick G.L. Rupp, and G.L. Bowie. 1982. Water Quality Assessment: A Screening Procedure for Toxic and Conventional Pollutants. U.S. EPA Environmental Research Laboratory, Athens, GA.

Misra, C., D.R. Nielsen, and J.W. Biggar. 1974. Nitrogen Transformations in Soil During Leaching: I. Theoretical Considerations. Soil Sci. Soc. Amer. Proc. 38:289–293.

Nash, R.G. 1980. Dissipation Rate of Pesticides from Soils. Chapter 17 in: CREAMS, A Field Scale Model for Chemicals, Runoff, and Erosion from Agricultural Management Systems. Vol. III. U.S. Department of Agriculture. Conservation Research Report No. 26.

Parker, C. A. et al. 1946. Fertilizers and Lime in the United States. USDA Misc. Publ. No. 586.

Parnarouskis, M.C., M.F. Flessner, and R.G. Potts. 1980. A Systems Approach to Chemical Spill Response Information Needs, In: Hazardous Chemicals – Spills and Waterborne Transportation. S.S. Weidenbaum (ed.) American Institute of Chemical Engineers.

Perrin, D.D. 1965. Dissociation Constants for Organic Bases in Aqueous Solution. Butterworths, London.

Perry, R.H. and C.H. Chilton (eds.). 1973. Chemical Engineer's Handbook. 5th Edition, McGraw-Hill, New York.

Pettyjohn, W.A., D.C. Kent, T.A. Prickett, H.E. LeGrand, and F.E. Witz. 1982. Methods for the Prediction of Leachate Plume Migration and Mixing, U.S. EPA Municipal Environmental Research Laboratory, Cincinnati, OH.

Pickens, J.F., J.A. Cherry, R.W. Gillham, and W.F. Merrit, 1977, Field Studies of Dispersion in A Shallow Sandy Aquifer, Proceedings of the Invitational Well-Testing Symposium, Berkeley, California.

Rao, P.S.C. 1982. Unpublished report prepared for Anderson-Nichols.

Rao, P.S.C. and J.M. Davidson. 1980. Estimation of Pesticide Retention and Transformation Parameters Required in Nonpoint Source Pollution Models. In: Environmental Impact of Nonpoint Source Pollution, M.R. Overcash and J.M. Davidson, Eds. Ann Arbor Science, Ann Arbor, MI. pp. 23-67.

Ritter, L.J., Jr. and R.J. Paquette. 1967. Highway Engineering, 3rd ed. The Ronald Press Co., New York.

Sax, N.I. 1979. Dangerous Properties of Industrial Materials, 4th ed., Van Nostrand Reinhold, New York.

Sergeant, E.P. and B. Dempsey. 1979. Ionization Constants of Organic Acids in Aqueous Solution, Pergammon Press, New York.

Smith, J.H., W.R. Mabey, N. Bohonos, B.R. Holt, S.S. Lee, T-W. Chou, D.C. Bomberger, and T. Mill. 1977. Environmental Pathways of Selected Chemicals in Freshwater Systems. Part I: Background and Experimental Procedures. U.S. EPA Environmental Research Laboratory, Athens, GA. EPA-600/7-77-113.

Stevenson, F.J. 1982. Humus Chemistry. John Wiley and Sons, New York.

Stewart, B.A., D.A. Woolhiser, H.W. Wischmeier, J.H. Caro, and M.H. Frere. 1976. Control of Water Pollution from Cropland, Vol. II - An Overview, U.S. Department of Agriculture, Hyattsville, MD. Prepared for U.S. EPA Environmental Research Laboratory, Athens, GA. EPA-600/2-75-026b.

Swann, R.L., P.J. McCall, and S.M. Unger. 1979. Volatility of Pesticides from Soil Surfaces, 177th National Meeting of the Am. Chem. Soc.

Thibodeaux, L.J. 1979. Chemodynamics. John Wiley and Sons, New York.

Thomas, R.B. and D.M. Whiting. 1977. Annual and Seasonal Precipitation Probabilities. U.S. EPA Environmental Research Laboratory, Ada, OK. EPA-600/2-77-182.

Thomas, R.G. 1982. Volatilization from Soil. Chapter 16 In: Lyman et al 1982.

U.S.C.G. 1974a. A Condensed Guide to Chemical Hazards, CG-446-1, U.S. Coast Guard, Washington, D.C.

U.S.C.G. 1974b. Hazardous Chemical Data, CG-446-2, U.S. Coast Guard, Washington, D.C.

U.S. SCS. 1964. Hydrology, National Engineering Handbook, Section 4, Pt. I,
    Watershed Planning, Washington, D.C.

U.S. SCS. 1971. SCS National Engineering Handbook, Section 4, Hydrology.
    U.S. Govt. Printing Office, Washington, DC.

U.V. Atlas of Organic Compounds Vols. 1-4. 1966-1971. Collaboration of
Photo-
    electric Spectrometry Group, London and Institut fur Spectrochemie and
    Angewandte Spectroskopie, Dortmund, Plenum Press, New York.

van Genuchten, M.Th. and W.J. Alves. 1982. Analytical Solutions of the One-
    Dimensional Convective-Dispersive Solute Transport Equation. U.S.
    Department of Agriculture, Tech. Bulletin No. 1661.

Verschueren, K. 1977. Handbook of Environmental Data on Organic Chemicals,
    Van Nostrand Reinhold Co., New York.

Wauchope, R.D. and R.A. Leonard. 1980. Pesticide Concentrations in Agricul-
    tural Runoff: Available Data and an Approximation Formula. Chapter 16
    in: CREAMS, A Field Scale Model for Chemicals, Runoff, and Erosion from
    Agricultural Management Systems. Vol. III. U.S. Department of
    Agriculture Conservation Research Report No. 26.

Weast, R.C. (ed). 1973. Handbook of Chemistry and Physics. 53rd ed., The
    Chemical Rubber Co., Cleveland, OH.

Weed, S.B. and J.B. Weber. 1974. Pesticide—Organic Matter Interactions.
    Chapter 3 in: Pesticides in Soil and Water, W.D. Guenzi, ed., Soil
    Science Society of America, Inc.

Windholz, M., ed. 1976. The Merck Index, 9th ed., Merck and Co., Inc.
    Rahway, NJ.

Appendix A

U.S. Soil Conservation Service
Runoff Estimation Period

(taken directly from: Stewart et al., 1976)

SIMULATION OF DAILY POTENTIAL DIRECT RUNOFF

INTRODUCTION

The amount and seasonal distribution of direct runoff was estimated to assess potential transport of pesticides and nutrients. The effects of some land management practices on direct runoff were also estimated. Hydrologists have developed several rainfall-runoff models of various degrees of complexity for making these estimates. The more physically realistic models are quite complicated and require a great deal of input informa-tion and computer time. The national scope of this report and the severe time constraints involved dictated the use of a rather simple method of estimating runoff from rainfall. Any input information required must also be readily available. After considering several possi-bilities, we decided to use the Soil Conservation Service procedure for estimating direct runoff from storm rainfall (4).

THE SOIL CONSERVATION SERVICE PROCEDURE FOR ESTIMATING DIRECT RUNOFF FROM STORM RAINFALL

The Soil Conservation Service procedure for estimat-ing direct runoff from storm rainfall (sometimes called the SCS curve number method) was designed to use the most generally available rainfall data: total daily rainfall. For this reason rainfall intensity is largely ignored. The basic relationship is the equation:

$$Q = \frac{(P-I_a)^2}{(P-I_a) + S} \; ; P \geq I_a \qquad (1)$$

where

Q = runoff in inches
P = rainfall in inches
I_a = initial abstraction in inches
S = potential maximum retention plus initial abstraction.

The initial abstraction before runoff begins is con-sidered to consist mainly of interception, infiltration and surface storage. Utilizing limited data from small experi-mental watersheds, the following empirical relationship was developed:

$$I_a = (0.2)S. \qquad (2)$$

Substituting this relationship into equation (1) gives

$$Q = \frac{(P-0.2S)^2}{P+0.8S} \; , P \geq (0.2)S \; , \qquad (3)$$

which is the rainfall-runoff relation used in the SCS method.

The parameter CN (runoff curve number of hydrol-ogic soil-cover complex number) is defined in terms of the parameter S as:

$$CN = \frac{1000}{S+10} \qquad (4)$$

Note that runoff equals rainfall when $S = 0$ and $CN = 100$.

The potential maximum retention, S, and therefore the runoff curve number are related to soil surface and profile properties, the vegetative cover, management practices, and the soil water content on the day of the storm. Solutions of equation (3) are shown as a family of curves in Fig. 1.

Soil water content on the day of the storm is accounted for by an Antecedent Moisture Condition (AMC) determined by the total rainfall in the 5-day period preceding the storm.

Three AMC groups have been established with the boundaries between groups dependent upon the time of year as shown in Table 1.

The seasonal difference in the AMC groupings is an attempt to account for the greater evapotranspiration between storms during the growing season.

The different infiltration characteristics of soils are accounted for by classifying soils into four groups based

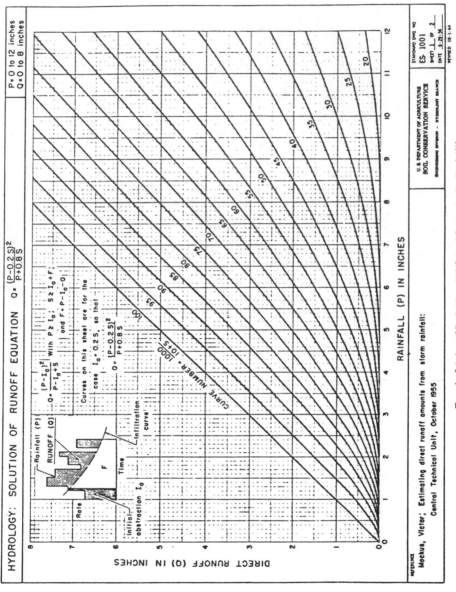

Figure 1.–Solutions of Eq. 3. [From SCS National Engineering Handbook (41)]

Table 1. Seasonal rainfall limits for antecedent moisture conditions[1]

AMC group	Total 5-day antecedent rainfall	
	Dormant season	Growing season
	inches	*inches*
I	<0.5	<1.4
II	0.5 - 1.1	1.4 - 2.1
III	>1.1	>2.1

[1] From SCS National Engineering Handbook (4).

upon the minimum rate of infiltration obtained for a bare soil after prolonged wetting. The influences of both the surface and the profile of a soil are included. The hydrologic soil groups as defined by SCS soil scientists in the National Engineering Handbook are:

A. (Low runoff potential). Soils having high infiltration rates even when thoroughly wetted and consisting chiefly of deep, well to excessively drained sands or gravels. These soils have a high rate of water transmission.

B. Soils having moderate infiltration rates when thoroughly wetted and consisting chiefly of moderately deep to deep, moderately well to well drained soils with moderately fine to moderately coarse textures. These soils have a moderate rate of water transmission.

C. Soils having slow infiltration rates when thoroughly wetted and consisting chiefly of soils with a layer that impedes downward movement of water, or soils with moderately fine to fine texture. These soils have a slow rate of water transmission.

D. (High runoff potential). Soils having very slow infiltration rates when thoroughly wetted and consisting chiefly of clay soils with a high swelling potential, soils with a permanent high water table, soils with a claypan or clay layer at or near the surface, and shallow soils over nearly impervious material. These soils have a very slow rate of water transmission.

The SCS has classified over 9,000 soils in the United States and Puerto Rico according to the above scheme. A sample from the extensive table in the SCS National Engineering Handbook is shown in Table 2. Rainfall-runoff data from small watersheds or infiltrometer plots were used to make the classifications where such data were available, but most are based on the judgement of soil scientists and correlators who used physical properties of the soils in making the assignments.

The interaction of hydrologic soil group (soil) and land use and treatment (cover) is accounted for by assigning a runoff curve number for average soil moisture condition (AMC II) to important soil cover complexes for the fallow period and the growing season. Rainfall-runoff data for single soil cover complex watersheds and plots were analyzed to provide a basis for making these assignments. Average runoff curve numbers for several soil-cover complexes are shown in Table 3. Average runoff curve numbers (AMC II) are for the average soil moisture conditions. AMC I has the lowest runoff potential. AMC III has the highest runoff potential. Under this condition the watershed is practically saturated from antecedent rains. Appropriate curve numbers for AMC I and III based upon the curve number for AMC II are shown in Table 4.

Curve numbers for a "good hydrologic condition" were used in the potential direct runoff simulations. "Hydrologic condition" refers to the runoff potential of a particular cropping practice. A row crop in good hydrologic condition will have higher infiltration rates and, consequently, less direct runoff than the same crop in poor hydrologic condition. Good hydrologic condition seemed an appropriate description of corn under modern management practices.

Seasonal variation not accounted for by the seasonal dependency of the AMC classes is included by varying the average moisture condition curve number according to the stages of growth of a particular crop. For the simulations reported here, with straight row corn as the index crop, the average (AMC II) curve number was set equal to that for fallow for the period from March 1 until the average emergence date for corn. Emergence dates were assumed to be 2 weeks after the average planting date reported by the USDA (5). During the growing season, AMC II curve numbers for each day were calculated by the following equation:

$$CN_i = F - \frac{C_i}{C_{ave}} (F - CN_{ave}) \tag{5}$$

where

CN_i = the curve number for the ith day for AMC II.

F = fallow curve number.

C_i = crop coefficient for the ith day. $C_i \leq 1$.

C_{ave} = average crop coefficient for the growing season.

CN_{ave}= average growing season curve number for AMC II.

The crop coefficients C_i are defined as the ratio of the crop evapotranspiration to potential evapotranspiration for a given day when soil water is not limiting. Crop

Table 2.—Soil names and hydrologic classifications[1] (Sample)

Soil	HSG	Soil	HSG	Soil	HSG	Soil	HSG	Soil	HSG
AABERG	C	AML	C	ALMY	B	AMLAUF	C	ARODSTOOK	
AASTAD	B	AMLSTROM	C	ALUNA	C	ANNABELLA	B	AROSA	C
ABAC	D	AMMER K	D	ALONSO	B	ANNANDALE	C	ARP	B
ABAJJ	C	AMGLT	D	ALOVAK	C	ANNISTON	B	ARRINGTON	B
ABBOTT	D	ANTANUM	C	ALPENA	B	ANORA	A	ARRITOLA	D
ABBOTTSTOWN	C	AMWAMNEE	C	ALPHA	C	ANOMES	C	ARROLIME	C
ABCAL	D	AIBLNITO	C	ALPUN	B	ANSARI	D	ARRON	D
ABEGG	B	AIKEN	B/C	ALPS	C	ANSEL	B	ARROW	B
ABELA	B	AIRMAN	D	ALSEA	A	ANSELMO	B	ARROWSMITH	B
ABELL	B	AILET	B	ALSPAUGH	C	ANSON	B	ARROYO SECO	B
ABERDEEN	D	AINAREA	B	ALSTAU	B	ANTELOPE SPRINGS	C	ARTA	C
ABES	D	AIRMOAT	D	ALSTOWN	b	ANTERO	C	ARTOIS	C
ABILENE	-	AIROTSA	B	ALTAMONT	D	ANT FLAT	C	ARVADA	D
ABINGTON	b	AIRPORT	D	ALTAMONT	D	ANTHO	B	ARVANA	B
ABIQUA	C	AITS	B	ALTAVISTA	C	ANTHONY	C	ARVESON	D
ABU	B/C	AJO	C	ALTDORF	D	ANTIGO	B	ARVILLA	B
ABUN	D	ARARA	A	ALTMAR	B	ANTIOCH	D	ARZELL	B
ABRA	C	AKASAKA	B	ALTU	C	ANTLER	C	ASA	C
ABRAHAM	B	ARELLA	C	ALTUGA	C	ANTOINE	C	ASBURY	B
ABSAROKEE	C	ALADDIN	B	ALTON	B	ANTROBUS	B	ASCALON	B
ABSCOTA	B	ALAE	A	ALTUS	B	ANTY	C	ASCHOFF	B
ABSHER	D	ALALLUM	B	ALTVAN	B	ANVIK	B	ASHBY	C
ABSTED	u	ALAGA	A	ALUM	B	ANWAY	B	ASHCROFT	B
ACACIO	C	ALAKAI	u	ALUSA	D	ANZA	B	ASHDALE	B
ACADEMY	u	ALAMA	B	ALVIN	B	ANZIANO	D	ASHE	B
ACADIA	D	ALAMANCE	B	ALVIRA	B	APACHE	D	ASHKUM	C
ACANA	D	ALAMC	C	ALVISO	D	APAKUIE	A	ASHLAR	B
ACASCO	D	ALAMOSA	C	ALVOR	C	APISHAPA	C	ASHLEY	A
ACE ITUNAS	B	ALAPAHA	D	AMADOR	D	APISON	D	ASH SPRINGS	C
ACEL	u	ALAPAI	A	AMAGON	D	APOPKA	A	ASHTON	B
ACKER	B	ALBAH	B	AMALU	D	APPIAN	C	ASNUE	B
ACKMEN	B	ALDANC	D	AMANA	B	APPLEGATE	C	ASHUELOT	C
ACME	C	ALBANY	C	AMARGOSA	D	APPLETON	C	ASHWOOD	C
ACU	B	ALBATON	u	AMARILLO	B	APPLING	B	ASKEW	C
ACULITA	B	ALBEE	C	AMASA	B	APRON	B	ASO	C
ACUNA	C	ALBEMARLE	B	AMBERSON	B	APT	C	ASOTIN	B
ACOVE	C	ALBERTVILLE	C	AMBOY	C	APTAKISIC	C	ASPEN	B
ACREE	C	ALBIA	C	AMBRAW	C	ARABY	A	ASPERMONT	B
ALKELANC	C	ALBION	B	AMEDEE	A	ARADA	C	ASSINNIBOINE	B
ACTON	B	ALBRIGHTS	C	AMELIA	C	ARANSAS	D	ASSUMPTION	B
ACUFF	B	ALCALDE	C	AMENIA	B	ARAPIEN	C	ASTATULA	A
ACWORTH	B	ALCESTER	B	AMERICUS	A	ARAVE	C	ASTOR	A/D
ALT	C	ALCOA	B	AMES	C	ARAVETON	B	ASTORIA	B
ADA	B	ALCONA	B	AMESHA	B	ARBELA	C	ATASCADERO	C
AUAIR	D	ALCOVA	B	AMHERST	C	ARBONE	B	ATASCOSA	D
ADAMS	A	ALDA	C	AMITY	C	ARBOR	B	ATCO	B
ADAMSON	B	ALDAX	D	AMMON	C	ARBUCKLE	C	ATENCIO	B
AJAMSTOWN		ALDEA	D	AMOLE	C	ARCATA	B	ATEPIC	D
AJAMSVILLE	C	ALDER	B	AMOR	B	ARCH	B	ATHELWOLD	B
ACATON	D	ALDERDALE	C	AMOS	C	ARCHABAL	B	ATHENA	B
AJAVEN	D	ALDERWOOD	C	AMSDEN	b	ARCHER	C	ATHENS	B
ADJIELUU	C	ALDING	C	AMSTERDAM	B	ARCHIN	C	ATHERLY	B
ADDISON	D	ALDWELL	C	ANTOFT	D	ARCO	B	ATHERTON	B/D
AJDY	C	ALEKNAGIR	B	AMY	B	ARCOLA	C	ATMAR	C
AUE	A	ALEMEDA	A	ANACAPA	B	ARD	C	ATMOL	B
ADEL	A	ALEX	A	ANAHUAC	D	ARDEN	B	ATKINSON	B
ADELAIDE	D	ALEXANDRIA	C	ANAMITE	D	ARDENVOIR	B	ATLAS	D
ADELANTU	B	ALEXIS	B	ANAPRA	B	ARDILLA	C	ATLEE	C
ADELINO	B	ALFORD	B	ANASAZI	B	AREDALE	B	ATORE	B/D
ADELPHIA	C	ALGANSEE	B	ANATONE	D	ARENA	C	ATOKA	C
ADENA	C	ALGERITA	B	ANAVERDE	D	ARENALES	C	ATON	B
ADGER	D	ALGIERS	C/D	ANAWALT	D	ARENDTSVILLE	B	ATRYPA	C
ADILIS	A	ALGOMA	B/D	ANCHO	B	ARENOSA	C	ATSION	C
ADIKUNDACK		ALHAMBRA	B	ANCHORAGE	A	ARENZVILLE	B	ATTERBERRY	B
ADIV	B	ALICE	A	ANCHOR BAY	D	ARGONAUT	D	ATTEWAN	A
AJJUNTAS	C	ALICEL	B	ANCHOR POINT	D	ARGUELLO	B	ATTICA	B
AJKINS	B	ALICIA	B	ANCLOTE	D	ARGYLE	B	ATTLEBORO	
ADLER	C	ALIDA	B	ANCO	C	ARIEL	C	ATWATER	B
ADOLPH	D	ALIRCHI	B	ANDEALY	C	ARIZO	C	ATWELL	C/D
ADRIAN	A/D	ALINE	A	ANDERS	C	ARKABUTLA	B	ATWOOD	B
AENEAS	B	ALKO	D	ANDERSON	B	ARKPORT	B	AUBBEENAUBBEE	B
AETNA	B	ALLAGASH	B	ANDES	C	ARLAND	C	AUBERRY	B
AFTON	D	ALLARD	B	ANDORINIA	C	ARLE	C	AUBURN	C/D
AGAR	B	ALLEGHENY	B	ANDOVER	D	ARLING	D	AUBURNDALE	D
AGASSIZ	D	ALLEMANDS	D	ANDREEN	C	ARLINGTON	C	AUDIAN	B
AGATE	D	ALLEN	B	ANDREES	B	ARLOVAL	B	AU GRES	C
AGAWAM	B	ALLENDALE	C	ANED	D	ARMAGH	D	AUGSBURG	B
AGENCY	C	ALLENS PARK	B	ANETH	A	ARMIJO	D	AUGUSTA	D
AWER	u	ALLENSVILLE	C	ANGELICA	B	ARMINGTON	D	AULD	D
AGHER	b	ALLENTINE	D	ANGELINA	B/D	ARMO	B	AURA	B
AGNED	B/C	ALLENWOOD	B	ANGELO	C	ARMOUR	B	AURORA	B
AGNUS	B	ALLESSIO	B	ANGIE	C	ARMSTER	D	AUSTIN	B
AGUA	B	ALLEY	C	ANGLE	A	ARMSTRONG	D	AUSTWELL	D
AGUADILLA	B	ALLIANCE	B	ANGLEN	B	ARMUCHEE	B	AURVASSE	D
AGUA DULCE	C	ALLIGATOR	D	ANGOLA	B	ARNEGARD	B	AUZQUI	C
AGUA FRIA	C	ALLIS	C	ANGOSTURA	C	ARNHART	C	AVA	C
AGUALT	B	ALLISON	C	ANMALT	D	ARNHEIM	B	AVALANCHE	B
AGUEDA	B	ALLOUEZ	C	ANIAK	B	ARNO	D	AVALON	B
AGUILITA	B	ALLOWAY	B	ANITA	D	ARNOLD	B	AVERY	B
AGUIRRE	u	ALMAC	B	ANKENY	A	ARMOT	C/D	AVON	B
AGUSTIN	B	ALMENA	B			ARNOLD	D	AVONBURG	D
AMATONE	D	ALMONT	D			ARNY	A	AVONDALE	E

NOTES   A BLANK HYDROLOGIC SOIL GROUP INDICATES THE SOIL GROUP HAS NOT BEEN DETERMINED
TWO SOIL GROUPS SUCH AS B/C INDICATES THE DRAINED/UNDRAINED SITUATION

[1] From SCS National Engineering Handbook (4).

Table 3. – Runoff curve numbers for hydrologic soil-cover complexes[1]

(Antecedent moisture condition II, and $I_a = 0.2 S$)

Cover			Hydrologic soil group			
Land use	Treatment or practice	Hydrologic condition	A	B	C	D
Fallow	Straight row	- - - -	77	86	91	94
Row crops	"	Poor	72	81	88	91
	"	Good	67	78	85	89
	Contoured	Poor	70	79	84	88
	"	Good	65	75	82	86
	" and terraced	Poor	66	74	80	82
	" " "	Good	62	71	78	81
Small grain	Straight row	Poor	65	76	84	88
	"	Good	63	75	83	87
	Contoured	Poor	63	74	82	85
	"	Good	61	73	81	84
	" and terraced	Poor	61	72	79	82
	" " "	Good	59	70	78	81
Close-seeded legumes[2] or rotation meadow	Straight row	Poor	66	77	85	89
	" "	Good	58	72	81	85
	Contoured	Poor	64	75	83	85
	"	Good	55	69	78	83
	" and terraced	Poor	63	73	80	83
	" " "	Good	51	67	76	80
Pasture or range		Poor	68	79	86	89
		Fair	49	69	79	84
		Good	39	61	74	80
	Contoured	Poor	47	67	81	88
	"	Fair	25	59	75	83
	"	Good	6	35	70	79
Meadow		Good	30	58	71	78
Woods		Poor	45	66	77	83
		Fair	36	60	73	79
		Good	25	55	70	77
Farmsteads		- - - -	59	74	82	86
Roads (dirt)[3]		- - - -	72	82	87	89
(hard surface)[3]		- - - -	74	84	90	92

[1] From SCS National Engineering Handbook (4).
[2] Close-drilled or broadcast.
[3] Including right-of-way.

Table 4.- Curve numbers (CN) and constants for the case $I_a = 0.2S$[1]

CN for condition II	CN for conditions I	III	S values[2]	Curve[2] starts where P =	CN for condition II	CN for conditions I	III	S values[2]	Curve[2] starts where P =
			(inches)	(inches)				(inches)	(inches)
100	100	100	0	0	60	40	78	6.67	1.33
99	97	100	.101	.02	59	39	77	6.95	1.39
98	94	99	.204	.04	58	38	76	7.24	1.45
97	91	99	.309	.06	57	37	75	7.54	1.51
96	89	99	.417	.08	56	36	75	7.86	1.57
95	87	98	.526	.11	55	35	74	8.18	1.64
94	85	98	.638	.13	54	34	73	8.52	1.70
93	83	98	.753	.15	53	33	72	8.87	1.77
92	81	97	.870	.17	52	32	71	9.23	1.85
91	80	97	.989	.20	51	31	70	9.61	1.92
90	78	96	1.11	.22	50	31	70	10.0	2.00
89	76	96	1.24	.25	49	30	69	10.4	2.08
88	75	95	1.36	.27	48	29	68	10.8	2.16
87	73	95	1.49	.30	47	28	67	11.3	2.26
86	72	94	1.63	.33	46	27	66	11.7	2.34
85	70	94	1.76	.35	45	26	65	12.2	2.44
84	68	93	1.90	.38	44	25	64	12.7	2.54
83	67	93	2.05	.41	43	25	63	13.2	2.64
82	66	92	2.20	.44	42	24	62	13.8	2.76
81	64	92	2.34	.47	41	23	61	14.4	2.88
80	63	91	2.50	.50	40	22	60	15.0	3.00
79	62	91	2.66	.53	39	21	59	15.6	3.12
78	60	90	2.82	.56	38	21	58	16.3	3.26
77	59	89	2.99	.60	37	20	57	17.0	3.40
76	58	89	3.16	.63	36	19	56	17.8	3.56
75	57	88	3.33	.67	35	18	55	18.6	3.72
74	55	88	3.51	.70	34	18	54	19.4	3.88
73	54	87	3.70	.74	33	17	53	20.3	4.06
72	53	86	3.89	.78	32	16	52	21.2	4.24
71	52	86	4.08	.82	31	16	51	22.2	4.44
70	51	85	4.28	.86	30	15	50	23.3	4.66
69	50	84	4.49	.90					
68	48	84	4.70	.94	25	12	43	30.0	6.00
67	47	83	4.92	.98	20	9	37	40.0	8.00
66	46	82	5.15	1.03	15	6	30	56.7	11.34
65	45	82	5.38	1.08	10	4	22	90.0	18.00
64	44	81	5.62	1.12	5	2	13	190.0	38.00
63	43	80	5.87	1.17	0	0	0	infinity	infinity
62	42	79	6.13	1.23					
61	41	78	6.39	1.28					

[1] From SCS National Engineering Handbook (4).
[2] For CN in Column 1.

Appendix B

Glossary

(Source: The Water Information Center, Port Washington, N.Y.)

Acidization - The process of forcing acid through a well screen or into the limestone, dolomite, or sandstone making up the wall of a borehole. The general objective of acidization is to clean incrustations from the well screen or to increase permeability of the aquifer materials surrounding a well by dissolving and removing a part of the rock constituents.

Anion - An atom or radical carrying a negative charge.

Annular Space (Annulus) - The space between casing or well screen and the wall of the drilled hole or between drill pipe and casing.

Aquiclude - A saturated, but poorly permeable bed, formation, or group of formations that impedes ground-water movement and does not yield water freely to a well or spring. However, an aquiclude may transmit appreciable water to or from adjacent aquifers, and where sufficiently thick, may constitute an important ground-water storage unit.

Aquifer - A geologic formation, group of formations, or part of a formation that is capable of yielding a significant amount of water to a well or spring.

Aquitard - Used synonymously with aquiclude.

Artesian - The occurrence of ground water under greater than atmospheric pressure.

Artesian (Confined) Aquifer - An aquifer bounded by aquicludes and containing water under artesian conditions.

Artificial Recharge - The addition of water to the ground-water reservoir by activities of man.

Backwashing - The surging effect or reversal of water flow in a well. Backwashing removes fine-grained material from the formation surrounding the borehole and, thus, can enhance well yield.

Barrier Well - A pumping well used to intercept a plume of contaminated ground water. Also a recharge well that delivers water to or in the vicinity of a zone of contamination under sufficient head to prevent the further spreading of the contaminant.

Base Flow - The flow of streams composed solely of ground-water discharge.

Biochemical Oxygen Demand (BOD) - A measure of the dissolved oxygen consumed by microbial life while assimilating and oxidizing the organic matter present in water.

Borehole - An uncased drilled hole.

Brine - A concentrated solution, especially of chloride salts.

Casing - Steel or plastic pipe or tubing that is welded or screwed together and lowered into a borehole to prevent entry of loose rock, gas, or liquid or to prevent loss of drilling fluid into porous, cavernous, or fractured strata.

Cation - An atom or radical carrying a positive charge.

Chemical Oxygen Demand (COD) - The amount of oxygen, expressed in parts per million, consumed under specified conditions in the oxidation of organic and oxidizable inorganic matter in waste water, corrected for the influence of chlorides.

Coliform Group - Group of several types of bacteria which are found in the alimentary tract of warm-blooded animals. The bacteria are often used as an indicator of animal and human fecal contamination of water.

Cone of Depression - The depression, approximately conical in shape, that is formed in a water-table or potentiometric surface when water is removed from an aquifer.

Connate Water - Water that was deposited simultaneously with the geologic formation in which it is contained.

Consumptive Use - That part of the water withdrawn that is no longer available because it has been either evaporated, transpired, incorporated into products and crops, or otherwise removed from the immediate water environment.

Contamination - The degradation of natural water quality as a result of man's activities, to the extent that its usefulness is impaired. There is no implication of any specific limits, since the degree of permissible contamination depends upon the intended end use, or uses, of the water.

Curie - The quantity of any radioactive material giving 3.7×10^{10} disintegrations per second. A picocurie is one trillionth of a curie, or a quantity of radioactive material giving 22.2 disintegrations per minute.

Drainage Well - A well that is installed for the purpose of draining swampy land or disposing of storm water, sewage, or other waste water at or near the land surface.

Dry Well - A borehole or well that does not extend into the zone of saturation.

Effluent - A waste liquid discharge from a manufacturing or treatment process, in its natural state, or partially or completely treated that discharges into the environment.

Eutrophication. - The reduction of dissolved oxygen in natural and man-made lakes and estuaries, leading to deterioration of the esthetic and life-supporting qualities.

Evapotranspiration - The combined processes of evaporation and transpiration.

Exfiltration - The leakage of effluent from sewage pipes into the surrounding soils.

Field Capacity - The moisture content of the soil after water has been removed by deep seepage through the force of gravity. It is the moisture retained largely by capillary forces.

Flow Path - The direction of movement of ground water and any contaminants that may be contained therein, as governed principally by the hydraulic gradient.

Fracture - A break in a rock formation due to structural stresses. Fractures may occur as faults, shears, joints, and planes of fracture cleavage.

Ground Water - Water beneath the land surface in the saturated zone that is under atmospheric or artesian pressure. The water that enters wells and issues from springs.

Ground-Water Reservoir - The earth materials and the intervening open spaces that contain ground water.

Hazardous Waste - Any waste or combination of wastes which pose a substantial present or potential hazard to human health or living organisms.

Head - The height above a standard datum of the surface of a column of water that can be supported by the static pressure at a given point.

Heavy Metals - Metallic elements, including the transition series, which include many elements required for plant and animal nutrition in trace concentrations, but which become toxic at higher concentrations. Examples are: mercury, chromium, cadmium, and lead.

Hydraulic Conductivity - The quantity of water that will flow through a unit cross-sectional area of a porous material per unit of time under a hydraulic gradient of 1.00 at a specified temperature.

Hydraulic Fracturing - The fracturing of a rock by pumping fluid under high pressure into a well for the purpose of increasing permeability.

Hydraulic Gradient - The change in static head per unit of distance along a flow path.

Infiltration - The flow of a liquid through pores or small openings.

Injection Well - A well used for injecting fluids into an underground stratum.

Intermittent Stream - A stream which flows only part of the time.

Ion Exchange - Reversible exchange of ions adsorbed on a mineral or synthetic polymer surface with ions in solution in contact with the surface. In the case of clay minerals, polyvalent ions tend to exchange for nonvalent ions.

Iron Bacteria - Bacteria which can oxidize or reduce iron as part of their metabolic process.

Irrigation Return Flow - Irrigation water which is not consumed in evaporation or plant growth, and which returns to a surface stream or ground-water reservoir.

Leachate - The liquid that has percolated through solid waste or other man-emplaced medium from which soluble components have been removed.

Loading Rate - The rate of application of a material to the land surface.

Mined Ground Water - Water removed from storage when pumpage exceeds ground-water recharge.

Mineralization - Increases in concentration of one or more constituents as the natural result of contact of ground water with geologic formations.

Monitoring (Observation) Well - A well used to measure ground-water levels, and in some cases, to obtain water samples for water-quality analysis.

Nonpoint Source - The contaminant enters the receiving water in an intermittent and/or diffuse manner.

Organic - Being, containing, or relating to carbon compounds, especially in which hydrogen is attached to carbon, whether derived from living organisms or not; usually distinguished from inorganic or mineral.

Overburden - All material (loose soil, sand, gravel, etc.) that lies above bedrock. In mining, any material, consolidated or unconsolidated, that overlies an ore body, especially deposits mined from the surface by open cuts.

Oxidation - A chemical reaction in which there is an increase in valence resulting from a loss of electrons; in contrast to reduction.

Percolate - The water moving by gravity or hydrostatic pressure through interstices of unsaturated rock or soil.

Percolation - Movement of percolate under gravity or hydrostatic pressure.

Perennial Stream - One which flows continuously. Perennial streams are generally fed in part by ground water.

Permeability - A measure of the capacity of a porous medium to transmit fluid.

Piezometric Surface - The surface defined by the levels to which ground water will rise in tightly cased wells that tap an artesian aquifer.

Plume - A body of contaminated ground water originating from a specific source and influenced by such factors as the local ground-water flow pattern, density of contaminant, and character of the aquifer.

Point Source - Any discernible, confined and discrete conveyance, including but not limited to any pipe, ditch, channel, tunnel, conduit, well, discrete fissure, container, rolling stock, or concentrated animal feeding operation from which contaminants are or may be discharged.

Potentiometric Surface - Used synonymously with piezometric surface.

Public Water Supply - A system in which there is a purveyor and customers: the purveyor may be a private company, a municipality, or other governmental agency.

Recharge - The addition of water to the ground-water system by natural or artificial processes.

Reduction - A chemical reaction in which there is a decrease in valence as a result of gaining of electrons.

Runoff - Direct or overland runoff is that portion of rainfall which is not absorbed by soil, evaporated or transpired by plants, but finds its way into streams as surface flow. That portion which is absorbed by soil and later discharged to surface streams is ground-water runoff.

Salaquifer - An aquifer which contains saline water.

Saline - Containing relatively high concentrations of salts.

Salt-Water Intrusion - Movement of salty ground water so that it replaces fresh ground water.

Saturated Zone - The zone in which interconnected interstices are saturated with water under pressure equal to or greater than atmospheric.

Self-Supplied Industrial and Commercial Water Supply - A system from which water is served to consumers free of charge, or from which water is supplied by the operator of the system for his own use.

Sludge - The solid residue resulting from a process or waste-water treatment which also produces a liquid stream (effluent).

Specific Conductance - The ability of a cubic centimetre of water to conduct electricity; varies directly with the amount of ionized minerals in
the water.

Storage (Aquifer) - The volume of water held in the interstices of the rock.

Strata - Beds, layers, or zones of rock.

Subsidence - Surface caving or distortion brought about by collapse of deep
mine workings or cavernous carbonate formations, or from overpumping of
certain types of aquifers.

Surface Resistivity (Electric Resistivity Surveying) - A geophysical prospecting operation in which the relative values of the earth's electrical
resistivity are interpreted to define subsurface geologic and hydrologic
conditions.

Surface Water - That portion of water that appears on the land surface,
i.e., oceans, lakes, rivers.

Toxicity - The ability of a material to produce injury or disease upon exposure, ingestion, inhalation, or assimilation by a living organism.

Transmissivity - The rate at which water is transmitted through a unit
width of an aquifer under a unit hydraulic gradient.

Unsaturated Zone (Zone of Artesian) - Consists of interstices occupied partially by water and partially by air, and is limited above by the land surface and below by the water table.

Upconing - The upward migration of ground water from underlying strata into
an aquifer caused by reduced hydrostatic pressure in the aquifer as a result of pumping.

Water Table - That surface in an unconfined ground-water body at which the
pressure is atmospheric. It defines the top of the saturated zone.

Water-Table Aquifer - An aquifer containing water under atmospheric conditions.

Well - An artificial excavation that derives fluid from the interstices of
the rocks or soils which it penetrates, except that the term is not applied
to ditches or tunnels that lead ground water to the surface by gravity.
With respect to the method of construction, wells may be divided into dug
wells, bored wells, drilled wells, and driven wells.

Well Capacity - The rate at which a well will yield water.

Withdrawal - The volume of water pumped from a well or wells.

Appendix C

Worksheets and Enlarged Nomographs for Rapid Assessment Procedures

Sheet _____ of _____
Calculated by _____ Date _____
Checked by _____ Date _____

WORKSHEET FOR RAPID ASSESSMENT NOMOGRAPH

ZONE: UNSATURATED _____

SATURATED _____

Site: _____ Date of Incident: _____

Location: _____

On Site Coordinator: _____ Agency: _____

Scientific Support
Coordinator: _____ Agency: _____

Compound Name: _____

Compound Characteristics: _____

REQUIRED PARAMETERS: **DATA SOURCES / COMMENTS**

C_o = _____   _____

V = _____   _____

D = _____   _____

k = _____   _____

R = $1 + \dfrac{B}{\theta} K_d$ = _____   _____

K_d = _____   _____

B = _____   _____

θ = _____   _____

PRELIMINARY CALCULATIONS:

1. $V^* = {}^{V}/_{R}$ = _____   3. $k^* = {}^{k}/_{R}$ = _____

2. $D^* = {}^{D}/_{R}$ = _____   4. $\sqrt{V^{*2} + 4D^*k^*}$ = _____

5	6	7	8	9				10	11	12
				See Footnote # 2				From Nomograph3		
x	t	${}^{x}/_{2D^*}$	$\sqrt{4D^*t}$	A_1	A_2	B_1	B_2	M_1	M_2	${}^{C}/_{C_o}$	C

Sheet _____ of _____
Calculated by _____ Date _____
Checked by _____ Date _____

NOMOGRAPH WORKSHEET (con't.)

ZONE: UNSATURATED _____
SATURATED _____

5	6	7	8	9				10	11	12
x	t	$x/2D*$	$\sqrt{4D*t}$	See Footnote # 2				From Nomograph[3]		C
				A_1	A_2	B_1	B_2	M_1	M_2	C/C_0

Footnotes: 1. Refer to Table 3.1 for definitions and units, and to Chapter 4 for estimation guidelines.

2. A_1 = Col.7 X (Item 1 - Item 4) = $\frac{x}{2D*}(V* - \sqrt{V*^2 + 4D*k*})$

A_2 = [Col.5 - Col.6 X Item 4] / Col.8 = $\frac{x - t\sqrt{V*^2 + 4D*k*}}{\sqrt{4D*t}}$

B_1 = Col.7 X (Item 1 + Item 4) = $\frac{x}{2D*}(V* + \sqrt{V*^2 + 4D*k*})$

B_2 = [Col.5 + (Col.6 X Item 4)] / Col.8 = $\frac{x + t\sqrt{V*^2 + 4D*k*}}{\sqrt{4D*t}}$

3. Figure 3.3 or Figure 3.4 (See Figure 3.3 for use of nomograph).

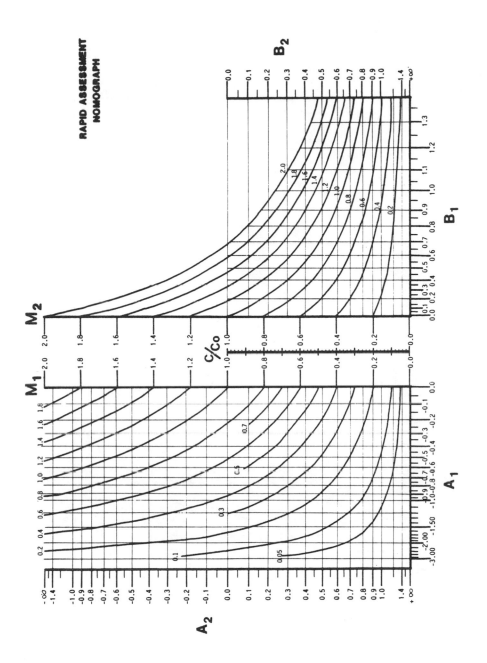

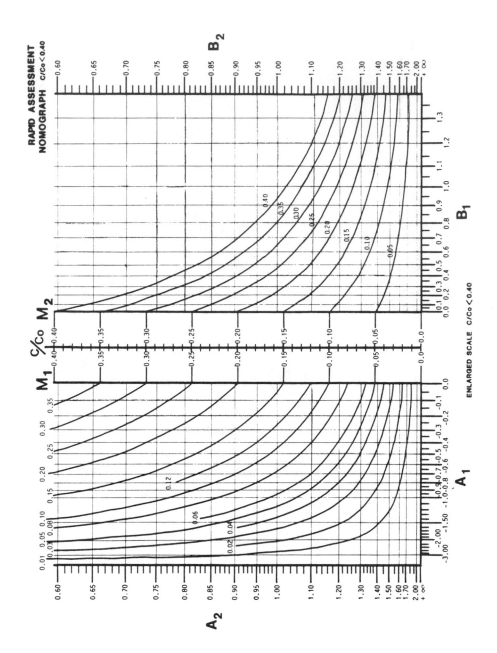

Addendum

The purpose of this addendum is to provide additional explanation on how to apply the procedures described in this manual to situations involving non-aqueous wastes. In this situation two phase flow may exist and the procedures described in the manual will not yield valid concentration predictions. Two phase flow may exist when both:

$$x < \frac{M}{RC_s A\theta} \quad \text{and} \quad t < \frac{M}{C_s AV\theta}$$

where x = distance

t = time

M = mass of contaminant

C_s = water solubility limit of contaminant

A = area of spill or discharge

V = velocity

R = retardation factor

θ = water fraction of soil

Under these conditions the procedures described in this manual should not be used.

Additionally, the user should understand that C_0 is the initial <u>water</u> phase concentration and can never exceed the water solubility limit. As explained in Section 4.1.2, lacking other information it is recommended that C_0 should be assumed equal to C_s.

Finally, for pulse input problems involving non-aqueous waste, the pulse duration (t_0) should be set equal to $M/(C_s AV\theta)$.

In summary, the following guidelines should be used for non-aqueous wastes:

o Constant Input Problems

1) set $C_0 = C_S$

2) apply only where $x > \dfrac{M}{RC_S A\Theta}$ and $t > \dfrac{M}{C_S AV\Theta}$

o Pulse Input Problems

1) set $C_0 = C_S$

2) set $t_0 = M/(C_S AV\Theta)$.

3) apply only where $x > \dfrac{M}{RC_S A\Theta}$ and $t > \dfrac{M}{C_S AV\Theta}$

EMERGING TECHNOLOGIES
FOR THE CONTROL OF HAZARDOUS WASTES

by

B.H. Edwards   J.N. Paullin   K. Coghlan-Jordan

Ebon Research Systems
Washington, DC

Pollution Technology Review No. 99

This book reviews and assesses emerging technologies or novel variations of established technologies for the control of hazardous wastes. Most of the hazardous wastes considered in the study are organic substances.

Current interest in hazardous waste handling methods and disposal practices is apparent almost daily in the news media. Methods which might reduce the ultimate disposal problems facing industry and/or decrease the actual quantity of wastes generated are actively being sought.

Three major technologies are covered in detail in the book—molten salt combustion, fluidized bed incineration, and ultraviolet (UV)/ozone destruction. Theory, unit operations, specific wastes treated, and economics are discussed for each of these.

Several other technologies in the developmental stages are also described. Included in this category are catalyzed wet oxidation, dehalogenation by treatment with UV irradiation and hydrogen, electron bombardment of trace toxic organic compounds, UV/chlorinolysis of aqueous organics, and catalytic hydrogenation-dechlorination of polychlorinated biphenyls (PCBs).

Among the wastes to be treated by these emerging technologies are various dioxins, PCBs, pesticides, herbicides, chemical warfare agents, explosives, propellants, nitrobenzene, plus hydrazine and its derivatives.

A **condensed table of contents** is listed below.

ISBN 0-8155-0943-X(1983)

146 pages

REMEDIAL ACTION TECHNOLOGY FOR WASTE DISPOSAL SITES

by

P. Rogoshewski   H. Bryson   K. Wagner

JRB Associates, Inc.
for the U.S. Environmental Protection Agency

Pollution Technology Review No. 101

The remedial actions which can be applied to control, contain, treat, or remove contaminants from uncontrolled hazardous waste sites and the nature of contamination at waste disposal sites are described in this comprehensive handbook. Improper disposal of industrial, commercial, and municipal solid and hazardous wastes is one of the nation's most pressing environmental problems. The quantity of wastes generated and disposed of annually is tremendous and growing. Cases of improper waste management have resulted in contamination of local groundwater, surface water, land, air, and food and forage crops.

As a result of clean-up operations that have already been conducted, and in anticipation of site clean-up activities that will result from recent regulatory action, many technologies have and are being developed. Those technologies specifically designed for clean-up of waste disposal sites are called "remedial actions." Remedial actions include surface, groundwater, leachate and gas migration controls; direct treatment methods; techniques for contaminated water and sewer lines; and processes for contaminated sediment removal.

The book details available technologies and describes how they may be selected and applied for the clean-up of disposal sites, with particular emphasis on hazardous waste sites. Information on each remedial action includes a general description; applications; design, construction, and/or operating considerations; advantages and disadvantages; and installation and annual operating costs, with examples where possible. **Chapter titles and selected subtitles** are given below.

ISBN 0-8155-0947-2

500 pages

ACID RAIN INFORMATION BOOK
Second Edition

Edited by

David V. Bubenick
GCA/Technology Division

This second edition discusses the major aspects of the acid rain problem which exists today with much new information; it points out the areas of uncertainty and summarizes current and projected research by various government agencies and other concerned organizations. This edition is a revised and greatly enlarged version of the original *Acid Rain Information Book* published in 1982. The wealth of information published in the two years since the first edition was completed made this revision both necessary and desirable—in order to provide a more complete picture of the acid rain situation.

Several recently released studies place the responsibility for acid rain on one industrial source or another. This book does not intend to point a finger; rather it attempts to present, simply, acid rain information.

Acid rain, caused by the emission of sulfur and nitrogen oxides to the atmosphere and their subsequent transformation to sulfates and nitrates, is one of the most widely publicized and emotional environmental issues of the day. The potential consequences of increasingly widespread acid rain demand that this phenomenon be carefully evaluated.

The book is organized in a logical progression from sources of pollutants affecting acid rain formation to the atmospheric transport and transformation of these pollutants and finally to the deposition of acid rain, the effects of that deposition, monitoring and modeling procedures, and possible mitigative measures and regulatory options. This information is followed by a discussion of uncertainties in the understanding of acid rain and a description of current and proposed research.

A condensed table of contents listing **chapter titles and selected subtitles** is given below.

ISBN 0-8155-0967-7 (1984)

397 pages

COSTS OF REMEDIAL RESPONSE ACTIONS
AT
UNCONTROLLED HAZARDOUS WASTE SITES

by

H.L. Rishel **T.M. Boston** **C.J. Schmidt**

SCS Engineers

Pollution Technology Review No. 105

This book presents conceptual design cost estimates for remedial response actions at uncontrolled hazardous waste sites. Thirty-five unit operations, covering uncontrolled landfill or surface impoundment disposal sites, were costed in mid-1980 dollars for the Newark, New Jersey area; and upper and lower cost averages for the contiguous 48 states were also prepared.

The data in the book are based on a review of pertinent literature with subsequent conversion to a consistent computational framework such that costs of remedial response options can be readily compared.

Listed below is a condensed table of contents including **chapter titles and selected subtitles plus the 35 unit operations** covered in the book.

ISBN 0-8155-0969-3 (1984)

144 pages

GENETIC ENGINEERING
AND
NEW POLLUTION CONTROL TECHNOLOGIES

by

James B. Johnston and Susan G. Robinson
Institute for Environmental Studies
University of Illinois at Urbana-Champaign

Pollution Technology Review No. 106
Biotechnology Review No. 3

This book documents the basis for the use of genetic technology to develop new pollution controls in the area of waste treatment; it describes the current state of the art and recommends a future research approach. The book is based on information emerging from expert panel discussions covering pollution problems, gene manipulations, and natural limits to biodegradation.

Spectacular recent advances in gene manipulation provide excellent opportunities to design new organisms with defined degradative capacities; however problems arise in bringing the organisms into contact with a polluted environment in suitable fashion. How to release "engineered" microorganisms to the environment, safely, is a fundamental concern, and a public review policy is suggested.

The study provides background information for scientists, engineers, research managers, and other decision-makers for the evaluation of opportunities for improving biological waste treatment. Policy issues, research problems, and scientific resources that bear on pollution abatement through genetic engineering are identified. The book is intended to promote an interdisciplinary approach to a promising new genetic technology.

A condensed table of contents listing **chapter titles and selected subtitles** is given below.

ISBN 0-8155-0973-1 (1984)

131 pages